Handbook of
Molecular Physics and
Quantum Chemistry

Handbook of Molecular Physics and Quantum Chemistry

Contributors

Mirzoaziz A. Khusenov and Ermuhammad B. Dushanov et al.

AURIS
Reference

www.aurisreference.com

Handbook of Molecular Physics and Quantum Chemistry

Contributors: Mirzoaziz A. Khusenov and Ermuhammad B. Dushanov et al.

Published by Auris Reference Limited

www.aurisreference.com

United Kingdom

Handbook of Molecular Physics and Quantum Chemistry

ISBN: 978-1-78154-889-9

British Library Cataloguing in Publication Data
A CIP record for this book is available from the British Library

Printed in the United Kingdom

Exclusively distributed by CBS Publishers & Distributors Pvt. Ltd.

Sales & Distribution Rights only for India, Pakistan, Bangladesh, Sri Lanka, Nepal and Bhutan.This book is not to be sold outside these territories.

Contents

List of Abbreviations

ANN	Artificial Neural Networks
ALMA	Atacama Large Millimeter Array
CNT	Carbon nanotube
DFT	Density functional theories
FCI	Full configuration interaction
GA	Genetic algorithm
ISM	Interstellar medium
LTE	Local thermodynamic equilibrium
ML	Machine learning
MWR	Method of weighted residuals
MD	Molecular dynamics
MSA	Multiple sequence alignment
PES	potential energy surface
RHF	Restricted Hartree-Fock
TBSH	Trotter-based surface-hopping

List of Contributors

Mirzoaziz A. Khusenov
Tajik Technical University Named after M.S. Osimi, Dushanbe, Tajikistan
Moscow State University of Technology "STANKIN", Moscow, Russia

Ermuhammad B. Dushanov
Institute of Nuclear Physics, Tashkent, Uzbekistan
Laboratory of Radiation Biology, JINR, Dubna, Moscow Region, Russia

Kholmirzo T. Kholmurodov
Moscow State University of Technology "STANKIN", Moscow, Russia
Laboratory of Radiation Biology, JINR, Dubna, Moscow Region, Russia
Dubna International University, Dubna, Moscow Region, Russia

Raul G. E. Morales
Centre for Environmental Sciences and Department of Chemistry Faculty of Sciences, Universidad de Chile, Santiago, Chile

Carlos Hernández
Department of Chemistry, Faculty of Basic Sciences Universidad Metropolitana de Ciencias de la Educación, Santiago, Chile

Masoud Saravi
Islamic Azad University, Nour Branch, Nour

Seyedeh-Razieh Mirrajei
Education Office of Amol, Amol, Iran

Nelson Henrique Morgon
Universidade Estadual de Campinas Brazil

Aline Thaís Bruni
Departamento de Química, Faculdade de Filosofia, Ciências e Letras de Ribeirão Preto, Universidade de São Paulo

Vitor Barbanti Pereira Leite
Departamento de Física, Instituto de Biociências, Letras e Ciências Exatas, Universidade Estadual Paulista, São José do Rio Preto Brazil

Irena Kratochvilova
Institute of Physics, Academy of Sciences of the Czech Republic, Prague, Czech Republic

Raghunathan Ramakrishnan
Department of Chemistry, Institute of Physical Chemistry, University of Basel, Klingelbergstrasse 80, CH-4056 Basel, Switzerland

Pavlo O. Dral
Max-Planck-Institut für Kohlenforschung, Kaiser-Wilhelm-Platz 1, 45470 Mülheim an der Ruhr, Germany
Computer-Chemie-Centrum, University of Erlangen-Nuremberg, Nägelsbachstr. 25, 91052 Erlangen, Germany

Matthias Rupp
Department of Chemistry, Institute of Physical Chemistry, University of Basel, Klingelbergstrasse 80, CH-4056 Basel, Switzerland

O. Anatole von Lilienfeld
Leadership Computing Facility, Argonne National Laboratory, 9700S. Cass Avenue, Lemont, Illinois, 60439, USA

Ryan Babbush
Department of Chemistry and Chemical Biology, Harvard University, Cambridge, MA 02138 USA

Peter J. Love
Department of Physics, Haverford College, Haverford, PA 19041, USA

Alán Aspuru-Guzik
Department of Chemistry and Chemical Biology, Harvard University, Cambridge, MA 02138 USA

M.-H. Yung
Center for Quantum Information, Institute for Interdisciplinary Information Sciences, Tsinghua University, Beijing, 100084, P. R. China
Department of Chemistry and Chemical Biology, Harvard University, Cambridge MA, 02138, USA

J. Casanova
Department of Physical Chemistry, University of the Basque Country UPV/EHU, Apartado 644, 48080 Bilbao, Spain

A. Mezzacapo
Department of Physical Chemistry, University of the Basque Country UPV/EHU, Apartado 644, 48080 Bilbao, Spain

McClean
Department of Chemistry and Chemical Biology, Harvard University, Cambridge MA, 02138, USA

J, L. Lamata
Department of Physical Chemistry, University of the Basque Country UPV/EHU, Apartado 644, 48080 Bilbao, Spain

A. Aspuru-Guzik
Department of Chemistry and Chemical Biology, Harvard University, Cambridge MA, 02138, USA

E. olano
Department of Physical Chemistry, University of the Basque Country UPV/EHU, Apartado 644, 48080 Bilbao, Spain
IKERBASQUE, Basque Foundation for Science, Alameda Urquijo 36, 48011 Bilbao, Spain

Troy Wymore
Pittsburgh Supercomputing Center, 300 South Craig Street, Pittsburgh, PA 15213 USA

Charles L. Brooks III
University of Michigan, Department of Chemistry and Biophysics, 930 North University Avenue, Ann Arbor, MI 48109 USA

Yulia Monakhova
Department of Chemistry, Saratov State University, Astrakhanskaya Street 83, Saratov 410012, Russia

Bernd Schneider
Max Planck Institute for Chemical Ecology, Hans Knöll-Str. 8, Jena 07745, Germany

Chang-Yu Hsieh
Chemical Physics Theory Group, Department of Chemistry, University of Toronto, Toronto, ON M5S 3H6, Canada

Raymond Kapral
Chemical Physics Theory Group, Department of Chemistry, University of Toronto, Toronto, ON M5S 3H6, Canada

Eno E. Ebenso
Department of Chemistry, North West University (Mafikeng Campus), Mmabatho 2735, South Africa

David A. Isabirye
Department of Chemistry, North West University (Mafikeng Campus), Mmabatho 2735, South Africa

Nnabuk O. Eddy
Department of Chemistry, Ahmadu Bello University, Zaria, Nigeria

P. Geerlings
Eenheid Algemene Chemie, Free University of Brussels (VUB), Pleinlaan 2,1050 Brussels,Belgium

F. De Proft
Eenheid Algemene Chemie, Free University of Brussels (VUB), Pleinlaan 2,1050 Brussels,Belgium

Preface

Molecular physics is the study of the physical properties of molecules, the chemical bonds between atoms as well as the molecular dynamics. Its most important experimental techniques are the various types of spectroscopy; scattering is also used. Quantum chemistry represents one of the most successful applications of quantum mechanics. It provides an excellent platform for understanding matter from atomic to molecular scales, and involves heavy interplay of experimental and theoretical methods. Handbook of Molecular Physics and Quantum Chemistry provides a detailed presentation of the most important theoretical concepts and methods for the study of molecules and molecular systems. In first chapter the quantum chemistry Tersoff potential in combination with classical trajectory calculations was used to investigate the interaction of the DNA molecule with a carbon nanotube (CNT). Second chapter focuses on cyanopolyynes as organic molecular wires in the interstellar medium. Third chapter highlights on numerical solution of linear ordinary differential equations in quantum chemistry by spectral method. Composite method employing pseudopotential at CCSD (T) level is presented in fourth chapter. In fifth chapter, we discuss theoretical methods for molecular conformational determination. The field that concerns ways to mimic the behavior of molecules and molecular systems is molecular modeling. It seeks a simplified or idealized description of molecular systems, making it possible to produce three-dimensional representations that provide insights into their behavior. The main goal of sixth chapter is to show how and why quantum chemistry modeling can /should be applied on class of organic materials with relatively high carrier mobility – phthalocyanines. Seventh chapter highlights about quantum chemistry structures and properties of 134 kilo molecules. In eighth chapter, we propose a radically different approach based on the quantum adiabatic algorithm. In this rapidly advancing paradigm of quantum computation, there is no need for Trotterization, phase estimation or logic gates. More generally, we show the first scalable quantum simulation scheme for fermionic systems using adiabatic quantum computing. In ninth chapter, we present an efficient toolkit for solving quantum chemistry problems based on the state-of-the-art in trapped-ion technologies. The aim of tenth chapter is on how both sequence-based bioinformatics and molecular dynamics (MD) simulations, particularly those using hybrid quantum chemical/molecular mechanical potential energy functions, can be employed to discover the most critical amino acids relating to the enzyme's specific function as well as the atomic details of an enzymatic mechanism. Eleventh chapter highlights on quantum chemical calculation of structural features favoring the formation of phenylphenalenones and Correlation functions in open quantum-classical systems have been described in twelfth chapter. The aim of thirteenth chapter is to find good theoretical parameters to characterize the inhibition property of the inhibitors, to establish correlations between inhibition efficiencies and some of the electronic properties of the studied molecules using different quantum chemical/ theoretical methods, quantitative structure activity relationship (QSAR) approach and local reactivity indices. Last chapter focuses on chemical reactivity as described by quantum chemical methods.

Chapter 1

MOLECULAR DYNAMICS SIMULATIONS OF THE DNA-CNT INTERACTION PROCESS: HYBRID QUANTUM CHEMISTRY POTENTIAL AND CLASSICAL TRAJECTORY APPROACH

Mirzoaziz A. Khusenov[1, 2], Ermuhammad B. Dushanov[3, 4], Kholmirzo T. Kholmurodov[2, 4, 5]

[1]Tajik Technical University Named after M.S. Osimi, Dushanbe, Tajikistan

[2]Moscow State University of Technology "STANKIN", Moscow, Russia

[3]Institute of Nuclear Physics, Tashkent, Uzbekistan

[4]Laboratory of Radiation Biology, JINR, Dubna, Moscow Region, Russia

[5]Dubna International University, Dubna, Moscow Region, Russia

ABSTRACT

In this work the quantum chemistry Tersoff potential in combination with classical trajectory calculations was used to investigate the interaction of the DNA molecule with a carbon nanotube (CNT). The so-called hybrid approach—the classical and quantum-chemical modeling, where the force fields and interaction between particles are based on a definite (but not unique) description method, has been outlined in some detail. In such approach the molecules are described as a set of spheres and springs, thereby the spheres imitate classical particles and the spring the interaction force fields provided by quantum chemistry laws. The Tersoff potential in hybrid molecular dynamics (MD) simulations correctly describes the nature of covalent bonding. The aim of the present work was to estimate the dynamical and structural behavior of the DNA-CNT system at ambient temperature conditions. The dynamical configurations were built up for the DNA molecule interacting with the CNT. The analysis of generated MD configurations for the DNA-CNT complex was carried out. For the DNA-CNT system the observations reveal an encapsulation-like behavior of the DNA chain inside the CNT chain. The discussions were made on possible use of the DNA-CNT complex as a candidate material in drug delivery and related systems.

INTRODUCTION

The novel computational approaches allow treating important problems in material fabrication and biomedical applications. Powerful modern computer-based molecular simulation methods became traditional tools in the industry of new materials and drugs. Herewith, the methods of computer molecular simulation involve conventional (classical based approach) or hybrid molecular dynamics, Monte-Carlo and ab initio quantum chemistry, and so on. Computer MD simulations of large molecular systems have quickly shown their power with the invention of modern supercomputers, as a powerful technique which allows one to investigate various physical or biological processes at the atomic/molecular level. X-ray or neutron measurements of physical, chemical, and biological structures inevitably involve the further use of computer models based on molecular and atomic simulation techniques [1-5].

Molecular simulation (conventional and hybrid MD) is based on classical Newtonian physics, modeling the particle interaction in molecules via the force fields defined in advance—empirically or calculated by other methods. Computer molecular simulation is a set of molecular simulation and quantum chemistry methods, or hybrids of these two kinds of methods exhibiting new possibilities.

The methods of computer molecular simulation (conventional MD, hybrid MD or MC (Monte Carlo)), which were first proposed more than 50 years ago, have been rapidly developing in the last 5 - 10 years with the invention of modern computing specialized clusters and supercomputers. In a MD study, the molecular systems are modeled deterministically by the integration of classical equations of motions; in MC, stochastically—with various ensembles. The MD methods are capable of modeling atomic molecular systems of up to thousands and millions of particles and simulating many system parameters and environmental configurations. MD simulation allows one to predict efficiently the ensemble properties and behavior, such as {P-V-T} relations, phase equilibrium, transport properties, structures of synthetic and biological macromolecules, docking of one molecule against another, etc. [1-8].

A scheme shown below (see Table 1) illustrates an example of hybrid MD approach used for a description of hydrogen molecule H_2 (two protons, (a, b) and two electrons, i = 1, 2; r_{12} and r_{ab}; r is interatomic distance).

Another example of a similar approach in quantum mechanics is an adiabatic (Born-Zommerfeldt's) approximation (see Table 2).

Thus, in so-called hybrid approach—the classical and quantum-chemical MD modeling—the molecules are described as a set of spheres and springs, where the force fields and interaction between particles are based on a definite (but not unique) description method and technique. In such technique the classical spheres and springs would imitate the interaction force fields which are actually provided by quantum chemistry laws (Figure 1). It is worth noting that for the development of multiscale models for complex chemical systems the Nobel Prize in Chemistry 2013 was awarded jointly to Martin Karplus, Michael Levitt and Arieh Warshel (Ref:http://www.nobelprize.org/nobel_prizes/chemistry/laureates/2013). So far, creating models of molecules as balls and sticks laid foundation for the powerful techniques and computer programs that are used to understand and predict protein structure conformations and dynamics, chemical reactions and related important processes [9-20].

Computational quantum chemistry research—ab initio, density functional theories (DFT), and others—in contrast to the conventional molecular simulations, is based on quantum physics. The computational quantum chemistry methods were first applied to the electronic structure of atoms or molecules, which yielded wave functions or a probability density functional describing the electron states.

Table 1: A scheme illustrating hybrid classical molecular dynamics (MD) and quantum chemistry molecular dynamics (qMD) approach

$E = E_a + E_b + U$	As results of quantum-mechanical calculations it's possible to define the wave function, and next, to calculate the total energy of system, E, at different interatomic distances, as well as the energies E_a and E_b of non-interacting atoms.
$U = E - E_a - E_b$	Next it's possible to calculate the interaction energy (pair potential) of hydrogen atoms as a function of a distance between them.
$E \sim \dfrac{\int \ldots \int \psi * \hat{H}\psi dr_{e1} \cdots dr_{12} dr_{12}}{\int \ldots \int \psi * \psi dr_{e1} \cdots dr_{12} dr_{12}}$	In the same way, it's possible to calculate the pair interaction potential of a complex many body system, thereby using semi empirical, strong coupling or density functional methods. With use of these methods one defines an approximate N electron wave function for calculation of system' total energy.
$E \sim \dfrac{\langle \psi * |\hat{H}|\psi \rangle}{\langle \psi * |\psi \rangle}$	Here the integration should be taken over all electronic degrees of freedom.
$U = \dfrac{\langle \psi * |\hat{H}|\psi \rangle}{\langle \psi * |\psi \rangle} - E_a - E_b$	So far, the pair interaction potentials of multi-atomic/multi-electronic systems could be defined on dependence of the distances.
$m_i \dfrac{d^2 r_i}{dt^2} = -\dfrac{dU(r_i)}{dr_i}$	The obtained in such way function to be used as pair interaction potential in classical molecular dynamics, thereby establishing the interrelation between quantum-mechanical calculations and molecular dynamics method.

Table 2: A scheme illustrating adiabatic (Born-Zommerfeldt's) approximation in hybrid classical MD and quantum chemisstry qMD approach

$y(r,R) = y_e(r,R)y_n(R)$	In quantum mechanics, in adiabatic (Born-Zommerfeldt's) approximation the total wave function $y(r,R)$ of a molecule could be expressed in the form of a product of functions of the electronic $y_e(r,R)$ and nuclei $y_n(R)$ states.
$\hat{H}y(r,R) = Ey(r,R)$	So far, in quantum-mechanical Schrödinger equation, the nuclei coordinate R in $y_e(r,R)$ to be as parameter, not as variable quantity.
$\hat{H}_e y(r,R) = E_e(R)y_e(r,R)$	This is a quantum-mechanical Schrödinger equation in the adiabatic (Born-Zommerfeldt's) approximation.
$m_i \dfrac{d^2 R_i}{dt^2} = -\nabla_i E(R)$	In this approach, we go on from the quantum mechanics up to the classical one, where the motions of nuclei are described by classical equation.
$E(R) = \sum_{bonded} E_i(R) + \sum_{non-bonded} E_i(R)$	Here the "empirical" force field includes the energies of chemical bonds (bonded interactions) as well as non-bonded interactions (Van der Waals, electrostatics, so on).

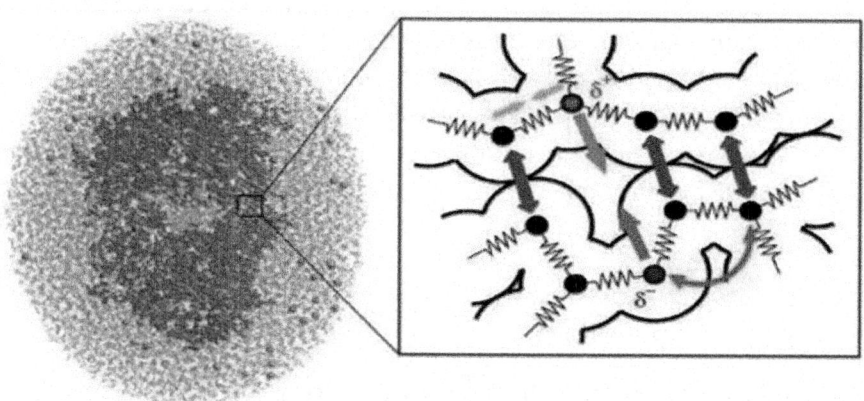

Figure 1: Molecules are described as a set of classical spheres & springs, which imitate the interaction force field provided by quantum chemistry laws. (Ref: a public domain of the World Wide Web).

The quantum chemistry methods provide greater accuracy but are restricted to a smaller molecular size because of their complexity and CPU costs. Quantum chemistry simulation is essential when chemical bonds are formed or broken. It is also used when force parameters are unknown or not applicable. The DFT methods are well established and used with increasing accuracy; the high-level wave function methods with large atomic orbital basis sets currently remain standard. As results of our quantum chemistry studies, we got force—field data, which enabled us to calculate the thermochemical, kinetic, and optical properties, NMR shifts, etc.

Thus, hybrid classical MD and quantum chemistry qMD facilitates a powerful multi-scale computational scheme (see Table 3).

In this work we have employed the quantum chemistry Tersoff potential in combination with classical trajectory calculations to investigate the interaction of the DNA molecule with a CNT. Tersoff potential is efficiently used for MD simulation of systems that contain carbon, silica, germanium, etc. alloys. In carbon nanotubes (CNTs) we have chemical bonding is hybridization sp^2(as graphite), which is stronger than sp^3 bond (of diamond). The nature of chemical bonding in CNTs is described by quantum chemistry, through the process of orbital hybridization. The Tersoff potential in hybrid MD simulations correctly describes the nature of covalent bonding. It's good for simulating systems that contain carbon, silica, germanium and alloys of these elements. The peculiarity of Tersoff potential is that it allows the breaking and formation of chemical bonds. That is associated with hybridization process. Tersoff potential is pair wise potential, but coefficient in attractive term depends on local environment. Thus, Tersoff potential possesses a many body nature. It is also worth noting that CNTs exhibit a unique electrical and chemical properties for organic materials, they possess a great interest for the material research and electronic applications. Depending on their chemical structure, CNTs can be used as an alternative to organic or inorganic semiconductors as well as conductors.

Table 3: A scheme illustrating multi-scale hybrid classical MD and quantum chemistry qMD approach

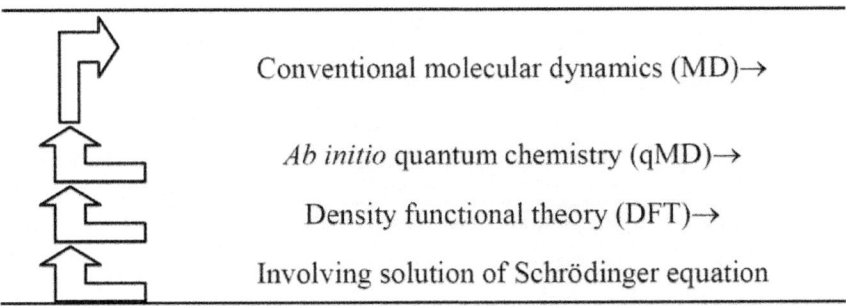

Conventional molecular dynamics (MD)$\rightarrow$

Ab initio quantum chemistry (qMD)$\rightarrow$

Density functional theory (DFT)$\rightarrow$

Involving solution of Schrödinger equation

The chemical bonding of nanotubes is composed entirely of sp^2 bonds, similar to those of graphite. This bonding structure, which is stronger than the sp^3 bonds found in diamonds, provides the molecules with their unique strength. Nanotubes naturally align themselves into "ropes" held together by Van der Waals forces. The nature of the bonding of a nanotube is described by quantum chemistry, specifically, orbital hybridization [21,22].

The physical and molecular mechanism of the interaction of DNA as the genetic material with CNT represents a great interest in today biophysics and biochemistry. Understanding the DNA-CNT interaction mechanism has crucially important for the purposes of a drug delivery in bio-medicine and nanotechnology applications. Along with the DNA-NP (nanoparticle) objects the DNA-CNT system represents a great interest in today biomedicine applications due to diagnostic and treatment of oncology diseases. Cancer, in which cells grow and divide abnormally, is one of the primary diseases with regard to how it responds to CNT drug delivery. Representing a revolutionarily potential for the biochemistry and medicine the use of CNTs in drug delivery has based on the enhancing of sufficient solubility and allowing of efficient tumor targeting. These aspects prevent CNTs from being cytotoxic and altering the function of immune cells. For today, cancer therapy involves surgery, radiation therapy, and chemotherapy. For example, recent experimental and simulation studies involve the interaction of DNA with highly localized high power beams and various nanoparticles (Ag, Au, etc.). These studies are aimed on targeted cancer therapy through the injection of metal microor nanoparticles into the tumor tissue with consequent local microwave or laser heating. Due to their good heat conductivities of NPs (Ag, Au, and so on) the experiments reveal that the only tumor cells to destroy, remaining normal cells undamaged. Nevertheless, such kind treatment methods are usually painful and kill normal cells in addition to producing adverse side effects. CNTs as drug delivery vehicles have shown a potential interest due to a targeting of specific cancer cells with a lower dosage rather than conventional drugs have [9-22].

The aim of this study is to perform the MD simulations to investigate the dynamical and structural behavior of the DNA-CNT model at ambient temperature conditions. The structural radial distribution functions and the dynamical configurations have built up for the DNA molecule interacting with the CNT. For the DNA-CNT system we have to observe an encapsulation-like behavior of the DNA chain inside the carbon nanotube to penetrate deep into. The discussions have made for possible use of the DNA-CNT formation as a candidate in drug delivery and related systems.

MATERIALS AND METHODS

A configuration snapshot of the DNA-CNT model is shown in Figure 2. The DNA strand was consisted of 1260 atoms and its structure was taken from the DL_ POLY database. Thereby specifying the DNA interaction potential, it contained 21 bonds, 1358 constraints, 2442 angles and 3525 dihedrals. The CNT (carbon nanotube) consists of 800 carbon atoms in a nanotube 41.7 angstrom in length (see Figure 3). In the description of the physical properties of CNT we employ the Tersoff potential. The Tersoff potential [23-33] is a special example of a density-dependent potential, which reproduces the properties of covalent bonding in systems containing carbon, silicon, germanium, etc., and alloys of these elements. A special feature of the potential is that it allows bond breaking and associated changes in bond hybridization.

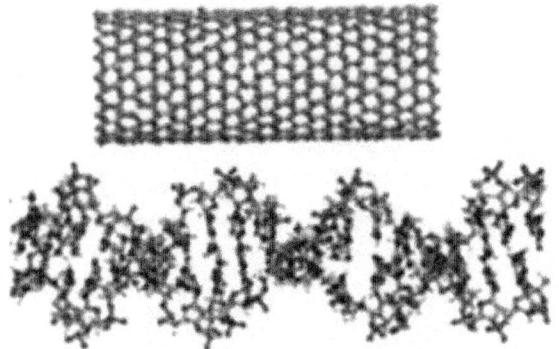

Figure 2: A configuration snapshot of the CNT (top) and DNA (bottom).

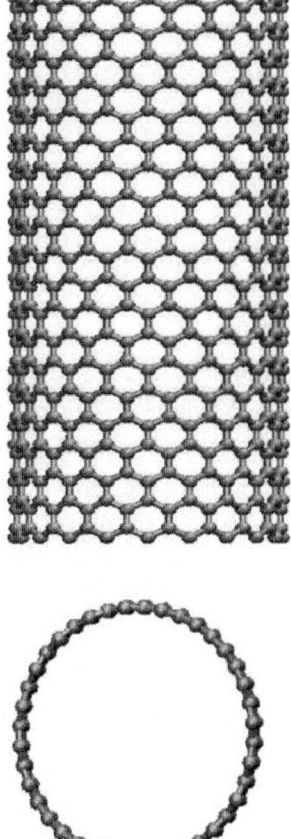

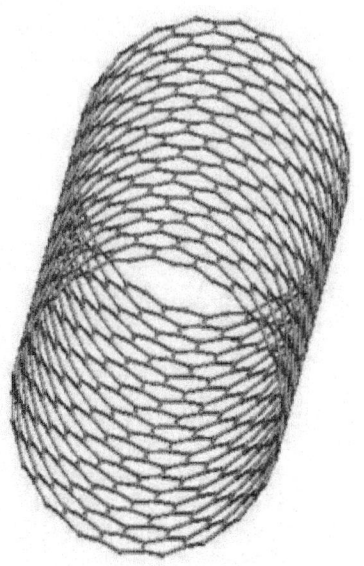

Figure 3: The CNT structural presentation (side and top views).

The energy is modelled as a sum of pair-like interactions where, however, the coefficient of the attractive term in the pair-like potential (which plays the role of a bond order) depends on the local environment giving a many-body potential.

The Tersoff potential has 11 atomic and 2 bi-atomic parameters and looks like:

$$U_{ij} = f_C(r_{ij})\left[f_R(r_{ij}) + \gamma_{ij} f_A(r_{ij}) \right] \tag{1}$$

where the potential parameters have the following forms:

$$f_R(r_{ij}) = A_{ij} \exp\left(-a_{ij} r_{ij}\right), \quad f_A(r_{ij}) = B_{ij} \exp\left(-b_{ij} r_{ij}\right),$$

$$f_C(r_{ij}) = 1/2 + \cos\left[\pi(r_{ij} - R_{ij})/(r_{ij} - S_{ij}) \right]/2,$$

$$\gamma_{ij} = \chi_{ij}\left(1 + \beta_i^{n_i} L_{ij}^{n_i}\right)^{-1/2 n_i}, \quad R_{ij} < r_{ij} < S_{ij},$$

$$L_{ij} = \sum_{k \neq i, j} f_C(r_{ik}) \omega_{ik} g(\theta_{ijk}), \quad f_C(r_{ij}) = 1, \text{ for } r_{ij} < R_{ij}$$

and $f_C(r_{ij}) = 0$, for $r_{ij} > S_{ij}$, $a_{ij} = (a_i + a_j)/2$,

$$b_{ij} = (b_i + b_j)/2,$$

$$g\left(\theta_{ijk}\right)=1+c_i^2\Big/d_i^2-c_i^2\Big/\left[d_i^2+\left(h_i-\cos\theta_{ijk}\right)^2\right],$$

$$A_{ij}=\left(A_iA_j\right)^{1/2},\quad B_{ij}=\left(B_iB_j\right)^{1/2},\quad R_{ij}=\left(R_iR_j\right)^{1/2},$$

$$S_{ij}=\left(S_iS_j\right)^{1/2}.$$

We have accepted the following values: $c_{ii}=1$, $c_{ij}=c_{ji}$, $w_{ii}=1$, $w_{ij}=w_{ji}$.

We simulated several DNA-CNT model systems using molecular dynamics (MD) simulation method. The MD simulation has been performed on the basis of the DL_POLY general-purpose code [32,33]. The elementary MD cell was a parallelipipped with a volume of V = (130, 130, 65) Å³. The integration algorithm was Berendsen NPT ensemble with the termostat and barostat relaxation times 2.0 ps.

For the CNT we used the Tersoff potential parameters of the DL_POLY software data base [24-26,33,34]: A = 1393.6, a = 3.4879, B = 346.74, b = 2.2119, R = 1.8, S = 2.1, b = 1.5724 × 10⁻⁷, h = 0.72751, c = 38049, d = 4.3484, h = −0.57058. The position of the CNT was fixed up using tethering potential as rigid body.

For the Van der Waals interactions between the DNA and CNT we used the Lennard-Jones (LJ), for the DNA used 12-6 potential:

$$U=4\varepsilon\left[\left(\frac{\sigma}{r}\right)^{12}-\left(\frac{\sigma}{r}\right)^6\right],\quad U=\frac{A}{r^{12}}-\frac{B}{r^6}\qquad(2)$$

All potential and parameters are shown in Tables 4-6, where C, O, N and P stands for carbon, oxygen, nitrogen and sulphur atoms of DNA and C_C denotes the carbon atom of CNT.

Table 4: The LJ potential parameters of the DNA-CNT model

Atomic pair/parameters	ε, kcal/mol	σ, Å
C_C-C	0.1	3.4
C_C-P	0.1	3.0
C_C-O	0.1	3.0
C_C-N	0.1	3.3

Table 5: The 12-6 potential parameters of the DNA model

Pair	A, Å^{12} kcal/mol	B, Å^{6} kcal/mol	Pair	A, Å^{12} kcal/mol	B, Å^{6} kcal/mol
C-C	1171340	667.5	P-P	8350780	3369.4
C-P	3145970	1481.6	P-O	1476420	1016.6
C-O	535958	452.2	P-N	1985180	1117.9
C-N	730530	500.7	N-N	450301	373.4
O-O	232116	198.1	N-O	325886	334.9

RESULTS AND DISCUSSIONS

Multiple configurations were generated for the DNACNT model with the MD simulation parameters described above. The DNA insertion into the CNT were relaxed at the temperatures around of $T = 310$ K. The output MD data were analyzed and visualized using different software facilities (VMD [35], DL_POLY utilities [32,33], etc.).

MD Simulated Structural DNA-CNT Configurations

In Figure 4 several sequential snapshots are shown to illustrate the DNA-CNT interaction process. In Figure 4 the top (left pictures) and side (right pictures) views of the DNA-CNT system along with the positions of the arbitrarily chosen DNA (P601 and P631) and CNT (CC1630) atoms are presented. From the stage of a close DNA-CNT contact formation, the DNA chain to start its penetration deeper into the CNT region. This is a fully spontaneous process; the only nature of the atomic interactions has governing the DNA-CNT complex formation. The penetration process is like as an initial stage of DNA encapsulation inside of a CNT. It is worth noting that this process is rather slow; for these MD simulations all processes are happening in vacuum. Obviously, the inclusion of water into consideration would accelerate the DNA encapsulation into the CNT. Therefore, CNTs have been proved to be an excellent transport candidate for many biomolecules, proteins or drugs [15-18].

The DNA-CNT Encapsulation

In Figure 5 a distance diagram is presented for the DNACNT close contact interaction. The positional changes of two distances, DNA (P601, P631)-CNT(C1630), illustrates a DNA penetration behavior deeper into a CNT region.

Table 6: The intermolecular potential parameters of the DNA model

C-C-C-N	2.5000	180	2
C-N-C-N	2.5000	180	2
H-C-N-C	11.250	180	2
N-C-N-C	11.250	180	2
N-C-C-H	5.6250	180	2
N-C-C-C	5.6250	180	2
H-C-C-H	5.6250	180	2
H-C-C-C	5.6250	180	2
C-N-C-N	0.1250	−180	6
N-C-N-H	0.1250	−180	6
H-N-C-N	0.1250	−180	6
C-C-N-H	0.1250	−180	6
C-C-N-C	0.1250	−180	6
O-C-N-H	0.1250	−180	6
O-C-N-C	0.1250	−180	6
C-N-C-N	0.1250	−180	6
C-N-C-H	0.2500	0	3
N-C-N-H	0.2500	0	3
C-N-C-N	0.2500	0	3
O-C-N-C	0.1667	180	2
H-C-N-C	0.1667	180	2
C-C-N-C	0.1667	180	2
H-C-C-H	0.1111	0	3
H-C-C-C	0.1111	0	3
O-C-C-H	0.1111	0	3
O-C-C-C	0.1111	0	3
C-C-C-C	0.1111	0	3

*Marked in italic elements have distinctive properties.

Harmonic bonds: $U(r) = k(r - r_0)^2/2$		
Bonds	k	r_0
O-H	300	1.724

Valence angles: $U(\theta) = k(\theta - \theta_0)^2/2$				
Bonds	k	θ_0	Bonds	k	θ_0
O-P-O	100.33	93.300	C-*N*-C	109.00	106.700
P-O-C	106.70	104.510	C-*N*-H	109.00	106.700
C-O-C	106.70	104.510	O-C-H	112.50	109.471
N-C-N	133.33	120.000	H-C-H	112.50	109.471
N-C-C	133.33	120.000	O-C-C	112.50	109.471
N-C-H	133.33	120.000	H-C-C	112.50	109.471
C-N-C	133.33	120.000	C-C-C	112.50	109.471
O-C-C	133.33	120.000	O-C-N	112.50	109.471
C-C-C	133.33	120.000	H-C-N	112.50	109.471
O-C-N	133.33	120.000	C-C-N	112.50	109.471

Dihedral angles: $U(\varphi) = A[1 + \cos(m\varphi - \delta)]$			
Bonds	A	δ	m
P-O-C-H	0.3333	0	3
P-O-C-C	0.3333	0	3
C-O-C-H	0.3333	0	3
C-O-C-N	0.3333	0	3
C-O-C-C	0.3333	0	3
C-O-P-O	0.5000	0	3
C-N-C-C	2.5000	180	2
N-C-C-O	2.5000	180	2
N-C-C-N	2.5000	180	2
C-C-C-O	2.5000	180	2

*Marked in italic elements have distinctive properties.

Some fluctuations of the distance between DNA and CNT take place, indicating the self-adjustment of the DNA positions before it begins to enter into the CNT. This process has seen to be quite similar with the encapsulation process, as indicated by other authors on biomolecule-CNT interactions [15-22]. Obviously, the Van der Waals and electrostatics interactions used in our simulation technique are both acceleration factors in the DNACND encapsulation process. It is well known that during the DNA insertion into CNT their interaction energies decrease.

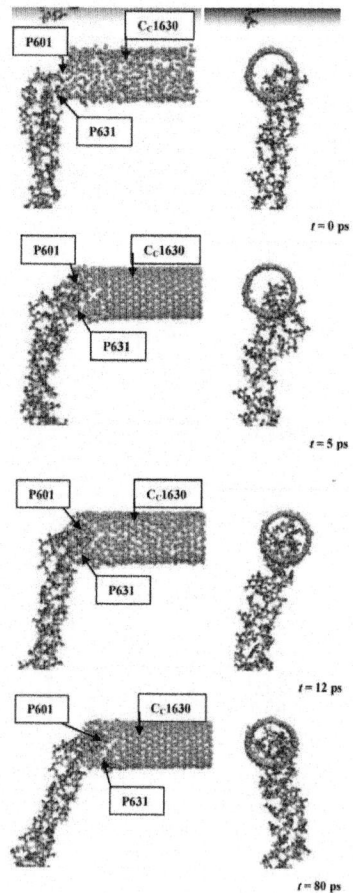

Figure 4: The DNA-CNT dynamics from their contact to penetrtion.

The continuous DNA encapsulation into CNT suggests a practical aspect of using such kind systems in an innovation technology, in material design for medicine in the future. Say, in manufacturing practical CNT-based devices for the delivery of genes, inhibitors, drugs and so on [15-22,34-38].

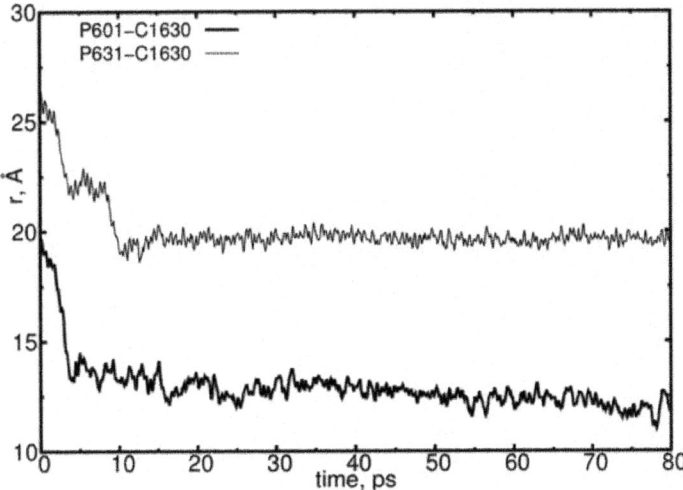

Figure 5: The DNA-CNT penetration dynamics from their contact. The positions of the arbitrarily chosen DNA (P601 and P631) and CNT (C1630) atoms are presented.

CONCLUSION

The quantum chemistry potentials in combination with classical trajectory calculations have to be an efficient approach in today computer molecular modeling. A lot of the physical, chemical and biological systems were under investigation in modern research using this kind hybrid approach. In the hybrid approach—the classical and quantum-chemical MD modeling, the molecules are described as a set of spheres and springs where the spheres imitate classical particles and the spring—the interaction force fields governed by quantum chemistry laws. In this work the interaction of DNA molecule with a carbon nanotube (CNT) was studied using Tersoff potential. The dynamical configurations were built up for the DNACNT model at ambient temperatures. The observations reveal an encapsulation-like behavior of the DNA chain inside the CNT. A possible use of the DNA-CNT complex as a candidate material in drug delivery and related systems was discussed.

ACKNOWLEDGEMENTS

This work has been performed as part of collaboration between JINR (Russia), RIKEN (Japan), and Keio University (Japan). This work was jointly supported by the JSPS (Japan Society for the Promotion of Science) and the RFBR (Russian Foundation for Basic Research); Grant No. 13-04-92100. The MD simulations have been performed using computer software, hardware facilities, and cluster machines at the CICC (JINR), RICC (RIKEN), and the Yasuoka Laboratory of Keio University (Japan).

REFERENCES

1. D. S. D. Larsson, Y. Wang and D. van der Spoel, Biochemistry, Vol. 48, 2009, pp. 1006-1015. http://dx.doi.org/10.1021/bi801952f

2. E. G. Marklund, D. S. D. Larsson, D. van der Spoel, A. Patriksson and C. Caleman, Physical Chemistry Chemical Physics, Vol. 11, 2009, pp. 8069-8078.http://dx.doi.org/10.1039/b903846a

3. R. Friemann, D. S. D. Larsson, Y. Wang and D. van der Spoel, Journal of the American Chemical Society, Vol. 131, 2009, pp. 16606-16607.http://dx.doi.org/10.1021/ja902962y

4. D. van der Spoel, E. Marklund, D. S. D. Larsson and C. Caleman, Macromolecular Bioscience, Vol. 11, 2011, pp. 50-59. http://dx.doi.org/10.1002/mabi.201000291

5. Kh. Kholmurodov, "Molecular Simulation Studies in Material and Biological Sciences," Nova Science Publishers Ltd., 2007, 196 p.

6. Kh. Kholmurodov, "Molecular Simulation in Material and Biological Research," Nova Science Publishers Ltd., 2009, 155 p.

7. Kh. Kholmurodov, "Molecular Dynamics of Nanobistructures," Nova Science Publishers Ltd., 2011, 210 p.

8. Kh. Kholmurodov, "Models in Bioscience and Materials Research: Molecular Dynamics and Related Techniques," Book of International Workshop MSSMBS'12, Nova Science Publishers Ltd., 2013, 208 p.

9. J. Santa Lucia Jr. and D. Hicks, Annual Review of Biophysics and Biomolecular Structure, Vol. 33, 2004, pp. 415-440.http://dx.doi.org/10.1146/annurev.biophys.32.110601.141800

10. K. J. Breslauert, R. Franks, H. Blockers and L. A. Markyt, Proceedings of the National Academy of Sciences of the United States of America, Vol. 83, 1986, pp. 3746-3750.http://dx.doi.org/10.1073/pnas.83.11.3746

11. V. Freyre-Fonseca, et al., Toxicology Letters, Vol. 202, 2011, pp. 111-119.http://dx.doi.org/10.1016/j.toxlet.2011.01.025

12. B. Trouiller, R. Reliene, A. Westbrook, P. Solaimani and R. H. Schiestl, Cancer Research, Vol. 69, 2009, pp. 8784- 8789. http://dx.doi.org/10.1158/0008-5472.CAN-09-2496

13. Q. Saquib, A. A. Al-Khedhairy, M. A. Siddiqui, F. M. Abou-Tarboush, A. Azam and J. Musarrat, Toxicology in Vitro, Vol. 26, 2012, pp. 351-361. http://dx.doi.org/10.1016/j.tiv.2011.12.011

14. R. Dunford, A. Salinaro, L. Cai, N. Serpone, S. Horikoshi, H. Hidaka and J. Knowland, FEBS Letters, Vol. 418, 1997, pp. 87-90. http://dx.doi.org/10.1016/S0014-5793(97)01356-2

15. C. Srinivasan, Current Science, Vol. 94, 2008, pp. 300- 301.

16. T. A. Hilder and J. M. Hill, Current Applied Physics, Vol. 8, 2008, pp. 258-261.http://dx.doi.org/10.1016/j.cap.2007.10.011

17. T. A. Hilder and J. M. Hill, Micro & Nano Letters, Vol. 3, 2008, pp. 41-49.http://dx.doi.org/10.1049/mnl:20080008

18. Q. Chen, Q. Wang, Y. C. Liu, T. Wu, Y. Kang, J. D. Moore and K. G. Gubbins, Journal of Chemical Physics, Vol. 131, 2009, Article ID: 015101.

19. I. Ali, D. Marenduzzo and J. M. Yeomans, Physical Review Letters, Vol. 96, 2006, Article ID: 208102. http://dx.doi.org/10.1103/PhysRevLett.96.208102

20. I. Ali, D. Marenduzzo and J. M. Yeomans, Biophysical Journal, Vol. 94, 2008, pp. 4159-4164. http://dx.doi.org/10.1529/biophysj.107.111963

21. J. V. Veetil and K. Ye, Biotechnology Progress, Vol. 23, 2007, pp. 517-531.http://dx.doi.org/10.1021/bp0602395

22. S. Dhar, Z. Liu, J. Thomale, H. J. Dai and S. J. Lippard, Journal of the American Chemical Society, Vol. 130, 2008, pp. 11467-11476. http://dx.doi.org/10.1021/ja803036e

23. J. Tersoff, Physical Review B, Vol. 39, 1989, pp. 5566- 5568.http://dx.doi.org/10.1103/PhysRevB.39.5566

24. Kh. Kholmurodov, G. Aru and K. Yasuoka, Natural Science, Vol. 2, 2010, pp. 902-910.http://dx.doi.org/10.4236/ns.2010.28111

25. Kh. Kholmurodov, et al., Chemical Physics, Vol. 402, 2012, pp. 41-47. http://dx.doi.org/10.1016/j.chemphys.2012.04.002

26. K. A. Affholter, S. J. Henderson, G. D. Wignall, G. J. Bunick, R. E. Haufler and R. N. Compton, The Journal of Chemical Physics, Vol. 99, 1993, pp. 9224-9229.http://dx.doi.org/10.1063/1.465538

27. Y. B. Melnichenko, G. D. Wignall, R. N. Compton and G. Bakale, The Journal of Chemical Physics, Vol. 111, 1999, pp. 4724-4728. http://dx.doi.org/10.1063/1.479234

28. H. E. Smorenburg, R. M. Crevecoeur, I. M. de Schepper and L. A. de Graaf, Physical Review E, Vol. 52, 1995, pp. 2742-2752. http://dx.doi.org/10.1103/PhysRevE.52.2742

29. F. Migliardo, V. Magazù and M. Migliardo, Journal of Molecular Liquids, Vol. 10, 2004, pp. 3-6. http://dx.doi.org/10.1016/j.molliq.2003.08.010

30. T. Tomiyama, S. Uchiyama and H. Shinohara, Chemical Physics Letters, Vol. 264, 1997, pp. 143-148. http://dx.doi.org/10.1016/S0009-2614(96)01290-0

31. A. D. Bokare and A. Patnaik, Journal of Chemical Physics, Vol. 119, 2003, pp. 4529-4538.http://dx.doi.org/10.1063/1.1594177

32. W. Smith and T. R. Forester, Journal of Molecular Graphics, Vol. 14, 1996, pp. 136-141.http://dx.doi.org/10.1016/S0263-7855(96)00043-4

33. W. Smith, T. R. Forester and I. T. Todorov, "The DL_POLY 2 User Manual," STFC Daresbury Laboratory Daresbury, Warrington WA4 4AD Cheshire, UK, Version 2.19, 2008.

34. Y. Kang, Y. C. Liu, Q. Wang, T. Wu and W. J. Guan, Biomaterials, Vol. 30, 2009, pp. 2807-2815. http://dx.doi.org/10.1016/j.biomaterials.2009.01.024

35. Y. H. Xie and A. K. Soh, Materials Letters, Vol. 59, 2005, pp. 971-975. http://dx.doi.org/10.1016/j.matlet.2004.10.079

36. D. S. D. Larsson, L. Liljas and D. van der Spoel, PLOS Computational Biology, Vol. 8, 2012, Article ID: e100 2502. http://dx.doi.org/10.1371/journal.pcbi.1002502

37. B.-D. Chen, C.-L. Yang, M.-S. Wang and X.-G. Ma, Chinese Physics B, Vol. 21, 2012, Article ID: 083103.

38. W. Humphrey, A. Dalke and K. Schulten, Journal of Molecular Graphics, Vol. 14, 1996, pp. 33-38. http://dx.doi.org/10.1016/0263-7855(96)00018-5

Chapter 2

CYANOPOLYYNES AS ORGANIC MOLECULAR WIRES IN THE INTERSTELLAR MEDIUM

Raul G. E. Morales[1], Carlos Hernández[2]

[1]Centre for Environmental Sciences and Department of Chemistry Faculty of Sciences, Universidad de Chile, Santiago, Chile

[2]Department of Chemistry, Faculty of Basic Sciences Universidad Metropolitana de Ciencias de la Educación, Santiago, Chile

ABSTRACT

Cyanopolyynes ($H[C{\equiv}C]_n$-CN or $HC_{2n+1}N$, where $n = 1, 2, 3, \cdots, n$) are commonly observed in the interstellar medium (ISM) as well as in the envelopes of carbon-rich stars. These linear molecular structures can be well described with a one-dimensional conduction model, which considers the scattering processes of electrons through the charge transfer conduction bridge of the $H[C{\equiv}C]_n$-molecular wire containing the CN group as an electron-acceptor terminal unit. Therefore, our results using this model enable a better understanding of the longest molecules observed in interstellar space and provide new insight into why these particular cyanopolyynes reach a maximum length, such as is observed from astronomical experimental spectral data and cosmological chemical models. Dipole moments and geometrical parameters of these cyanopolyynes were obtained from ab initio molecular orbital calculations using the restricted Hartree-Fock approach and $6\text{-}311G^*$ basis set, in order to obtain the inner resistance as a new parameter of chemical reaction feasibility for this molecular series. Using this last molecular parameter, we have been able to analyze the possibility of identifying long molecular species that can be found under local thermodynamic equilibrium in some ISM such us $HC_{25}H$, $HC_{27}H$, and $HC_{29}N$, which have not been observed at present.

INTRODUCTION

New chemical models applied to the circumstellar envelope surrounding carbon-rich stars or particular interstellar mediums (ISM) are being developed to determine the presence and abundance of different molecular species [1-5]. Furthermore, the production of large molecules in heterogeneous astronomical environments is one of several theoretical astrochemical conjectures used to explain how complex molecular systems such as fullerenes in the ISM can originate from long linear molecular compounds such as cyanopolyynes.

All of the smaller members of the series have been detected in the ISM, particularly in circumstellar envelopes surrounding carbon-rich stars, corresponding to cold circumstellar shells, and they all have a conjugated, unsaturated carbon chain terminated at one end by an H atom and at the other by the CN group.

The smaller members of these series are well known at the laboratory, but under hard experimental conditions and extreme detection sensitivity, only its well known the rotational spectra up to $HC_{17}N$, as the longest members of the series [6]. However, HC_7N to $HC_{11}N$ were first discovered in the ISM by means of theoretical predictions of their rotational spectra and the subsequent detection of these spectra by radio telescopes. While $HC_{11}N$, the largest linear interstellar molecule was the last-discovered member of these cyanopolyynes series [7,8].

Since 1978, several spectral observations from various interstellar regions have reported these types of oligomeric structures, particularly polyynic wires involving the CN group [9,10]. However, cosmological and theoretical conjectures derived from observational considerations have systematically demonstrated the absence of molecular wires longer than five units.

These molecular structures, from a physicochemical point of view, can be seen as a special case of molecular systems characterized by a basic triple bond unit that is repeated in a linear sequence generating several oligomeric structures, generally between one and five units as a maximum length with a CN terminal group. Therefore, the molecular structure can be defined by an electron-donor (D) terminal group, an intermediate group constituted by an unsaturated carbon molecular bridge (W) and an electron-acceptor (A) terminal group, where an electronic charge transfer process from D to A through the linear molecular structure determines several particular characteristics of these D-W-A systems.

Whenever this kind of molecular wires has been a subject of long data research, particularly associated to the concept of molecular electronic device [11-14], previous studies on polyenic systems in our laboratory were based on the electronic conduction properties of conjugated oligomeric compounds of the D-W-A type [14-17]. These studies have permitted us to focus our attention on these interstellar cyanopolyynes.

Therefore, in the present work, we have started a physicochemical model that allows us to derive electronic conduction properties from the dipole moments, a classical parameter used in structural molecular spectroscopy studies, in order to understand why long molecular organic wire concentrations are not favored from a structural point of view. Thus, our molecular results are coherent to observational evidence based on the radial column density distribution of cyanopolyynes observed in the circumstellar envelope of carbon-rich stars such as the IRC+10216 system [4] analyzed in the present work. Furthermore, our analysis extended to the ISM under local thermodynamic equilibrium (LTE) anticipates eventual observational evidence of very low ranges of radial column densities of long species such as $HC_{25}H$, $HC_{27}H$, and $HC_{29}N$, which have not been observed at present.

THE ONE-DIMENSIONAL CONDUCTION MODEL

A few years ago, we developed a one-dimensional conduction model for D-W-A molecular systems [15,17]. Our model consider a scattering process of electrons through the internal charge transfer conduction bridge (W) to determine the molecular resistance and the molecular resistivity by means of the linear and non-linear contributions derived from the dipole moments and the p-molecular orbital bridge length (L) as fundamental physicochemical parameters.

This one-dimensional conduction model is based on a novel comprehension of the role of these oligomeric compounds, where the p-conduction channel of the molecular wires can be seen as a one-dimensional charge migration channel induced by the electron-acceptor group, assuming a typical scattering process of electrons as found in a metal wire system [18]. Thus, the inner conductance of this molecular wire determines the molecular resistance, and the final charge distribution determines the dipole moment. This model can be applied to any oligomeric molecular system that preserves the orientation of the dipole moment through the main axis of the molecular wire [17]. The molecular wire length and the dipole moment due to the ground-state chargetransfer process induced by the CN group are two fundamental parameters of the ground electronic state of these cyanopolyynes.

Therefore, the dipole moment of every nth oligomer can be represented as [17]

$$\mu_n = \mu_o + \mu_\infty \left\{ 1 - e^{-\gamma L(n)} \right\} \tag{1}$$

where μ_o is the dipole moment of the first compound of the oligomeric series (HCN) without a bridge unit (n = 0), μ_∞ is a molecular constant of the oligomeric series at the limit value for L®¥, $L(n)$ is the molecular wire length of the nth oligomer (-[C≡C]$_n$-), and g is the wire conduction constant of this oligomeric molecular series.

By considering a scattering process of the electronic flow through the molecular wire length(L), we obtain the atomic scale conductance $G=(2e^2/h)$ $f(T,R)$, where e is the electron charge, h is Planck's constant, T and R are the transmission and reflection probability factors, respectively, and $f(T,R)$ is a function of T and R given by $f(T,R)=T/R$, where T + R = 1.

If under this approach [17] we consider an electronic scattering transmission of the type $T=T_0e^{-\gamma L}$, with T_0 = 1, we obtain the molecular resistance $(R=1/G)$ according to:

$$R(L)=12.91\left[e^{gL}-1\right](k\Omega) \tag{2}$$

where $(h/2e^2)=12{,}910\,\Omega$ or 12.91 kW. However, to determine the linear molecular resistivity (ρ), this last equation can be expanded as a Maclaurin series,

$$R(L)=12.91\left[\gamma L+\frac{1}{2}!\gamma^2 L^2+\frac{1}{3}!\gamma^3 L^3+\cdots+\frac{1}{m}!\gamma^m L^m\right] \tag{3}$$

where the first term defines the linear contribution to the molecular resistance, given by $R_l(L)=12.91[\gamma L]\ (k\Omega)$, and the linear molecular resistivity follows as $\rho=R_l(S/L)$, where S is the molecular wire cross-section estimated to be 4.5 Å^2 [14].

RESULTS AND DISCUSSION

It is well known fact by experimentalist the inherent difficulties that these cyanopolyynes present in order to reach dipole moment measurements in vacuum. Additionally, molecular instability or hard experimental conditions have been observed in the laboratory synthesis of the larger members of the cyanopolyynes. However, these oligomeric compounds, from a geometrical and charge distribution perspective, can be well described by means of density-functional molecular-orbital calculations. Therefore, we used the SPARTAN'06 Quantum Mechanics Program (PC/x86), which employs the restricted Hartree-Fock (RHF) approach and the 6-311G* basis set, to determine the best molecular geometry optimization and the electronic charge distribution as a function of the length of the molecular wire for the oligomeric series under study. The molecular wire lengths and dipole moments of the cyanopolyynes, calculated using this theoretical approach are shown in Table 1. Figure 1 shows a plot of the dipole moment versus the molecular wire length for the cyanopolyynes.

Table 1: Molecular wire length and dipole moments of the cyanopolyynes

Molecular series	Formula	Molecular wire length[*] (Å)	Dipole moment[*] (debye)
H-C≡N	HCN	0.00	3.215
H-(C≡C)$_1$-C≡N	HC$_3$N	2.570	4.061
H-(C≡C)$_2$-C≡N	HC$_5$N	5.135	4.673
H-(C≡C)$_3$-C≡N	HC$_7$N	7.699	5.117
H-(C≡C)$_4$-C≡N	HC$_9$N	10.26	5.439
H-(C≡C)$_5$-C≡N	HC$_{11}$N	12.83	5.673
H-(C≡C)$_6$-C≡N	HC$_{13}$N	15.39	5.843
H-(C≡C)$_7$-C≡N	HC$_{15}$N	17.96	5.966
H-(C≡C)$_8$-C≡N	HC$_{17}$N	20.52	6.056
H-(C≡C)$_9$-C≡N	HC$_{19}$N	23.08	6.120
H-(C≡C)$_{10}$-C≡N	HC$_{21}$N	25.65	6.168
H-(C≡C)$_{11}$-C≡N	HC$_{23}$N	28.21	6.202
H-(C≡C)$_{12}$-C≡N	HC$_{25}$N	30.74	6.226
H-(C≡C)$_{13}$-C≡N	HC$_{27}$N	33.34	6.244
H-(C≡C)$_{14}$-C≡N	HC$_{29}$N	35.90	6.258

[*]RHF molecular orbital calculations in the 6.311G* basis set.

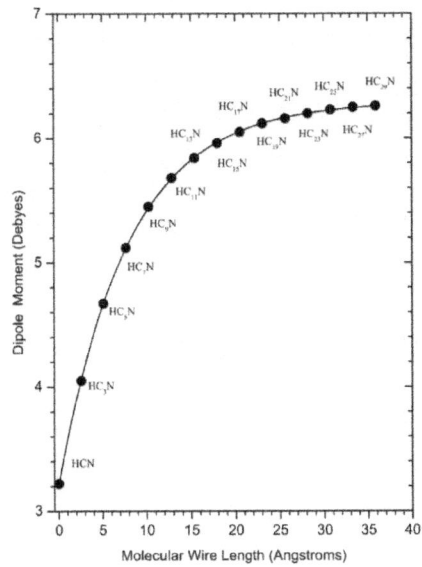

Figure 1: Dipole moments of the cyanopolyynes (H-[C≡C]$_n$- CN), n = 0 to 29, in the RHF approach using a 6-311G* basis set.

We can observe a functional dependence for both parameters according to the behavior expected in Equation (1). Therefore, the expected functional dependence between m_n and L has permitted us to make a best fit for the curve, where the final parameters of Equation (1) are the following: g = 0.125 ± 0.001 (Å^{-1}), with m_0 = 3.215 ± 0.006 (Debye), m_* = 3.077 ± 0.006 (Debye), and r^2 = 0.99996.

First, one can see how the dipole moment changes as a function of the molecular length. The dipole moments of the long species converge to 6.292 ± 0.012 Debye. Only insignificant changes are expected after $HC_{29}N$. This phenomenon shows the inherent and limited property of the CN group to be an electron acceptor group in this molecular wire. Reflection due to electronic charge transfer becomes more important than transmission in terms of the wire length. In other words, the CN group is not strong enough to stabilize a long molecular array.

The molecular wire g-conduction constant exhibits a behavior similar to that of other oligomeric compounds [17]. Now, based on the g-conduction constant derived from Figure 1, we can determine the electronic resistance as a function of the length of the molecular wire according to Equation (2). Table 2 depicts the molecular resistance of these cyanopolyynes under study. Furthermore, if we use the linear component of the molecular resistance, as can be seen in Equation (3), these cyanopolyynes have a linear resistivity of 72.6 mW cm, which is similar to other molecular wires previously reported [17] (see Table 3).

Table 2: Molecular resistances of the cyanopolyynes

Cyanopolyynes	Molecular resistance (kΩ)
HC_3N	4.9
HC_5N	11.6
HC_7N	20.9
HC_9N	33.6
$HC_{11}N$	51.3
$HC_{13}N$	75.5
$HC_{15}N$	109.0
$HC_{17}N$	154.9
$HC_{19}N$	218.2
$HC_{21}N$	305.8
$HC_{23}N$	426.0
$HC_{25}N$	589.2
$HC_{27}N$	820.5
$HC_{29}N$	1134.8

From our results, we can appreciate how this molecular wire series follows the one-dimensional conduction model very well. The molecular parameters, such as the geometry and dipole moments, appear to be quite accurate with the RHF approach and the 6-311G* basis set.

Table 3: Molecular wires linear resistivities

Material	$\rho\,[\mu\Omega\,\text{cm}]$
Interstellar molecular wire[a]	
$H[C\equiv C]_nCN$	72.6 ± 0.6
Molecular wires[b]	
$Me_2N(CH=CH)_nCHO$	104 ± 4
$Me_2N(C\equiv C)_nCHO$	75 ± 2
$Me_2NHC(C=C)_nCHCHO$	9.0 ± 0.2
Macroscopic wires[c]	
Cu (electrolytic)	1.7
Constantan	49
Hastelloy C	125
Polyacetylene (doped)[d]	100.0

[a]This work; [b]Morales & González [17]; [c]Lide [19]; [d]Kanatzidis [20].

Therefore, the molecular resistance can be used as a new criterion for chemical reactions feasibility of these linear molecular species. This is due to the fact that all species are synthesized from the same original reactants under the same conditions of LTE in bulk delimited by low-temperature dense-cloud regions. Thus, the molecular resistance to the internal charge transfer emerges as an indicator of the polarity strength of the ground state during the molecular formation process. Consequently, every new extension of the molecular wire of the cyanopolyynes due to -C≡Cunit gradually presents an additional resistance to the internal charge transfer process and, subsequently, their reaction feasibility necessarily decreases, due to the weak attraction force of the cyano electron-acceptor group.

Over 50 different molecules, including several oligomeric species, have been detected in IRC+10216 or CW Leonis, which is a carbon-rich star that is embedded in a thick dust envelope. This stellar source has been well studied both observationally and theoretically, and chemical models have carefully reproduced detailed radial distributions of these cyanopolyynes.

Thus, the report by Millar et al. [4] is an extension of the new standard model adapted to large molecules in a carbon-rich environment, including 3851 reactions and 407 gas-phase species. According to these authors the positive ion-molecule and neutral-neutral chemistry leading to the production of cyanopolyyne species was extended from that of the interstellar model so that cyanopolyynes as complex as $HC_{23}N$ were included. Reactions involving the radical CN and hydrocarbons were involved in the formation of cyanopolyynes, as they are in dense clouds, but reactions between the radical C_2H and smaller cyanopolyynes were far more important in the IRC+10216 envelope chemistry. Therefore, the role of the CN group in the chemical reactions associated with cyanopolyyne synthesis is determined by the electronic feasibility of the charge transfer between the molecular wire and the electron acceptor group. Thus, we can expect a relationship between the oligomeric species density distribution and the molecular resistance to the internal charge transfer of the molecular wires that determines the final probability of the molecular array density under LTE.

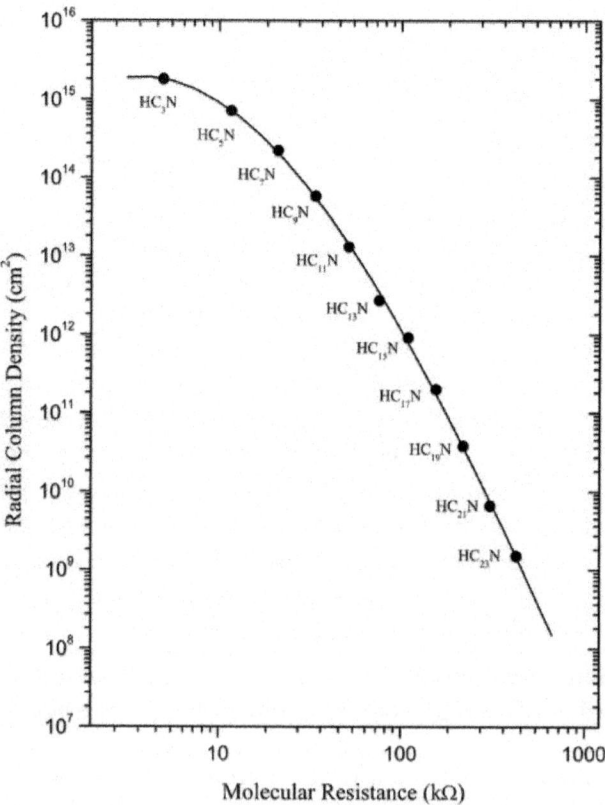

Figure 2: Theoretical radial column densities for cyanopolyynes (H-[C≡C]n-CN) in the circumstellar envelope of the carbon-rich star IRC + 10216 [4] versus molecular resistances calculated in the present work.

In particular, observed and theoretical radial column densities for the cyanopolyynes in IRC + 10216 are well represented by different complex astrophysical models. Therefore, we have used the best theoretical data determined at present by the Millar et al. model [4], which include HC_3N to $HC_{23}N$, and we have correlated these density data to our molecular resistances as can be seen in Figure 2. A very good logarithmic correlation between both data sets can be observed, given an appropriate basement to our hypothesis respect to the chemical reaction feasibility in terms of the molecular wire length.

Other significant fact emerges from the broad scale of cyanopolyyne densities and the molecular wire lengths observed, where this linear correlation presents a new practical method for predicting new species at very low concentrations. Thus, we have extended the observed trend by means of an extrapolation to the $HC_{25}N$, $HC_{27}N$, and $HC_{29}N$ molecular resistances in order to obtain the radial column densities expected for these molecules.

The extrapolated values are 2.8×10^8, 5.07×10^7, and 9.02×10^6 (cm^{-2}), respectively. These expected values can give new inside to observational registration based on large exposition time required for low concentrations, only possible to obtain from Atacama Large Millimeter Array (ALMA), the new international observatory project.

CONCLUSIONS

Our results present a new line of analysis respect to long molecular wires when chemical models are being used in a complex network of reaction schemes under LTE as well as in radial column density estimations of ISM. Effectively, our predictions about how the radial column densities of long molecular wires decline with increasing molecular resistance introduces a new criterion for radio searches of long molecules, which is important given the large amount of integration time that will be necessary for observing these particular molecular systems.

Our work also shows how the dipole moments of an oligomeric series increase with chain length up to reach a certain saturation point, as can be seen in this cyanopolyynes series. Therefore it is not possible to distinguish between long molecular wires which dipole moments are similar enough according to the error measurements. In our case, the $HC_{29}N$ define a natural detection limit of the present series.

This work motivates studies of other molecular wires in the ISM that are currently in progress in our laboratory.

ACKNOWLEDGEMENTS

The authors would like to acknowledge the Centre for Environmental Sciences of the University of Chile for financial support.

REFERENCES

1.   E. Herbst and E. F. van Dishoeck, "Complex Organic Interstellar Molecules," Annual Review of Astronomy and Astrophysics, Vol. 47, No. 1, 2009, pp. 427-480.doi:10.1146/annurev-astro-082708-101654

2.   J. R. Pardo, J. Cernicharo and J. R. Goicoechea, "Observational Evidence of the Formation of Cyanopolyynes in CRL 618 through the Polymerization of HCN," Astrophysical Journal, Vol. 628, No. 1, 2005, pp. 275-282. doi:10.1086/430774

3.   P. M. Woods, T. J. Millar, E. Herbst and A. A. Zijlstra, "The Chemistry of Protoplanetary Nebulae," Astronomy and Astrophysics, Vol. 402, No. 1, 2003, pp. 189-199.doi:10.1051/0004-6361:20030215

4.   T. J. Millar, E. Herbst and R. P. A. Bettens, "Large Molecules in the Envelope Surrounding IRC+10216," Monthly Notices of the Royal Astronomical Society, Vol. 316, No. 1, 2000, pp. 195-203.

5.   M. A. Cordiner and T. J. Millar, "Density-Enhanced Gas and Dust Shells in a New Chemical Model for IRC+ 10216," Astrophysical Journal, Vol. 697, No. 1, 2009, pp. 68-78.doi:10.1088/0004-637X/697/1/68

6.   M. C. McCarthy, W. Chen, M. J. Travers and P. Thaddeus, "Microwave Spectra of 11 Polyyne Carbon Chains," Astrophysical Journal, Vol. 129, No. 2, 2000, pp. 611-623.

7.   M. B. Bell, P. A. Feldman, S. Kwok and H. E. Matthews, "Detection of $HC_{11}N$ in IRC+10216," Nature, Vol. 295, 1982, pp. 389-391. doi:10.1038/295389a0

8.   M. B. Bell, P. A. Feldman, M. J. Travers, M. C. McCarthy, C. A. Gottlieband and P. Thaddeus, "Detection of HC11N in the Cold Dust Cloud TMC-1," Astrophysical Journal Letters, Vol. 483, No. 1, 1997, pp. L61-L64. doi:10.1086/310732

9.   H. W. Kroto, C. Kirby, R. M. Walton, L. W. Avery, N. W. Broten, J. M. MacLeod and T. Oka, "The Detection of Cyanohexatriyne, $H(CC_3)CN$, in Heiles' Cloud 2," Astrophysical Journal Letters, Vol. 219, No. 3, 1978, pp. L133-L137.

10.   D. Smith, "The Ion Chemistry of Interstellar Clouds," Chemical Reviews, Vol. 92, No. 7, 1992, pp. 1473-1485. doi:10.1021/cr00015a001

11.   F. L. Carter, "Molecular Electronic Devices," Marcel Dekker Inc., New

York, 1986.

12. C. Joachim and S. Roth, "Atomic and Molecular Wires," Kluwer Academic Publishers, Dordrecht, 1997. doi:10.1007/978-94-011-5882-4

13. D. K. James and J. M. Tour, "Molecular Wires," Topics in Current Chemistry, Vol. 257, 2005, pp. 33-62. doi:10.1007/b136066

14. C. Hernández and R. G. E. Morales, "Bridge Effect in Charge-Transfer Photoconduction Channels. 1. Aromatic Carbonyl Compounds," Journal of Physical Chemistry, Vol. 97, No. 45, 1993, pp. 11649-11651. doi:10.1021/j100147a016

15. R. G. E. Morales and C. González-Rojas, "Dipole Moments of Polyenic Oligomeric Systems. Part I. A OneDimensional Molecular Wire Model," Journal of Physical Organic Chemistry, Vol. 11, No. 12, 1998, pp. 853-   856.   doi:10.1002/(SICI)1099-1395(199812)11:12<853::AID-POC74>3.0.CO;2-Y

16. C. González and R. G. E. Morales, "Molecular Resistivities in Organic Polyenic Wires. I. A One-Dimensional Photoconduction Charge Transfer Model," Chemical Physics, Vol. 250, No. 3, 1999, pp. 279-284. doi:10.1016/S0301-0104(99)00335-3

17. R. G. E. Morales and C. González-Rojas, "Dipole Moments of Polyenic Oligomeric Systems. Part II. Molecular Organic Wire Resistivities: Polyacetylenes, Allenes and Polylines," Journal of Physical Organic Chemistry, Vol. 18, No. 9, 2005, pp. 941-944. doi:10.1002/poc.931

18. R. Landauer, "Electrical Resistance of Disordered OneDimensional Lattices," Philosophical Magazine, Vol. 21, No. 172, 1970, pp. 863-867. doi:10.1080/14786437008238472

19. D. R., Lide, "CRC Handbook of Chemistry and Physics," 85th Edition, CRC Press Inc., Boca Ranton, 2004, pp. 12-234.

20. M. G. Kanatzidis, "Polymeric Electrical Conductors," Chemical Engineering News, Vol. 68, No. 49, 1990, pp. 36-54. doi:10.1021/cen-v068n049.p036

Chapter 3

NUMERICAL SOLUTION OF LINEAR ORDINARY DIFFERENTIAL EQUATIONS IN QUANTUM CHEMISTRY BY SPECTRAL METHOD

Masoud Saravi[1] and Seyedeh-Razieh Mirrajei[2]

[1]Islamic Azad University, Nour Branch, Nour

[2]Education Office of Amol, Amol, Iran

INTRODUCTION

The problem of the structure of hydrogen atom is the most important problem in the field of atomic and molecular structure. Bahr's treatment of the hydrogen atom marked the beginning of the old quantum theory of atomic structure, and wave mechanics had its inception in Schrodinger 's first paper, in which he gave the solution of the wave equation for the hydrogen atom. Since the most differential equations concerning physical phenomenon could not be solved by analytical method hence, the solutions of the wave equation are based on polynomial (series) methods. Even if we use series method, some times we need an appropriate change of variable, and even when we can, their closed form solution may be so complicated that using it to obtain an image or to examine the structure of the system is impossible. For example, if we consider Schrodinger equation, i.e.,

$$\varphi'' + (2mEh^{-2} - \alpha^2 x^2)\varphi = 0,$$

we come to a three-term recursion relation, which work with it takes, at least, a little bit time to get a series solution. For this reason we use a change of variable such as

$$\varphi = e^{-\alpha x^2/2} f(x),$$

or when we consider the orbital angular momentum, it will be necessary to solve

$$\frac{d^2s}{d\theta^2} + cot\theta \frac{ds}{d\theta} + \left(\frac{c}{h^2} - \frac{m^2}{sin^2\theta}\right)s = 0.$$

As we can observe, working with this equation is tedious. Another two equations which occur in the hydrogen atom wave equations, are Legendre and Laguerre equations, which can be solved only by power series methods.

In next section, after a historical review of spectral methods we introduce Clenshaw method, which is a kind of spectral method, and then solve such equations in last section. But, first of all, we put in mind that this method can not be applied to atoms with more electrons. With the increasing complexity of the atom, the labour of making calculations increases tremendously. In these cases, one can use variation or perturbation methods for overcoming such problems.

HISTORICAL REVIEW

Spectral methods arise from the fundamental problem of approximation of a function by interpolation on an interval, and are very much successful for the numerical solution of ordinary or partial differential equations. Since the time of Fourier (1882), spectral representations in the analytic study of differential equations have been used and their applications for numerical solution of ordinary differential equations refer, at least, to the time of Lanczos.

Spectral methods have become increasingly popular, especially, since the development of Fast transform methods, with applications in problems where high accuracy is desired. Spectral methods may be viewed as an extreme development of the class of discretization schemes for differential equations known generally as the method of weighted residuals (MWR) (Finlayson and Scriven (1966)). The key elements of the MWR are the trial functions (also called expansion approximating functions) which are used as basis functions for a truncated series expansion of the solution, and the test functions (also known as weight functions) which are used to ensure that the differential equation is satisfied as closely as possible by the truncated series expansion. The choice of such functions distinguishes between the three most commonly used spectral schemes, namely, Galerkin, Collocation(also called Pseudospectral) and Tau version. The Tau approach is a modification of Galerkin method that is applicable to problems with non-periodic boundary conditions. In broad terms, Galerkin and Tau methods are implemented in terms of the expansion coefficients, where as Collocation methods are implemented in terms of physical space values of the unknown function.

The basis of spectral methods to solve differential equations is to expand the solution function as a finite series of very smooth basis functions, as follows

$$y_N(x) = \sum_{n=0}^{N} a_n \phi_n(x),$$

(1)

in which, one of our choice of ϕ_n, is the eigenfunctions of a singular Sturm-Liouville problem. If the solution is infinitely smooth, the convergence of spectral method is more rapid than any finite power of $1/N$. That is the produced error of approximation (1), when $N ® ¥$, approaches zero with exponential rate. This phenomenon is usually referred to as "spectral accuracy". The accuracy of derivatives obtained by direct, term by term differentiation of such truncated expansion naturally deteriorates. Although there will be problem but for high order derivatives truncation and round off errors may deteriorate, but for low order derivatives and sufficiently high-order truncations this deterioration is negligible. So, if the solution function and coefficient functions of the differential equation are analytic on[,] a b , spectral methods will be very efficient and suitable. We call function y is analytic on [,] a b if is infinitely differentiable and with all its derivatives on this interval are bounded variation.

CLENSHAW METHOD

In this section, we are going to introduce Clenshaw method. For this reason, first we consider the following differential equation:

$$Ly = \sum_{0}^{M} f_{M-i}(x)D^i y = f(x), \; x \in [-1,1],$$

(2)

$$By = C,$$

(3)

where $L = \sum_{0}^{M} f_{M-i}(x)D^i$, and f_i , $i = 0,1,...,M, f$, are known real functions of x, D^i

denotes i^{th} order of differentiation with respect to x ,B is a linear functional of rank M and $C \in \hat{A}^M$. Here (3) can be initial, boundary or mixed conditions. The basis of spectral methods to solve this class of equations is to expand the solution function, y , in (2) and (3) as a finite series of very smooth basis functions, as given below

$$y_N(x) = \sum_{n=0}^{N} a_n T_n(x),$$

(4)

where, $\{T_n(x)\}_0^N$ is sequence of Chebyshev polynomials of the first kind. By replacing y_N in (2), we define the residual term by $r_N(x)$ as follows

$$r_N(x) = Ly_N - f .$$

$$(5)$$

In spectral methods, the main target is to minimize $r_N(x)$, throughout the domain as much as possible with regard to (3), and in the sense of pointwise convergence. Implementation of these methods leads to a system of linear equations with N +1 equations and N +1 unknowns $a_0, a_1, ..., a_N$.

The Tau method was invented by Lanczos in 1938. The expansion functions $\phi_n (n = 1, 2, 3, ...)$ are assumed to be elements of a complete set of orthonormal functions. The approximate solution is assumed to be expanded in terms of those functions as $u_N = \sum_{n=1}^{N+m} a_n \phi_n$, where m is the number of independent boundary constraints $Bu_N = 0$ that must be applied. Here we are going to use a Tau method developed by Clenshaw for the solution of linear ODE in terms of a Chebyshev series expansion.

Consider the following differential equation:

$$P(x)y'' + Q(x)y' + R(x)y = S(x), x \in (-1,1),$$
$$y(-1) = \alpha , y(1) = \beta .$$

$$(6)$$

First, for an arbitrary natural number N, we suppose that the approximate solution of equations (6) is given by (4). Our target is to find $\underline{a} = (a_0, a_1, ..., a_N)^t$. For this reason, we put

$$P(x) \cong \sum_{i=0}^{N} \xi_i T_i(x),$$

$$Q(x) \cong \sum_{i=0}^{N} \gamma_i T_i(x),$$

$$R(x) \cong \sum_{i=0}^{N} \lambda_i T_i(x).$$

$$(7)$$

Using this fact that the Chebyshev expansion of a function $u \in L_w^2(-1,1)$ is

$u(x) = \sum_{k=0}^{\infty} \hat{u}_k T_k(x); \hat{u}_k = \frac{2}{\pi c_k} \int_{-1}^{1} u(x) T_k(x) w(x) dx$, we can find coefficients ξ_i, γ_i and λ_i as follows:

$$\xi_i = \frac{2}{\pi c_i} \int_{-1}^{1} \frac{P(x)T_i(x)}{\sqrt{1-x^2}} dx$$

$$\gamma_i = \frac{2}{\pi c_i} \int_{-1}^{1} \frac{Q(x)T_i(x)}{\sqrt{1-x^2}} dx$$

$$\lambda_i = \frac{2}{\pi c_i} \int_{-1}^{1} \frac{R(x)T_i(x)}{\sqrt{1-x^2}} dx,$$

$$(8)$$

where, $c_0 = 2$ and $c_i = 1$ for $i \geq 1$.

To compute the right-hand side of (8) it is sufficient to use an appropriate numerical integration method. Here, we use $(N+1)$- point Gauss-Chebyshev-Lobatto quadrature

$$x_j = \cos\frac{\pi j}{N}, w_j = \frac{\pi}{\tilde{c}_j N}, 0 \leq j \leq N,$$

where $\tilde{c}_0 = \tilde{c}_N = 2$ and $\tilde{c}_j = 1$ for $j = 1,2,...,N-1$.

Note that, for simplicity of the notation, these points are arranged in descending order, namely, $x_N < x_{N-1} < ... < x_1 < x_0$, with weights

$$w_k = \frac{\pi}{N}, \quad 1 \leq k \leq N-1,$$

$$= \frac{\pi}{2N}, \quad k = 0, k = N,$$

and nodes $x_k = \cos\frac{\pi k}{N}$, $k = 0,1,...,N$. That is, we put:

$$\xi_i \cong \frac{\pi}{N} \sum_{k=0}^{N''} P(\cos(\frac{k\pi}{N}))T_i(\cos(\frac{k\pi}{N})),$$

and using $T_i(x) = \cos(i\cos^{-1}x)$, we get

$$\xi_i \cong \frac{\pi}{N} \sum_{k=0}^{N''} P(\cos(\frac{k\pi}{N}))\cos(\frac{\pi ik}{N}),$$

where, notation å " means first and last terms become half .Therefore, we will have :

$$\xi_i \cong \frac{\pi}{N} \sum_{k=0}^{N''} P(\cos(\frac{k\pi}{N})) \cos(\frac{\pi ik}{N}),$$

$$\gamma_i \cong \frac{\pi}{N} \sum_{k=0}^{N''} Q(\cos(\frac{k\pi}{N})) \cos(\frac{\pi ik}{N}),$$

$$\lambda_i \cong \frac{\pi}{N} \sum_{k=0}^{N''} R(\cos(\frac{k\pi}{N})) \cos(\frac{\pi ik}{N}).$$

(9)

Now, substituting (4) and (9) in equations (6), and using the fact that

$$y'(x) \cong \sum_{m=0}^{N} a_m^{(1)} T_m(x), \ a_m^{(1)} = \frac{2}{c_m} \sum_{p=m+1}^{N} p a_p, \ m = 0,1,\dots, N-1, \ a_N^{(1)} = 0,$$

$$m+p = odd$$

$$y''(x) \approx \sum a_m^{(2)} T_m(x), \ a_m^{(2)} = \frac{1}{c_m} \sum_{p=m+2}^{N} p(p^2 - m^2) a_p, \ m = 0,1,\dots, N-2, \ a_{N-1}^{(2)} = a_N^{(2)} = 0,$$

$$m+p = even$$

we get

$$\sum_{i=0}^{N} \sum_{m=0}^{N} \xi_i a_m^{(2)} T_i(x) T_m(x) + \sum_{i=0}^{N} \sum_{m=0}^{N} \gamma_i a_m^{(1)} T_i(x) T_m(x) + \sum_{i=0}^{N} \sum_{m=0}^{N} \lambda_i a_m T_i(x) T_m(x) = S(x),$$

(10)

$$\sum_{i=0}^{N} a_i T_i(-1) = \alpha,$$

$$\sum_{i=0}^{N} a_i T_i(1) = \beta.$$

(11)

Now, we multiply both sides of (10) by $\dfrac{2}{\pi c_j} \dfrac{T_j(x)}{\sqrt{1-x^2}}$, and integrate from -1 to 1, we obtain

$$\frac{2}{\pi c_j} \sum_{i=0}^{N} \sum_{m=0}^{N} [\xi_i a_m^{(2)} + \gamma_i a_m^{(1)} + \lambda_i a_m] \int_{-1}^{1} \frac{T_i(x) T_m(x) T_j(x)}{\sqrt{1-x^2}} dx$$

$$= \frac{2}{\pi c_j} \int_{-1}^{1} \frac{S(x) T_j(x)}{\sqrt{1-x^2}} dx, j = 0, 1, ..., N-2,$$

(12)

where,

$$\int_{-1}^{1} \frac{T_i(x) T_m(x) T_j(x)}{\sqrt{1-x^2}} dx = \begin{cases} \pi & , i = m = j = 0 , \\ \frac{\pi}{2} \delta_{i,m} & , i+m>0, j=0, \\ \frac{\pi}{4}(\delta_{j,i+m} + \delta_{j,|i-m|}), & j>0, \end{cases}$$

(13)

with, $\delta_{i,j} = 1$, when $i = j$, and zero when $i \neq j$.

We can also compute the integrals in the right-hand side of (12) by the method of numerical integration using $N+1$-point Gauss-Chebyshev-Lobatto quadrature. Therefore, substituting (13) in (12) and using the fact that $T_i(\pm 1) = (\pm 1)^i$ equations (12) and (11) make a system of $N+1$ equations for $N+1$ unknowns, $a_0, a_1, ..., a_N$, hence we can find $(a_0, a_1, ..., a_N)^t$ from this system.

NUMERICAL EXAMPLES

As we mentioned the important problem in the field of atomic and molecular structure, is solution of wave equation for hydrogen atom. In this section we will solve Schrodinger, Legendre and Laguerre equations, which occur in the hydrogen atom wave equations, by Clenshaw method and observe the power of this method comparing with usual numerical methods such as Euler's or Runge-Kutta's methods. We start with Schrodinger's equation.

Example 1. Let us consider

$$\varphi'' + (2mEh^{-2} - \alpha^2 x^2)\varphi = 0.$$

Assume $\alpha = 2, mEh^{-2} = -1$, with $\varphi(0) = 1, \varphi(1) = e$. The exact solution is $\varphi(x) = e^{-x^2}$.

Here interval is chosen as [0,1], but using change of variable such as $t = \frac{x+1}{2}$ we can transfer interval [0,1] to [-1,1].

We solve this equation by Clenshaw method and compare the results for different values of N. The results for N=4, 7, 10, 13, respectively, were:

$1.660 \times 10^{-2}, 4.469 \times 10^{-5}, 5.901 \times 10^{-8}, 7.730 \times 10^{-11}$.

As we expected when N increases, errors decrease.

Example 2. Consider Legendre's equation given by

$(1 - x^2)y'' - 2xy' + \lambda(\lambda + 1)y = 0$.

As we know, this equation for $\lambda = 2$, and boundary conditions $y(\pm 1) = -2$ has solution $y(x) = 1 - 3x^2$. The results for N=4, 6, 10 were:

$5.5511 \times 10^{-17}, 2.2204 \times 10^{-16}, 2.7756 \times 10^{-17}$.

Since our solution is a polynomial then for N>3, we come to a solution with error very closed to zero. If such cases you find the error is not zero but closed to it, is because of rounding error. We must put in our mind that the results by this method will be good if the exact solution is a polynomial.

We end this section by solving Laguerre's equation.

Example 3. Consider

$xy'' + (1 - x)y' + \lambda y = 0$.

Suppose $\lambda = 2$ and boundary conditions are given by $y(-1) = \frac{7}{2}$, $y(1) = -\frac{1}{2}$.

The exact solution is $(x) = 1 - 2x + x^2/2$.

Here we have again a polynomial solution, so we expect a solution with very small error. We examined for different values of N such as N=2, 3 and get the results 0 and 3×10^{-17}, respectively.

Results in these examples show the efficiency of Clenshaw method for obtaining a good numerical result. In case of singularity, one can use pseudo-spectral method. Some papers also modified pseudo-spectral method and overcome the problem of singularity even if the solution function was singular.

REFERENCES

1.   Babolian. E, Bromilow. T. M, England. R, Saravi. M, 'A modification of pseudo-spectral method for olving linear ODEs with singularity', AMC 188 (2007) 1260-1266.

2.   Babolian. E, Delves. L .M, A fast Galerkin scheme for linear integro-differential equations, IMAJ. Numer. Anal, Vol.1, pp. 193-213, 1981.

3.   Canuto. C, Hussaini. M. Y, Quarteroni. A, Zang. T. A, Spectral Methods in Fluid Dynamics, Springer- Verlag,NewYork,1988.

4.   Delves. L. M, Mohamed. J. L, Computational methods for integral equations, Cambridge University Press, 1985.

5.   Gottlieb. D, Orszag. S. A, Numerical Analysis of Spectral Methods, Theory and Applications, SIAM,Philadelphia,1982.

6.   Lanczos. C, Trigonometric interpolation of empirical and analytical functions, J. Math. Phys. 17 (1938) 123-129.

7.   Levine. Ira N, Quantum Chemistry, 5th ed, City University of NewYork, Prentice-Hall Publication, 2000.

8.   Pauling. L, Wilson. E.B, Quantum Mechanic, McGraw-Hill Book Company, 1981.

Chapter 4

COMPOSITE METHOD EMPLOYING PSEUDOPOTENTIAL AT CCSD (T) LEVEL

Nelson Henrique Morgon

Universidade Estadual de Campinas Brazil

INTRODUCTION

Thermochemical data are among the most fundamental and useful information of chemical species which can be used to predict chemical reactivity and relative stability. Thus, it is not surprising that an important goal of computational chemistry is to predict thermochemical parameters with reasonable accuracy (Morgon, 1995a). Reliability is a critical feature of any theoretical model, and for practical purposes the model should be efficient in order to be widely applicable in estimating the structure, energy and other properties of systems, as isolated ions, atoms, molecules(Ochterski et al., 1995), or gas phase reactions(Morgon, 2008a).

What is the Importance of These Studies?

For instance gas phase reactions between molecules and ions, and molecules and electrons are known to be important in many scientifically and technologically environments. On the cosmic scale, the chemistry that produces molecules in interstellar clouds is dominated by ion-molecule reactions. Shrinking down to our own planet, the upper atmosphere is a plasma, and contains electrons and various positive ions. Certain anthropogenic chemical compounds (including SF_6 and perfluorocarbons) can probably not be destroyed within the troposhere or stratosphere, but may be removed by reactions with ions or electrons in the ionosphere. Recent years have seen a massive growth in the industrial use of plasmas, particularly in the fabrication of microelectronic devices and components. The chemistry within the plasma, much of which involves ion-molecule and electron-molecule reactions, determines the species that etch the surface, and hence the outcome and rate of an etching process. Much of the chemistry that is often labelled as 'organic' or 'inorganic' involves ion-molecule reactions, usually carried out in the presence of a solvent. For instance, S_N2 reactions, such as $OH^- + CH_3Cl$, fall into this category. To gain a clearer picture of how these (gas phase) reactions occur, it is advantageous to study them removed from the (very great) perturbations due to the solvent.

So, the need for thorough studies of ion-molecule and electron-molecule reactions are thus well established, ranging from the astrophysical origins of molecules, through the survival of the earth's atmosphere, to modelling the plasmas that underpin many advanced processing technologies. There is intrinsic interest too in the studies, as they help to explore the nature and progress of binary encounters between molecules and ions, and molecules and electrons. At the most basic level answers are needed to the following questions - how fast does a reaction proceed? and what are the products of the reaction? What determines which reactions occur? and what products are formed? Beyond these may come questions about the detailed dynamics of the reaction, such as how changing the energy of the reactants may influence the progress and outcome of the reaction.

Many powerful experimental techniques have been developed to give the basic data of reaction rate coefficients and products (usually just the identification of the ion product). These results are part of the raw data needed to understand and model the complex chemistry occurring in the diverse environments identified above. There is much information that is not directly available from the experimental data. This includes identification of the neutral products of a reaction, knowledge of the thermochemistry of the reaction, and characterization of the pathway that connects reactants to products. By invoking some general rules, the experimental observations can be used to provide partial answers. Thus the fast flow techniques that are used to provide much of the experimental data on ion-molecule reactions can only detect the occurrence of very rapid reactions ($k >= 10^{-12}$ cm^3 s^{-1}), which places an upper bound on the exothermicity of the reaction of +20 kJ mol^{-1}. In cases where there are existing reliable enthalpies of formation of each of the species in a proposed pathway to an observed ion product, this rule can test whether the suggested neutral products may be correct. In other cases, where the enthalpy of formation of just one of the species involved in a reaction (usually the product ion) is unknown, the observation of a specific reaction pathway can be used to place a bound on the previously unkown enthalpy of formation. Finally for reactions which are known to be exothermic, if the experimenal rate coefficient is observed to be less than the capture theory rate coefficient, then it is usual to conclude that there must be some bottleneck or barrier to the reaction.

What can Theoretical Calculations Add to the Experimental Data?

Three important and fundamental gas-phase thermochemical properties from a theoretical and experimental point of view are the standard heat of formation $(\Delta_f H^o_{gas})$, the electron (EA) and proton (PA) affinities. Thus, it is not surprising that an important goal of computational chemistry is to predict such thermochemical parameters with reasonable accuracy, which can be useful in the gas phase reaction studies. Proton transfer reactions are also of great importance in chemistry and in biomolecular processes of living organisms(Ervin, 2001). Absolute values of proton affinities are not always easy to obtain and are often derived from relative measurements with respect to reference molecules. Relative proton affinities are usually measured by means of high pressure mass spectrometry, with triple quadrupole and ion trap mass spectrometers (Mezzache et al., 2005) or using ion mobility spectrometry (Tabrizchi & Shooshtari, 2003). The importance and utility of the EA extend well beyond the regime of gas-phase ion chemistry. A survey of examples illustrates the diversity of areas in which electron affinities play a role: silicon, germanium clusters, interstellar chemistry, microelectronics, and so on.

The standard heat of formation, which measures the thermodynamic stability, is useful in the interpretation of the mechanisms of chemical reactions (Badenes et al., 2000). On the other hand, theoretical calculations represent one attempt to study absolute values of electron or proton affinity and other thermochemical properties (Smith & Radom, 1991). However, accurate calculations of these properties require sophisticated and high level methods, and great amount of computational resources. This is particularly true for atoms of the 2nd, 3rd, ..., periods and for calculating properties like the proton and electron affinity of anions. Gaussian-n theories (G1, G2, G3, and G4) (Curtiss et al., 1997; 1998; 2000; 2007) have given good results for properties like proton and electron affinities, enthalpies of formation, atomization energies, and ionization potentials. These theories are a composite technique in which a sequence of well-defined ab initio molecular orbital calculations is performed to arrive at a total energy of a given molecular species. There are other techniques that have been demonstrated to predict accurate thermochemical properties of chemical species, and are alternative to the Gaussian-n methods: the Correlation Consistent Composite Approach (ccCA)(DeYonker et al., 2006), the Multireference Correlation Consistent Composite Approach (MR-ccCA)(Oyedepo & Wilson, 2010), the Complete Basis Set Methods (CBS) and its versions: CBS-4M, CBS-Lq, CBS-Q, CBS-QB3, CBS-APNO(Montgomery Jr. et al., 2000; Nyden & Petersson, 1981; Ochterski et al., 1996; Peterson et al., 1991), and Weizmann Theories (W1 to W4)(Boese et al., 2004; Karton et al.,

2006; Martin & De Oliveira, 1999; Parthiban & Martin, 2001). Recently, we have implemented and tested a pseudopotential to be used with the G3 theory for molecules containing first-, second-, and non-transition third-row atoms (G3CEP) (Pereira et al., 2011). The final average total absolute deviation using this methodology and the all-electron G3 were 5.39 kJ mol^{-1} and 4.85 kJ mol^{-1}, respectively. Depending on the size of the molecules and the type of atoms considered, the CPU time was drastically decreased.

COMPUTATIONAL METHODS

In this chapter we have developed a computational model similar to version of the G2(MP2, SVP) theory (Curtiss et al., 1996). Both theories are based on the additivity approximations to estimate the high level energy for the extended function basis set. While G2(MP2,SVP) is based on the additivity approximation to estimate the QCISD(T) energy for the extended 6-311+G (3df,2p) basis set: E[QCISD(T)/ 6-311+G(3df,2p)] E[QCISD(T)/6-31G(d)] + E[MP2/6-311+G(3df,2p)] - E[MP2/6-31G(d)]. Our methodology employs CCSD(T) energies in addition to the the the valence basis sets adapted for pseudopotential (ECP) (Stevens et al., 1984) using the Generator Coordinate Method (GCM) procedure (Mohallem & Dreizler, 1986; Mohallem & Trsic, 1985). The present methodology which relies on small basis sets (representation of the core electrons by ECP) and an easier and simpler way for correcting the valence region (mainly of anionic systems) appears as an interesting alternative for the calculation of thermochemical data such as electron and proton affinities or heat of formation for larger systems.

Development of Basis Sets

The GCM has been very useful in the study of basis sets(Morgon, 1995a;b; 2006; 2008b; 2011; Morgon et al., 1997). It considers the monoelectronic functions $\psi(1)$ as an integral transform,

$$\psi(1) = \int_0^\infty f(\alpha)\, \phi(\alpha, 1)\, d\alpha$$

(1)

where f(α) and φ(α, 1) are the weight and generator functions respectively (gaussian functions are used in this work), and α is the generator. The existence of the weight functions (graphical display of the linear combination of basis functions) is an essential condition for the use of GCM. Analysis of the behavior of the weight functions by the GCM permits the atomic basis set to be adapted in such a way as to yield a better description of the core electrons (represented by ECP) and the valence orbitals (corrected by addition of the extra diffuse functions), in the molecular environment. With the exception of some simple systems the analytical expression of the weight functions is unknown. Thus, an analytical solution of the integral transform (Eq. 1) is not viable in most cases, and suggests the need of numerical techniques to solve Eq. 1 (Custodio, Giordan, Morgon & Goddard, 1992; Custodio, Goddard, Giordan & Morgon, 1992). The solution can be carried out by an appropriate choice of discrete points on the generate coordinate, represented by:

$$\alpha_{i,(k)} = exp[\Omega_{o,(k)} + (i-1) \cdot \Delta\Omega_{(k)}], \quad i = 1, 2, 3, \ldots N_{(k)} \tag{2}$$

The discretization of the set is defined by the following parameters: an initial value (Ω_o), an increment $(\Delta\Omega)$, and by the number of primitives used (N) for a given orbital k (s, p, d, ...). The search for the best representation is obtained using the total energy of the electronic ground state as the minimization criterion. The SIMPLEX search method (Nelder & Mead, 1965) can be adapted to the any electronic structure program to provide the minimum energy of the ground state of the atom corresponding to the optimized discretization parameters.

The basic procedure consists of the following steps:

(a)　search of the optimum discretized parameter set for the atoms using the GCM for variation on the generator coordinate space. The core electrons are represented by a pseudopotential. The discretization parameters $(\Omega_o$ and $\Delta\Omega)$ are defined with conjunction with this ECP;

(b)　the minimum energy criterion is observed and the characteristics of the atomic orbital weight functions are analyzed;

(c)　extra functions (polarizaton or diffuse funcions - s, p, d and f) are obtained by observing the convergent behavior of the weight functions of the outer atomic orbitals (s and p). These extra functions are needed for the correct description of the electronic distribution in an anion (diffuse character of electronic cloud).

To the heavy atoms f type polarization functions are not available in these valence basis sets for this kind of ECP(Stevens et al., 1984). So, it was need to define the value of these f functions for Br and I atoms. The determination of the best value was carried out considering the smaller difference between the PAs (experimental and theoretical) values considering the Br− and I− anions. The f exponent values found are 0.7 and 0.3 for Br and I atoms, respectively. In fact two sets of basis functions are used, a small basis and a larger basis, with extra diffuse and polarization functions, B0 and B1, respectively. Calculations with basis B1 are naturally much more expensive than those employing basis B0, so it is important to have computational schemes that perform the minimum number of calculations using basis B1.

For instance, the B0 basis set is defined as: (31) for H; (311/311) for C, O, F, S, and Cl; and (411/411) Br and I. For more refined energy calculations (B1 basis set are used), this set was augmented with additional diffuse and polarization functions (p for H and s, p, d, and f for heavy atoms) to yield a (311/11) set for H; (311/311/11/1) for C, O, F, S, and Cl; and (411/411/11/1) for Br and I atoms.

Molecular Calculations

In many problems to be addressed by electronic structure methodology, high accuracy is of crucial importance. In order to obtain energies that may approach chemical accuracy ($\approx$10 kJ.mol−1) calculations must take account of electron correlation. The 'ideal' methodology would have been a multi-reference configuration interaction all electron calculation with several large, flexible basis sets to enable extrapolation to the complete basis set limit. Depend on the size of the systems, performing accurate calculations (methodology and basis sets) represents a significant challenge. Morgon et al. (Morgon, 1998; Morgon et al., 1997; Morgon & Riveros, 1998) have been developing techniques to tackle such problems. These are centered around the use of effective core potentials, in which the inner electrons are represented by an effective potential derived from calculations on atoms. The electronic wavefunction itself then only contains the outer electrons.

The procedure to the molecular calculations employing this methodology is:

(a)   optimization of the molecular geometries and vibrational analysis are carried out at HF/B0 level. The harmonic frequencies confirm that the stationary points correspond to minima and are used to compute the zero-point energies;

(b)   further optimization is carried out at MP2/B0 level;

(c)   at the MP2 equilibrium geometry corrections to the total energies are performed at higher level of theory. First, this is carried out at CR-CCSD[T]/B0 level (Completely Renormalized Coupled-Cluster with Single and Double and Perturbative Triple excitation) (Kowalski & Piecuch, 2000) (or at CCSD(T)/B0 level for EA calculations), and later by addition of extra functions (s, p, d, and f) at MP2/B1 level.

Thus, these results coupled to additive approximations for the energy yield an effective calculation at a high level of theory,

$$\approx E[CR-CCSD[T]/(B1)] = E[CR-CCSD[T]/(B0)] +$$

$$+ E[MP2/(B1)] - E[MP2/(B0)] + ZPE[HF/(B0)] * scal \tag{3}$$

where scal (0.89) is the scaling factor on the vibrational frequencies. The CR-CCSD[T] (Completely Renormalized Coupled-Cluster with Single and Double and Perturbative Triple excitation) methodology refers to size-extensive left eigenstate completely renormalized (CR) coupled-cluster (CC) singles (S), doubles (D), and noniterative triples (T). This approach is abbreviated as CR-CCL and is appropriately described by Piecuch(Piecuch et al., 2002) and Ge(Ge et al., 2007).

An alternative model was developed for the study of heat of formation. This model employs valence basis sets aug-CCpVnZ (n = 2, 3, and 4) (Dunning, 1989). These basis sets were adapted to ECP using the GCM and are identified by ECP+ACCpVnZm (m = modified). The energies are obtained through the extrapolations to the complete basis set limit (CBS) using Peterson mixed exponential/Gaussian function extrapolation scheme (Feller & Peterson, 1999).

$$E(MP2) = E_{CBS} + B\,exp[-(x-1)] + C\,exp[-(x-1)^2] \tag{4}$$

where x = 2, 3, and 4 come from ECP+ACCpV2Zm, ECP+ACCpV3Zm and ECP+ACCpV4Zm energies, respectively.

For this electronic property, molecular calculations consist of:

(a)   optimization of the molecular geometries and vibrational analysis are carried out at HF/ECP+ACC2Zm level. The harmonic frequencies are employed to characterize the local minima and to compute the zero-point energies;

(b)   further optimization is carried out at MP2/ECP+ACC2Zm level;

(c)   at the MP2 equilibrium geometry corrections to the total energies are performed at higher level of theory. First, at CR-CCL/ECP+ACC2Zm level, and calculations by addition of extra functions (s, p, d, and f) at MP2/ECP+ACC3Zm and MP2/ECP+ACC4Zm levels. The E[MP2/ECP+ACC5Zm] is estimated throught the Eq. 4.

(d)   Finally, the results are coupled through additive approximations, and the energy corresponds to an effective calculation at a high level of theory,

$$\approx E[CR-CCL/ECP+ACC5Zm] = E[CR-CCL/ECP+ACC2Zm] + $$
$$+ E[MP2/ECP+ACC5Zm] - E[MP2/ECP+ACC2Zm] - $$
$$+ ZPE[HF/ECP+ACC2Zm]*scal + E(HLC) \tag{5}$$

where scal is the scaling factor on ZPE. The method also includes an empirical higher-level correction (HLC) term. This term is given by either Eq. 6 or Eq. 7 depending on whether the species is a molecule or an atom:

$$HLC_{molec.} = -C\cdot n_\beta - D\cdot(n_\alpha - n_\beta) \tag{6}$$

$$HLC_{atom} = -A \cdot n_\beta - B \cdot (n_\alpha - n_\beta)$$

(7)

In these equations, n_α and n_β are the numbers of α and β electrons, respectively. The parameters A (4.567mH), B (2.363mH), C (4.544mH), and D (2.337mH) were obtained by fitting to the experimental data of heats of formation. It is important to note that since it only depends on the number of α and β valence electrons, the HLC cancels entirely from most reaction energies, except when the reactions involve a mixture of atoms and molecules (as in heats of formation and bond dissociation energies) and/or when spin is not conserved (Lin et al., 2009). It should be also noted that the absolute values of the calculated energies have no real significance, as no common energy zero for different atoms has been used. This arises from how the effective core potentials are constructed. When differences are formed, the differences between the zeros cancel, to leave for instance the difference in energy between the products and reactants of a reaction.

Additionally, an alternative approach for molecular geometry optimization and harmonic frequencies calculation can be considered through the use of the DFT (B3LYP, M06, ...).

RESULTS AND DISCUSSION

Basis Sets

The existence of the weight functions (graphic representations of the linear combination of the atomic orbitals) is the fundamental condition to use the GCM. The analysis of the behavior of the weight functions by the GCM allows the fitting of the atomic basis sets in order to get a better description of the electrons in the molecular environment. The analysis of the weight functions of the outermost atomic orbitals suggested the need for improvements of the basis sets for the heavy atoms. Observing the plots of the weight functions of the outermost orbitals it is possible to establish the best fit of the basis sets. Using the atom of Cl as an example, the representation of the weight function of the atomic orbital 3s is shown in Fig. 1. The continuous line represents the plot of the weight function of the original primitive basis set for the all electron system. The dashed line represents the same weight function obtained using the pseudopotential. This figure shows that the 10 inner functions (large α) have a contribution close to zero towards the description of the weight function. The vertical solide line cutoff of the basis set indicates precisely where the pseudopotential starts to represent the core atomic region.

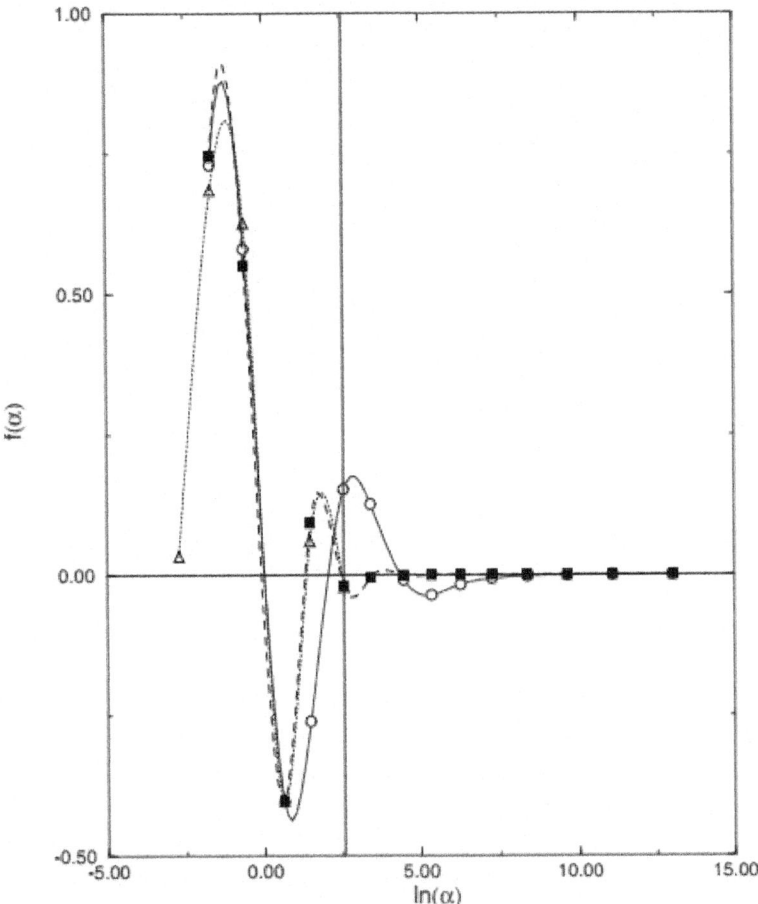

Figure 1: Weight functions for the 3s atomic orbital of Cl in systems with all electrons (ae, continuous line), with the pseudopotential (pp, dashed line), with the addition of one diffuse function plus pseudopotential (dif+pp, dot-dashed line) and the cutoff point line represented by the vertical solid line.

Proton Affinity

Proton affinity is a very sensitive property of the electronic structure and it is appropriate to test our methodology. Table 1 shows the results of the acidity (in kJ.mol^{-1}) for a set of anions systems. These results were obtained using the B0 and B1 basis sets and Eq. 3. A comparison is also presented with experimental results. One can observe that our theoretical results are very close to the experimental errors (well within 5 kJ.mol^{-1}) with root mean square deviation of 4.14 kJ mol^{-1} .

Table 1: Proton Affinity (kJ.mol^{-1}) calculated with the method given by Eq. 3 and comparison with experimental values

System	PA$_{Calc.}$	PA$_{Exp.}^a$
F$^-$	1556.17 (-2.17)	1554.0
CH$_2$F$^-$	1710.21 (0.79)	1711 ± 17
Cl$^-$	1395.56 (-0.56)	1395.0
CH$_2$Cl$^-$	1658.18 (-1.18)	1657 ± 13
Br$^-$	1354.80 (-1.80)^b	1353.0 ± 8.80
CHClBr$^-$	1563.4 (0.82)	1560.0 ± 13.00
CH$_2$Br$^-$	1640.54 (2.46)	1643 ± 13
I$^-$	1315.10	1315.0
CH$_2$I$^-$	1617.47 (-1.47)	1616 ± 21
Acetamide,2,2,2-trichloride	1493,6	1436 ± 8.8
Formamide-N,N-dimethyl	1671.34 (-1.34)	1670.0 ± 17.00
Formamide	1510.13 (-5.13)	1505.0 ± 8.80
methyl cyclopentanol	(2.88)	1559.00 ± 8.40
2,methyl cyclopentanol	(2.99)	1558.00 ± 8.40
CH$_2$CHCHI$^-$	1555.24 (-4.24)	1551.0 ± 8.8
δ_{rmsd}	4.14	

^{a}Experimental values from NIST Webbase - http://webbook.nist.gov/chemistry/. ^{b} (PA$_{Exp}$. - PA$_{Calc}$.).

Electron Affinity

In the Table 2 are the electron affinities calculated with the CCSD(T)/B1 energy from Eq. 3. It also shows a comparison between our results and the experimental values, where experimental data are available. The root mean square deviation (δ_{rmsd}) calculated is 0.15 eV. The use of pseudopotential is competitive, mainly in systems containing S, Cl, and Br atoms. The computational time is almost constant for analogous systems with Cl and Br atoms, because in these cases we have an equal number of outer electrons. In calculations involving all electrons the computational performance is totally different and increases with the number of electrons. The CR-CCSD[T]/ECP computational demand is decreased by 10% when compared with all-electron calculations. For molecules containing Cl, Br or I atom the time is drastically decreased (Morgon, 2006).

Heat of Formation

In the Table 3 are the heats of formation calculated with the CCR-CL/ECP+ACC5Zm from Eq. 6. It also shows a comparison between our results and the experimental values, where experimental data are available. The average error using this methodology with respect to experimental results is closer to 10 kJ.mol^{-1} .

Table 2: Electron Affinity (eV) calculated with the method given by Eq. 3, and comparison with experimental values

System	$EA^a_{Calc.}$	$EA^b_{Exp.}$
Br	3.36 (-)[c]	-
Br_2	2.53 (-0.11)	2.42
CBr_3	2.49 (0.08)	2.57 ± 0.12
CCl_3	2.22 (-0.05)	2.17 ± 0.10
CF_3	1.74 (0.08)	1.82 ± 0.05
CF_3COO	4.56 (-0.10)	4.46 ± 0.18
$CH_2=CH$	0.71 (-0.04)	0.67 ± 0.02
$CH_2=CHCHBr$	0.96 (-)	-
$CH_2=CHCHF$	0.50 (-)	-
$CH_2=CCH_3$	0.69 (-0.13)	0.56 ± 0.19
$CH_2=NO_2$	2.40 (0.07)	2.48 ± 0.01
CH_2Br	0.91 (-0.12)	0.79 ± 0.14
CH_2BrCOO	3.99 (-0.01)	3.98 ± 0.16
CH_2Cl	0.70 (0.04)	0.74 ± 0.16
CH_2ClCOO	4.02 (-0.11)	3.91 ± 0.16
CH_2F	0.17 (0.08)	0.25 ± 0.18
CH_2FCOO	4.55 (-0.76)	3.79 ± 0.16
CH_3	0.08 (0.00)	0.08 ± 0.03
CH_3CH_2O	1.75 (-0.03)	1.71
CH_3CH_2S	2.02 (-0.07)	1.95
CH_3O	1.53 (0.05)	1.57
CH_3S	1.75 (0.12)	1.87
$CHBr_2$	1.74 (-0.03)	1.71 ± 0.08
$CHBr_2COO$	4.37 (-0.11)	4.26 ± 0.16
$CHClBr$	1.64 (-0.17)	1.47 ± 0.04
CHF_2COO	4.25 (-0.10)	4.15 ± 0.16
Cl	3.60 (0.01)	3.61
Cl_2	2.55 (-0.15)	2.40 ± 0.20
F	3.34 (0.06)	3.40
F_2	2.92 (0.08)	3.00 ± 0.07
Formamide-N,N-dimethyl	0.65 (-)	-
Formamide	3.34 (-)	-
HS	2.27 (0.05)	2.32
NH_2	0.70 (0.07)	0.77 ± 0.01
OH	1.77 (0.06)	1.83
iPrO	1.87 (0.01)	1.87
nPrO	1.74 (0.05)	1.79 ± 0.03
δ_{rmsd}	0.15	

[a]The higher level correction (HLC) from G3 theory was added to the final energy.

[b]Experimental values from NIST Webbase - http://webbook.nist.gov/chemistry/. [c] ($EA_{Exp.}$ - $EA_{Calc.}$).

Table 3: Heats of formation (in kJ mol^{-1}) with the method given by Eq. 6, and comparison with experimental values

Molecule	Point Group	Ground State	$\Delta_f H^o_{gas}$ calc	$\Delta_f H^o_{gas}$ exp[a]
OF	$C_{\infty v}$	$^2\Pi$	110.999 (2.21)[c]	108.78
OF$_2$	C_{2v}	$^1A_1$	28.582 (4.06)	24.52
OCl	$C_{\infty v}$	$^2\Pi$	103.482 (2.26)	101.22
OCl$_2$	C_{2v}	$^1A_1$	92.341 (4.48)	87.86
SF	$C_{\infty v}$	$^2\Pi$	8.661 (-4.31)	12.97
SF$_2$	C_{2v}	$^1A_1$	-289.573 (7.08)	-296.65
SF$_3$	C_s	$^2A'$	-512.605 (-9.57)	-503.03
SF$_4$	C_{2v}	$^1A_1$	-775.362 (-12.20)	-763.16
SF$_5$	C_{4v}	$^2A_1$	-887.770 (20.68)	-908.45
SF$_6$	O_h	$^1A_{1g}$	-1218.986 (1.48)	-1220.47
SCl	$C_{\infty v}$	$^2\Pi$	153.843 (-2.62)	156.47
SCl$_2$	C_{2v}	$^1A_1$	-19.262 (-1.69)	-17.57
SCl$_3$	C_s	$^2A'$	22.346 (-14.47)	36.82[b]
SCl$_4$	C_{2v}	$^1A_1$	-15.301 (-12.38)	-2.92[b]
SCl$_5$	C_{4v}	$^2A_1$	-38.76 (-2.78)	-35.98[b]
SCl$_6$	O_h	$^1A_{1g}$	-87.54 (-4.7)	-82.84[b]

[a]Experimental values from NIST Webbase - http://webbook.nist.gov/chemistry/.

[b] Ref. (Ditter & Niemann, 1982).

[c] ($\Delta_f H_{gas}$ Calc. - $\Delta_f H_{gas}$ Exp.).

CONCLUSIONS

The proton and electron affinities and the heats of formation of some simple systems obtained by the procedure outlined in this paper are in very good agreement with experimental values. These results can be compared with those obtained by sophisticated and computationally more expensive calculations. ECP-based methods have been shown to be powerful, and of affordable computational cost for the systems addressed in this work. This is due to three features:

1)   the number of steps employed during the calculations,

2)   the smaller basis sets used in our methodology, and

3)   the use of ECP.

The use of adapted basis functions for atoms by the Generator Coordinate Method along with the use of the pseudopotential allows a high quality calculation at a lower computational cost. The present methodology - Eqs. 3 and 6, which relies on small basis sets (representation of the core electrons by ECP) and an easier and simpler way for correcting the valence region (mainly of anionic systems) appears as an interesting alternative for the calculation of thermochemical data such as electron and proton affinities and enthalpies of formation for larger systems. The CCSD(T)/B1 method have been shown to be powerful, and of affordable computational cost for the systems containing atoms of the 2nd and 3rd periods.

ACKNOWLEDGMENTS

I would like to thank the computational facilities of Chemistry Institute at UNICAMP and the financial support from Conselho Nacional de Desenvolvimento Científico e Tecnológico (CNPq) and Fundação de Amparo à Pesquisa de São Paulo (FAPESP).

REFERENCES

1. Badenes, M. P., Tucceri, M. E. & Cobos, C. J. (2000). Zeitschrift für Physikalische Chemie 214: 1193.

2. Boese, A. D., Oren, M., Atasoylu, O., Martin, J. M. L., Kállay, M. & Gauss, J. (2004). J. Chem. Phys. 120: 4129–4141.

3. Curtiss, L. A., Raghavachari, K., Redfern, P. C. & Pople, J. A. (1997). J. Chem. Phys. 106: 1063.

4. Curtiss, L. A., Raghavachari, K., Redfern, P. C. & Pople, J. A. (1998). J. Chem. Phys. 109: 7764–7776.

5. Curtiss, L. A., Raghavachari, K., Redfern, P. C. & Pople, J. A. (2000). J. Chem. Phys. 112: 7374.

6. Curtiss, L. A., Redfern, P. C. & Raghavachari, K. (2007). J. Chem. Phys. 126: 084108.

7. Curtiss, L. A., Redfern, P. C., Smith, B. J. & Radom, L. (1996). J. Chem. Phys. 104: 5148.

8. Custodio, R., Giordan, M., Morgon, N. H. & Goddard, J. D. (1992). Int. J. Quantum Chem. 42: 411.

9. Custodio, R., Goddard, J. D., Giordan, M. & Morgon, N. H. (1992). Can. J. Chem. 70: 580.

10. DeYonker, N. J., Cundari, T. R. & Wilson, A. K. (2006). J. Chem. Phys. 124: 114104.

11. Ditter, G. & Niemann, U. (1982). Phillips J. Res. 37: 1.

12. Dunning, T. H. (1989). J. Chem. Phys. 90: 1007.

13. Ervin, M. K. (2001). Chem. Rev. 101: 391.

14. Feller, D. & Peterson, K. A. (1999). J. Chem. Phys. 110: 8384.

15. Ge, Y., Gordon, M. S. & Piecuch, P. (2007). J. Chem. Phys. 174: 174106.

16. Karton, A., Rabinovich, E., Martin, J. M. L. & Ruscic, B. (2006). J. Chem. Phys. 125(14): 144108. Kowalski, K. & Piecuch, P. (2000). J. Chem. Phys. 18: 113.

17. Lin, C. Y., Hodgson, J. L., Namazian, M. & Coote, M. L. (2009). J. Phys. Chem. A 113: 3690. Martin, J. M. L. & De Oliveira, G. (1999). J. Chem. Phys. 111: 1843–1856.

18. Mezzache, S., Bruneleau, N., Vekey, K., Afonso, C., Karoyan, P. Fournier, F. & Tabet, J.-C. (2005). J. Mass Spectrom. 40: 1300.

19. Mohallem, J. R. & Dreizler, R. M. Trsic, M. (1986). Int. J. Quantum Chem. - Symp. 20: 45.

20. Mohallem, J. R. & Trsic, M. (1985). Z. Phys. A - Atoms and Nuclei 322: 538.

21. Montgomery Jr., J. A., Frisch, M. J., Ochterski, J. W. & Petersson, G. A. (2000). J. Chem. Phys. 112: 6532.

22. Morgon, N. H. (1995a). J. Phys. Chem. A 99: 17832.

23. Morgon, N. H. (1995b). J. Phys. Chem. 99: 17832.

24. Morgon, N. H. (1998). J. Phys. Chem. A 102: 2050.

25. Morgon, N. H. (2006). Int. J. Quantum Chem. 106: 2658.

26. Morgon, N. H. (2008a). J. Braz. Chem. Soc. 19: 74.

27. Morgon, N. H. (2008b). Int. J. Quantum Chem. 108: 2454.

28. Morgon, N. H. (2011). Int. J. Quantum Chem. 111: 1555–1561.

29. Morgon, N. H., Argenton, A. B., Silva, M. L. P. & Riveros, J. M. (1997). J. Am. Chem. Soc. 119: 1708.

30. Morgon, N. H. & Riveros, J. M. (1998). J. Phys. Chem. A 102: 10399.

31. Nelder, J. A. & Mead, R. (1965). Computer J. 7: 308.

32. Nyden, M. R. & Petersson, G. A. (1981). J. Chem. Phys. 75: 1843.

33. Ochterski, J. W., Petersson, G. A. & Montgomery Jr., J. A. (1996). J. Chem. Phys. 104: 2598.

34. Ochterski, J. W., Petersson, G. A. & Wiberg, K. B. (1995). J. Amer. Chem. Soc. 117: 11299.

35. Oyedepo, G. A. & Wilson, A. K. (2010). J. Chem. Phys. 114: 8806.

36. Parthiban, S. & Martin, J. M. L. (2001). J. Chem. Phys. 114: 6014–6029.

37. Pereira, D. H., Ramos, A. F., Morgon, N. H. & Custodio, R. (2011). J. Chem. Phys. 135: 034106.

38. Peterson, K. A., Tensfeldt, T. G. & Montgomery Jr., J. A. (1991). J. Chem.

Phys. 94: 6091.

39. Piecuch, P., Kucharski, S. A., Kowalski, K. & Musial, M. (2002). Comp. Phys. Commun. 149: 71.

40. Smith, B. J. & Radom, L. (1991). J. Phys. Chem. 95: 10549.

41. Stevens, W. J., Basch, H. & Krauss, M. (1984). J. Chem. Phys. 81: 6026.

42. Tabrizchi, M. & Shooshtari, S. (2003). J. Chem. Thermodyn. 35: 863.

Chapter 5

QUANTUM CHEMISTRY AND CHEMOMETRICS APPLIED TO CONFORMATIONAL ANALYSIS

Aline Thaís Bruni[1] and Vitor Barbanti Pereira Leite[2]

[1]Departamento de Química, Faculdade de Filosofia, Ciências e Letras de Ribeirão Preto, Universidade de São Paulo

[2]Departamento de Física, Instituto de Biociências, Letras e Ciências Exatas, Universidade Estadual Paulista, São José do Rio Preto Brazil

INTRODUCTION

Conformational Analysis: Early History and Importance

Molecular structure plays a special role in science. Knowledge of the atomic arrangement is essential in order to be able to elucidate chemical properties and processes. The first advances in determining molecular structure occurred in nineteenth century. Around 1812, Jean-Baptiste Biot, a French physicist, discovered optical activity by observing polarized light shifting when crossing a quartz crystal. He observed that the light was displaced to the right in some cases and to the left in others. The conclusion was that rotation of polarized light by quartz is an inherent property of the crystal. Interested in the phenomenon, Biot noticed in further studies that similar effects were found when polarized light passed through certain liquids such as natural oils (lemon extract and laurel), alcoholic solutions of camphor, some sugars and tartaric acid. (Drayer, 1993; Cintas, 2007; Gal, 2011) Biot's observations were very important in laying foundation for the concept of optical activity. In 1948, Louis Pasteur discovered molecular chirality when studying a mixture of tartaric acid crystals.(Gal, 2007) He patiently performed the manual separation of tartarate enantiomer crystals (Cintas, 2007) and observed that each solution made with them was able to displace polarized light in one direction. He concluded that compounds with nonsuperimposable molecular asymmetry have identical chemical properties despite the inverse behavior related to polarized light.

Pasteur argued that the optical activity of organic solutions is related to molecular geometry. This insight was far ahead of the organic structural theory of the time.(Drayer, 1993) Although Pasteur was the first to show a relationship between optical activity and molecular symmetry, he was not able to say exactly how a molecule could be right- or left-handed. The main advances in this idea occurred in 1874 when a theory of organic structure in three dimensions was independently and simultaneously developed by Jacobus Henricus van't Hoff in Holland, and Joseph Achille Le Bel in France. (Drayer, 1993; Cintas, 2007) In 1865, August Kekulé proposed his theory of the benzene molecular structure and proposed that the carbon atom has valence 4.(Brush, 1999) His principal idea was that the carbon atom is tetravalent and can form valence bonds with other carbon atoms yielding to chains. These carbon chains can sometimes have closed arrangements, forming rings. (Drayer, 1993) Van't Hoff and Le Bel proposed that the four valences of the carbon atom were not planar, but directed into three-dimensional space. Van't Hoff specifically proposed that the spatial arrangement was tetrahedral. Later, he used the tetrahedron as a graphic representation of the valence arrangement around the carbon atom and also used this model to explain the physical property of optical activity.(Ramberg & Somsen, 2001) A compound containing a four different substituted carbon – described by Van't Hoff as asymmetric carbon - would be capable of existing in two distinctly different nonsuperimposable forms. Finally, he stated that the asymmetric carbon atom was the cause of molecular asymmetry and optical activity.(Drayer, 1993) Le Bel, in turn, also published his stereochemical ideas in 1874, but with a different approach to the problem from that presented by Van't Hoff. His hypothesis was not based on the tetrahedral model for the carbon atom and the fixed valences between the atoms. His investigation was into the asymmetry as a whole, without evaluating the individual atoms. The full system was considered in his evaluation, and his interpretation could be inserted into the field that is currently understood as molecular asymmetry. He mentions the tetrahedral carbon atom only in special cases, and not as a general principle. Many molecules confirm Le Bel's concepts of molecular asymmetry. Allenes, spiranes, and biphenyls are some examples of asymmetric molecules that do not contain any asymmetric carbons. Van't Hoff's and Le Bel's different approaches can be explained by the origin of their formation. Van't Hoff, based on Kekulé tetrahedron models, suggested the concept of the asymmetric carbon atom. On the other hand, Le Bel based his investigations on Pasteur's considerations of the connections between optical rotation and molecular structure.(Drayer, 1993) The historical development of conformational search does not end here and has many other important aspects and particularities. Our goal was just to give a basic outline of the initial concepts and how they influence current conformational understanding. Despite the historical progress in conformational studies, the advances in structure determination has been relatively recent and have been made possible by the development of analytical instruments and computational tools. Early structural studies were applied only to small molecules or substructures that could be expressed in terms of a few settings.(Allen et al., 2010) Currently, a great evolution is occurring in mechanisms for determining and understanding molecular structures. The relationship between geometry and energy is experimentally measurable and gives an idea of the balance between energy factors involved in each structure. (Pietropaolo et al., 2011) Reactivity and other properties are directly linked to the conformational arrangement of molecules.(Hunger & Huttner, 1999) Every chemical property must be understood according to its molecular structure and atomic connections. (Pietropaolo et al., 2011) Indeed, knowledge of structural arrangement is important since it underlies studies in chemical reactions and other molecular behaviors. There are experimental techniques for the structural determination, such as X-ray, magnetic resonance, infrared, mass spectroscopy and others. In this chapter we will discuss theoretical methods for molecular conformational determination. The field that concerns ways to mimic the behavior of molecules and molecular systems is molecular modeling. It seeks a simplified or idealized description of molecular systems, making it possible to produce three-dimensional representations that provide insights into their behavior. As computer tools have enjoyed a spectacular increase in last decades, theoretical methods are invariably associated with computer modeling. This has become a powerful tool for evaluating molecular structure, from which special chemical information about molecular behavior can be inferred. (Pietropaolo et al., 2011)

STATEMENT OF THE PROBLEM

For the theoretical and computational determination of molecular properties it is necessary to previously determine the minimum energy structure of the system being studied. A central issue is to probe the equilibrium configuration of the molecular system. The way that energy varies with the coordinates is usually referred to as the potential energy surface. At the atomic level, the interaction energy between atoms is essentially ruled by quantum mechanics, which provides the basic elements and methods used in molecular modeling. However, the potential energy surface can be addressed with different degrees of approximation, i.e., ab initio, effective potentials or even more coarse-grained potentials. Irrespective of the details with which the system is considered, one usually faces the problem of a highly dimensional system with the occurrence of multiple minima. Low energy minima play an important role in determining molecular properties, and the determination of these minima conformational states is a non-trivial task, usually referred to as energy minimization method for exploring the energy surface. If four or more atoms are connected in chain by single bonds we can suppose that there is considerable flexibility in the molecule. The existence of hindered rotation about a single bond is one of the fundamental concepts in conformational analysis.(Mo & Gao, 2007) The understanding of the connections between the atoms is related to the internal coordinate parameters, i.e., bond length, bond angle and dihedral angle, and is essential in designing molecular models.

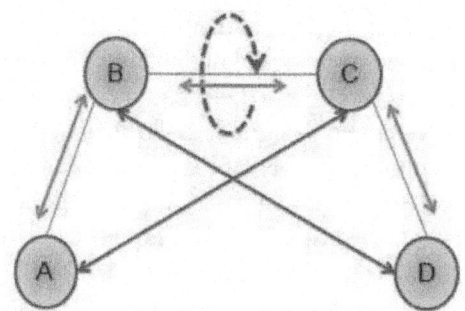

Figure 1: Model for a generic molecule with four different atoms.

For instance, let us consider a molecule composed of four different atoms which are singlebond linked (Figure 1). The green arrows represent the bond stretch and the average value is the bond length; the red arrows correspond to the angle formed by three sequential atoms, i.e., the angle bond; the curved blue arrow indicates the free rotation around the only single bond able to perform changes in the molecule conformation, as shown in Figure 2:

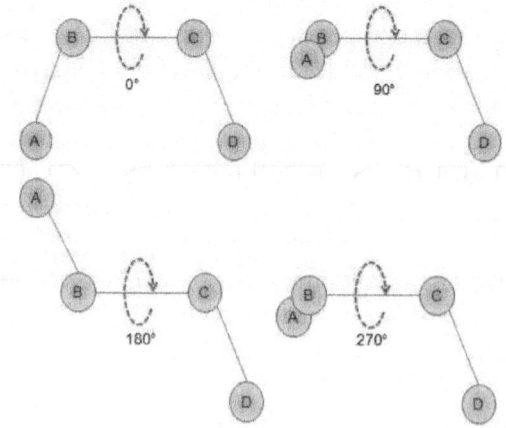

Figure 2: 90° rotation around the single bond.

In other words, different conformations are obtained when a dihedral angle is rotated. A dihedral angle is that composed by the planes formed by the sequence of three atoms (Figure 3):

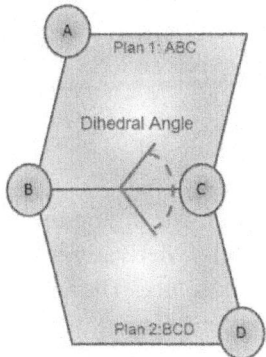

Figure 3: Dihedral angle representation for the molecule ABCD.

The main questions on molecular modeling are concerned with a good way of finding the global minimum energy structure. Important information is also concerned with the behavior of the conformational space. It is not only necessary to know which is the global minimum, but also the whole shape of the potential energy surface (PES) The main characteristic of this conformational phase space is that it is exponentially large, and a computationally hard problem. (Fraenkel, 1993) One problem that illustrates this difficulty is the protein folding, in which one searches for the global energy minimum structure associated with its functional conformation. If a method can be used to describe the relevant potential energy surface of a given molecule, it can also accurately elucidate its behavior against many interest situations. Many techniques have been presented, and it is not the goal of this work to make a deep study on them. A good discussion of practical methods is given by Leach.(Leach, 2001) We intend to give a brief idea of the most popular techniques used for investigating the molecular structure computationally. For conformational sampling, one can imagine a hierarchy of methods with different computational costs.(Seabra et al., 2009) However, there is no sovereign truth about what is the best method for performing a conformational analysis. Each situation must be evaluated. The best method is one that has the best fit to the problem studied; in practical terms it will provide the answer as quickly as possible, using the least amount of computer resources.

Stochastic Methods of Conformational Analysis

The literature reports many methods for trying to solve multiconformational problems, and most of them are based on stochastic approaches. Put simply, stochastic methods work with random variables, such as initial conformations for the search or the steps probing the configuration phase space. The simple criterion for establishing a minimum energy conformation is that the first derivatives of the energy E with respect to each variable (x_i) is zero and the second derivatives are all positive:

$$\frac{\partial E}{\partial x_i} = 0 \quad \text{and} \quad \frac{\partial^2 E}{\partial x_i^2} > 0.$$

The algorithms that search for minimum energy states can be classified into two groups: those which use derivatives of the energy with respect to the coordinates, and those which do not. The most used derivative minimization methods are the steepest descent, line search in one dimension and conjugate gradient methods.(Leach, 2001) These algorithms are very useful for conducting local (restricted) searches of minima, or downhill searches to the nearest minimum, since they are not able to overcome energy barriers. They are often used in combination with other stochastic methods. In the remainder of this section we discuss examples of stochastic methods. Havel, Kuntz and Crippen described distance geometry algorithms in conformational analysis.(Havel et al., 1983b) Given the impossibility of examining all possible conformations, they introduced a method which is capable of finding global optima without considering all possible solutions by means of combinatorial optimization. The method is known as brunch and bound and involves logical tests that allow whole classes of solutions to be eliminated without examining them one by one. The method converts a set of distance ranges (or "bounds") into a set of Cartesian coordinates that are consistent with those bounds. (Spellmeyer et al., 1997) The efficiency of a branch and bound algorithm depends on how effective these tests are compared to the time required to perform them. (Havel et al., 1983a; Havel et al., 1983b) In another study, Havel et al presented the basic theorems of distance geometry in Euclidean space. They proposed new algorithms and described refinements to the existing ones. All these algorithms were similar because they utilize geometric principles in order to interpret structural relationships. (Havel et al., 1983b) According to Leach and Smellie,(Leach & Smellie, 1992) distance geometry is a method for searching conformational space in which a structure is initially formulated in terms of interatomic distances. Any molecular system can be described as the set of minimum and maximum interatomic distances between all pairs of atoms in the molecule. The complete conformational space of the molecule is contained within this space. In distance geometry, a matrix is defined as the set of minimum and maximum distances, and then used to create a series of conformers that are consistent with those distances.(Spellmeyer et al., 1997) Another tool for performing conformational searches is the genetic algorithm, a stochastic method first introduced by Holland in 1975. Genetic algorithm (GA) is a method applied to solve problems using a natural evolution process simulation. It is a stochastic method developed in analogy to Darwin's theory of evolution in order to perform the optimization.(Brodmeier & Pretsch, 1994; Lucasius, 1993; Nair & Goodman, 1998) Genetic algorithm is commonly used for studying a large-scale space of possible solutions. The goal is to identify the best solutions within that space without the need to evaluate all possibilities.(Yanmaz et al., 2011) The GA is the optimization of a large number of possible solutions using a randomly generated population. When applied to conformational analysis, the population of interest consists of different conformations. The biological evolution of this population is simulated. A population of trial solutions is iteratively manipulated by a series of genetic operators to satisfy an objective function. The adjustment is calculated, and a new population is generated according to operators, such as selective reproduction, recombination and mutation. The process is repeated until the minimum energy structures are obtained.(Lucasius & Kateman, 1994; Beckers et al., 1996; Beckers et al., 1997) Artificial Neural Networks (ANN) are another example of stochastic methods used in conformational analysis. This method is based on concepts of the behavior of the human brain. Although artificial neural networks are primitive compared to their biological counterparts, they exhibit some interesting properties which make them useful as multivariate tools in various fields of research. During the last decade, ANN have been successfully applied in non-linear modeling, classification, signal processing and process control.(Derks & Buydens, 1996) The properties of a molecule are intimately linked to the conformations that it adopts and so an understanding of the conformational space is important in rationalizing and predicting its behavior.(Jordan et al., 1995) Among the most popular stochastic methods for covering the conformational space are Monte Carlo (MC) and Molecular Dynamics (MD). They are similar in the sense that both procedures include the same representation of molecules and use classical force fields for the potential energy terms, under periodic boundary conditions. The main purpose of these methods is to sample the phase space and to use the force fields ability to represent the conformational space near minima and connecting transition structures.(Jorgensen & TiradoRives, 1996; Grouleff & Jensen, 2011) However, large differences are found in sampling and configuring space available to the system. For MC, a new configuration is generated by selecting a random molecule or part of it, rotating it, translating it, and performing an internal structural variation. These changes do not necessarily need to follow a realistic physical trajectory. The acceptance of the new configuration is, however, determined by the Metropolis sampling algorithm. The sampling criterion is set in a way that enhances the likelihood of probing low energy conformations. Application over enough configurations yields properly Boltzmann-weighted averages for structure and thermodynamic properties. For MD, given a set of initial conditions (position and velocities of all atoms), new configurations are generated by application of Newton's equations of motion, so that the new atomic positions and velocities of all atoms are determined simultaneously over a small time step. In both cases, the force field controls the total energy (MC) and forces (MD), which determines the evolution of the systems. (Jorgensen & TiradoRives, 1996) Examples of problems related to large systems are the interaction between drug and the receptor, and protein behavior and folding. Molecular docking procedures are capable of predicting the three-dimensional structure of macromolecular complexes and their binding affinity. The information required is simple and corresponds to the structures of the receptor and ligand and the presumable interfacing region between them. Besides the simplicity of these docking procedures, they have low computational costs. However, molecular plasticity and solvation effects are not, or are only approximately, taken into account in these approaches. Free energy simulations may be then used to investigate the molecular association process and to predict binding affinity. (Biarnés et al., 2011) It is important to realize that sometimes the probing of PES addresses singular questions, which involve association of several methods, also called hybrid methods. A particular wellknown tailored one is the quantum mechanics/molecular dynamics approach, also known as QM/MM approach. This is a molecular simulation method that combines the strength of both QM (high accuracy in specific regions) and MD fast calculations (in not so crucial regions), in such a way that it efficiently allows the study of chemical processes in solution and in proteins. When stochastic methods are used to find minimum energy conformations, asymptotic states in restricted regions of the phase space are probed. This means that there is no end point in the search, and the convergence cannot be assured.

SYSTEMATIC SEARCH IN CONFORMATIONAL ANALYSIS

As seen before, stochastic techniques use different heuristics to randomly cover the conformational space. These algorithms apply a perturbation to the initial conformer and minimum energy conformation is associated with the lowest energy state that is found through out this procedure. They provide a sampling of energy minima structures and the shape of the PES is obtained in an indirect way. Beyond the stochastic methods there are procedures that do not work with random choice to cover conformational space. These classes of methods are described as deterministic and are capable of searching the conformational map in a systematic way, providing a direct knowledge of PES shape. These searches divide conformational space into quantized units and apply algorithms to search this discrete space or define a set of heuristic rules that are used to drive the search.(Smellie et al., 2003) Systematic methods are those that explore all conformational space at some fixed degree of resolution. To perform the systematic search, a molecule must be numerically described by its atoms' internal coordinates. The internal coordinates are bond length, angle bond and dihedral (torsion) angle. For a given initial structure the systematic conformational search is conducted by regular variation in dihedral angles (Figure 2). Although a systematic search can obtain the morphology of a molecule's energetic behavior directly, this method is not feasible for evaluating complex systems. (Beusen et al., 1996) Systematic search is most usefully applied for molecules with few degrees of freedom.(Li Manni et al., 2009) According to literature (Beusen et al., 1996), to cover the PES corresponding to the conformational space, different molecular structures must be systematically generated by rotating the torsion angles around the single bonds between 0° and 360°. The number of conformations is given by:

$$\text{Number of conformations} = s^N \tag{1}$$

where N is the number of free rotation angles, and s is the number of defining steps according to the angle increment:

$$s = \frac{360°}{\theta_i} \tag{2}$$

with θ_i being the dihedral increment of angle i. An examination of equation (1) reveals that the number of conformations generated will exponentially increase in proportion to the number of bonds with free rotation in the molecule under study. A problem arises if the number of steps is large, i.e., when a very refined surface is required by small angle increments. This problematic behavior of the systematic study of PES, described as combinatorial explosion, is the major restriction involved in this kind of search. Figures 4 and 5 illustrate how combinatorial explosion works. In Figure 4, we have a representation of the system growth where many single bonds can be rotated. The combinatorial explosion problem is represented by Figure 5. The number of branches to be considered is shown by the ramification achieved according the number of angles (A, B, C, D...) and will depend on the dihedral increment chosen. Due to the problem involved in combinatorial explosion, systematic search becomes nonviable for studying large molecules, since the number of degrees of freedom increases. A useful strategy for reducing the dimensionality of the conformational space is to perform systematic conformational searches on small portions of the molecule (either as isolated fragments or in situ). Using these optimal parts, one builds the conformation of the whole molecule with only limited additional searching of the relative conformations of the fragments. Approaches that incorporate this principle are known as "build-up" methods. (Beusen et al., 1996; Izgorodina et al., 2007) There are some strategies for overcoming the combinatorial explosion. We will focus our discussion on procedures that involve chemometrical approaches.

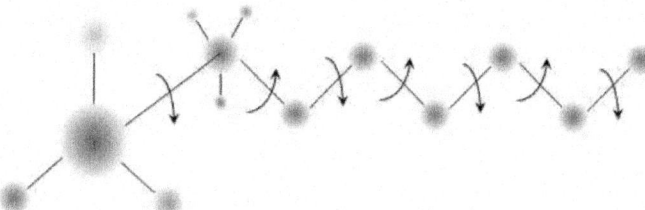

Figure 4: A general structure with many single bonds.

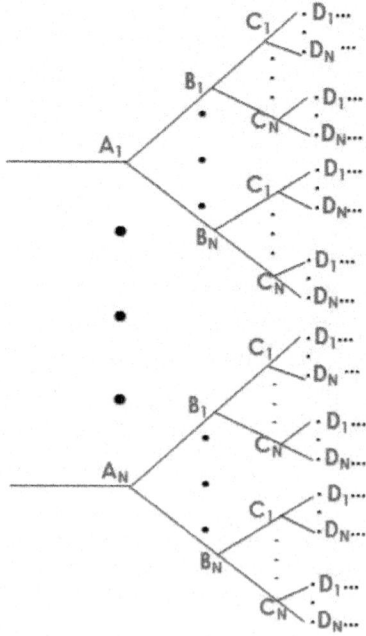

Figure 5: Branches generated by the dihedral angles A, B, C and D.

CHEMOMETRICS AND STRUCTURE DETERMINATION

A conformational search, independently of the method chosen, usually involves large amounts of data. Sometimes, data achieved from a given methodology must be explored by an additional technique. According to Geladi (2003) "data exploration means taking a look at the data to find interesting phenomena, often without prior expectations. As a result, outliers, clustering of objects and gradients between clusters may be detected."(Geladi, 2003) Chemometrics has been used extensively in recent years for exploring chemical problems by means of computer tools and statistical observations. The literature presents many definitions for chemometrics. For our purposes, this field of knowledge is better defined as a combination of two definitions found in the literature:

- According to Wold (1995), chemometrics can be understood as a way "to get chemically relevant information out of measured chemical data, how to represent and display this information, and how to get such information into data";(Wold, 1995)

- For Beeb (1998), chemometrics corresponds to "the entire process whereby data (e.g., numbers in a table) are transformed into information used for decision making." (Beebe, 1998)

The above definitions indicate that chemometrics offers a broad approach to chemical measurement sciences. It is not restricted to the actual experimental analysis but also considers what happens before and after it. (Massart et al., 2004) It is the goal of chemometrics to extract the information from the data. (Ramos et al., 1986) Chemometrical approaches have been applied to conformational analysis for handling special difficulties of large amounts of data generated both by stochastic and by systematic searches. Among the various chemometrical techniques, Principal Component Analysis (PCA) is the most commonly used for conformational problems. In many ways, it forms the basis for multivariate data analysis. PCA is a multivariate method of analysis whose main concern is to reduce the dimensions needed to portray accurately the characteristics of a large dimensional data matrix.(Beebe, 1998; Wold et al., 1987) This mathematical procedure consists of eliminating a large number of correlated variables without changing the characteristics of the original data-set that contribute most to its variance. For an easy graphical representation, consider a two-dimensional set of variables as shown in Figure 6 (a). PCA can be performed on the original variables as shown in Figure 6 (b) and new axes, called Principal Components, arise to account for the maximum variation. A subsequent rotation (Figure 6(c)) is made on these new PC axes in order to rewrite the original variables in terms of this new axes-system. Each PC is constructed as a linear combination of variables:

$$P_i = \sum_{j=1}^{v} c_{i,j} x_j$$

(3)

where P_i is the ith principal component and $c_{i,j}$ is the coefficient of the variable $x_{i,j}$. (Leach, 2001) There are v such variables. The first principal component PC1 is chosen in order to maximize the data variance of the axis. The second and subsequent ones are chosen to be orthogonal to each other and account for the maximum variance in the data not yet described by previous principal components. A variety of algorithms can be used to calculate the principal components. The most commonly employed approach is singular value decomposition SVD. (Golub & Loan, 1996)

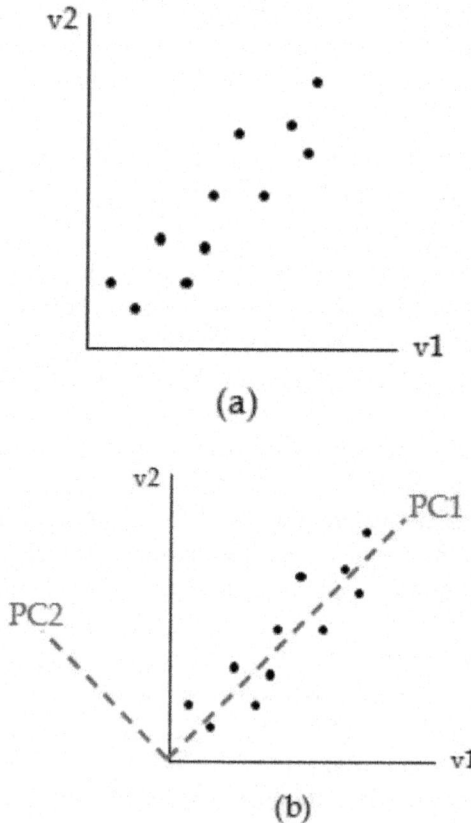

(a)

(b)

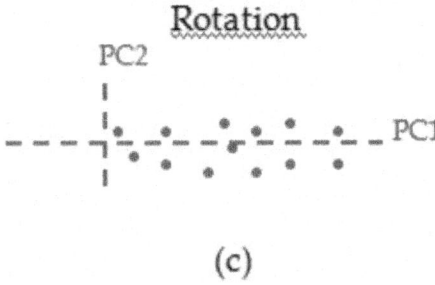

(c)

Figure 6: PCA procedure: (a) original data set; (b) PCA on original data set and (c) Variables according to the new PC coordinates.(Beebe, 1998).

A matrix of arbitrary size can be decomposed into the product of three matrices in such a way that:

$$X = USV_t$$

(4)

where U and V are square orthogonal matrices. The matrix U (whose columns are the eigenvectors of XXt) contains the coordinates of samples along the PC axes. The V matrix (which contains the eigenvectors of the correlation matrix XtX) contains the information about how the original variables were used to make the new axes[$c_{i,j}$ coefficients in eq. (3)]. The S matrix is a diagonal matrix that contains the eigenvalues of the correlation matrix (standard deviations) or singular values of each of the new PCs. The diagonalization of symmetric matrices (such as XXt and XtX) and SVD are fundamental problems in linear algebra (Golub & Loan, 1996), for which computationally efficient software has been developed and can be used on a routine basis (Hanselma et al., 1997) for very large-size matrices.

In chemistry, PCA was introduced by Malinowski around 1960 under the name Principal Factor Analysis, and further developed after 1970.(Malinowski, 2003) Principal Component Analysis can be used for crystallographic structure data; in its general form, conformational analysis is applied to multivariate numerical problems. (Allen et al., 2010) Many studies report on the use of PCA for handling Molecular Dynamics data. Among them, we highlight the application that uses PCA for mapping potential energy surfaces, by the quantitative visualization of a macromolecular energy funnel. (Becker, 1998) Other examples where PCA can be applied in molecular structure determination can also be found in recent studies. (Das et al., 2011; AraujoAndrade et al., 2010; Kiralj et al. 2007; Oblinsky et al., 2009; Silva et al., 2011)

PAIRS OF DIHEDRAL ANGLES-SYSTEMATIC ANALYSIS

There is a variety of theoretical methods that are capable of locating minimum energy structures in the potential energy surface. The problem of stochastic methods is that there is no natural end point for the conformational search. In some cases, only a small subset of conformational space is explored and the convergence of the system is not guaranteed. Only Systematic Conformational Analysis maps the conformational space completely. We stress the principal difficulty inherent in this method is the combinatorial explosion. In a previous study (Bruni et al., 2002), a new methodology was introduced for controlling the combinatorial explosion through a systematic reduction in the size of the system by means of chemometrics. This method consists of a small systematic conformational analysis, in which the conformational space is studied by rotation of the important free rotation in pairs, described as Pairs of Dihedral Angles-Systematic Analysis – PDA-SA. The main objective is to reduce the dimension of the investigated system. The idea is to address the conformational space in small portions, evaluating PES in combinations of angles in pairs. If the problem of combinatorial explosion is controlled, the conformational space can be sufficiently refined in the regions of minimum energy, taking care to minimize the information lost. The energy surfaces are obtained for each pair of angles and the number of conformations is given by Equation 5:

$$\text{Number of conformations} = s^2 \frac{N(N-1)}{2}$$

$$(5)$$

where s and N have the same meaning as in Equation 1. The number of conformations, in this case, is given by the combinatory arrangement of the N dihedrals in pairs. The main observation of the comparison between equations (1) and (5) is that the number of conformations as given by Eq. (1) increases exponentially with the number of bonds with free rotation, while from Eq. (5), the number of studied conformations increases quadratically with N. As the number of free rotation angles increases, the difference in the number of conformers generated by these two equations becomes more evident.

The computational procedure for PDA-SA can be organized in five basic steps:

- Molecular Building: The interest molecule must be defined in terms of its internal coordinates: bond length, angle bond and dihedral angles. There are many softwares able to define this molecular initial structure. A quantum chemistry optimization is required at this step in order to adjust internal parameters. The best method must be chosen according to the system under study.

- Dihedral Pair Rotation: The PDA-SA conformational search begins and the combination of the existing pairs of angles is taking account. Sometimes it is only possible to choose a dihedral increment with a less refined value. A rough PES is obtained in this case. The matrix to be analyzed consists of energy values from potential surfaces for angle combinations, and they are grouped according to Figure 7 for N angles. Appendix A shows the energy values for omprazole basic structure. The idea is to perform a cyclical permutation on the data, and this matrix form ensures that no information about the total PES is lost. The energy values obtained for each angle rotation as a function of the others allow the conformational space to be completely mapped. The major advantage is that the shape of these small portions can be visually observed, since we have a 3-D fitting. (see Figures 9 and 10)

- PCA application on data matrix: After the energy matrix statement, PCA is performed on the data. The regions with minimum energy points on the grid search can be easily selected. The number of selected regions will depend on the nature of the studied system.

- Refinement with a short dihedral increment: The regions initially obtained in step 3 can be refined with a small angle increment. It is important to emphasize that this step is not obligatory, since a small dihedral increment can be used in step 3, depending on the studied system. However, previous experience in this methodology (Bruni et al., 2002; Bruni & Ferreira, 2008) shows that this is the easiest procedure, i.e., firstly make rotations with a large dihedral increment and subsequently refine the minimum energy regions selected by PCA with small dihedral increments.

- Optimization of the final structure: the procedure described above provides angle values for the conformational search with a good level of accuracy. When these values are combined, we obtain all the possible minimum structures. Those structures constrained by the angle values obtained by PCA analysis are submitted to final optimization and the resulting structures are considered to be those of minimum energy.

In the study that introduced this method, the approach was successfully tested in the analysis of omeprazole and its derivatives, in which the results were in agreement with the experimental ones.(Bruni et al., 2002) In a second study, the technique was used to find minimum energy conformations of omeprazole derivative molecules in a QSAR study. (Bruni & Ferreira, 2002) It was shown that conformational analysis is crucial when establishing SAR/QSAR models using theoretically calculated descriptors, and they are strongly dependent on the details of molecular structure. Though all minima conformation have similar energetic values, some calculated properties are very sensitive to the structural variation, which is understandable since electronic properties are intrinsically dependent on molecular conformation.(Bruni & Ferreira, 2002) Omeprazole's racemization barrier and decomposition reaction was also studied. Quantum chemistry coupled to PDA-SA chemometric method was used to find all omeprazole minimum energy structures. To obtain the racemization barriers it was essential that the starting structure was in a global energy minimum. In that work, for all the studied structures, there was no change in the values of the racemization barriers, which confirmed the identification of the most stable structures for omeprazole.(Bruni & Ferreira, 2008)

$$
\begin{array}{c|ccccc}
i\backslash j & 1 & 2 & 3 & \dots & N \\
\hline
1 & - & E_{12} & E_{13} & \dots & E_{1N} \\
2 & E_{21} & - & E_{23} & \dots & E_{2N} \\
3 & E_{31} & E_{32} & - & \dots & E_{3N} \\
\vdots & \vdots & \vdots & \vdots & \ddots & \vdots \\
N & E_{N1} & E_{N2} & E_{N3} & \dots & -
\end{array}
$$

Figure 7: Matrix scheme for N angles: the discrete energy values for each rotation angle must be evaluated. E_{ij} are the energy matrices with elements E_{ij}^{km}, in which k and m are the angle increment indices for the angles i and j, respectively.

This approach is straightforward and in principle would have no size limits for its application. However, it presents limitation due to some initial condition dependence. Given a system with N degrees of freedom, for each pair of angles there are N-2 parameters that can interfere in the method. For example, in Figure 4 the potential energy surface for first and last dihedral angle combinations depends on the dihedral angles conformation between them. When the dihedral angles are too far from each other along the chain of atoms, the method may not become feasible. In this case the method may need to be repeated with different initial conditions to improve the sampling of the configurational phase space, and moreover we cannot be sure that we have reached the global minimum. When the correlations between the pairs of angles do not depend strongly on these initial conditions the method is very useful. Such system corresponds to small molecules, not so flexible, in which there are few large potential basins, such as omeprazol and its derivatives. (Bruni et al., 2002; Bruni & Ferreira, 2002; Bruni & Ferreira, 2008) The limit of validity for this method is under investigation. We are applying this method to study the IAN peptide, which is a tetrapeptide isobutyryl-(ala)3-NH-methyl. (Nascimento et al., 2009) This is the smallest polypeptide that can have secondary-like structure (an helix) (Becker & Karplus, 1997) and it has 11 free rotation bonds. For a flexible system, such as this, the initial condition dependence in the calculation of the minimum energy conformations is expected to increase with the size of the system. Since the system is more flexible and expected to be more rugged, we partially overcome this problem by using small angle increments steps, in order to probe all local minima of the system and compare them.

NUMERICAL RESULTS

Study of Basic Structure for Omeprazole and Derivatives

Initially, the basic structure of omeprazole and derivatives was evaluated. This structure has three bonds with free rotation. To validate the proposed methodology, two different approaches were performed. In the first approach, pairs of angles were taken account ((1,2), (1,3) e (2,3) in Fig. 8) and the number of conformations is given according to Equation 5. The resulting matrix analyzed was composed by the energy values from the potential energy surface for each angle combination (see matrix example in Figure 7). A matrix with discrete energy values for the basic structures with 30° angle increment in Equation 2 is showed in Appendix A.

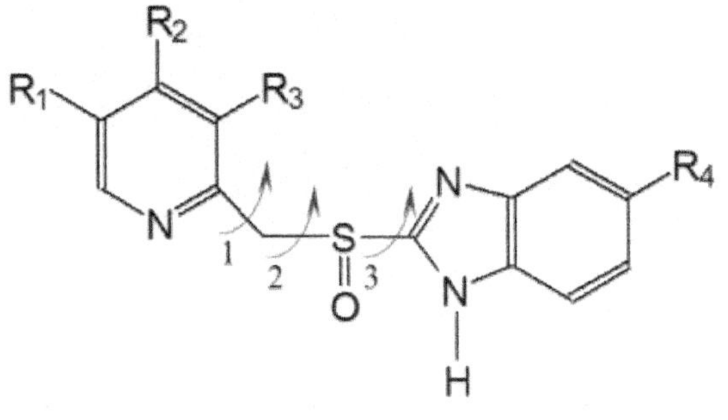

Figure 8: Basic structure for omeprazole and derivatives.

Three PES were obtained for a 30° dihedral increment and are showed in Figures 9 and 10. Figure 9 shows the original energy values and Figure 10 shows the same surfaces, but with a 0,12 hartrees cut off for better visualization. PCA was performed on autoscaled original data and the results are shown in Figure 11. 64% of the whole information is cumulated in first and second Factors (or Principal Components-PCs). The convergence of the points for one region is observed. Figure 12 shows the PCA results for the leveled data in 0,12 hartrees. Factor 1 and Factor 2 now cumulate 73% of the entire information.

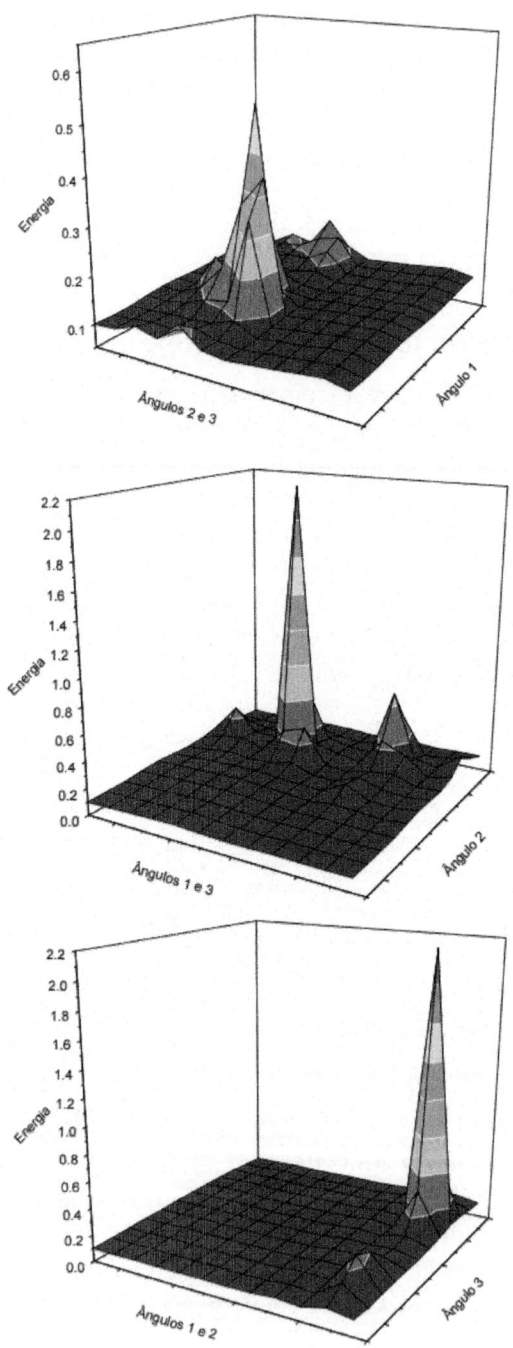

Figure 9: Orignal PES obtained from PDA-SA method for structure from Fig. 8.

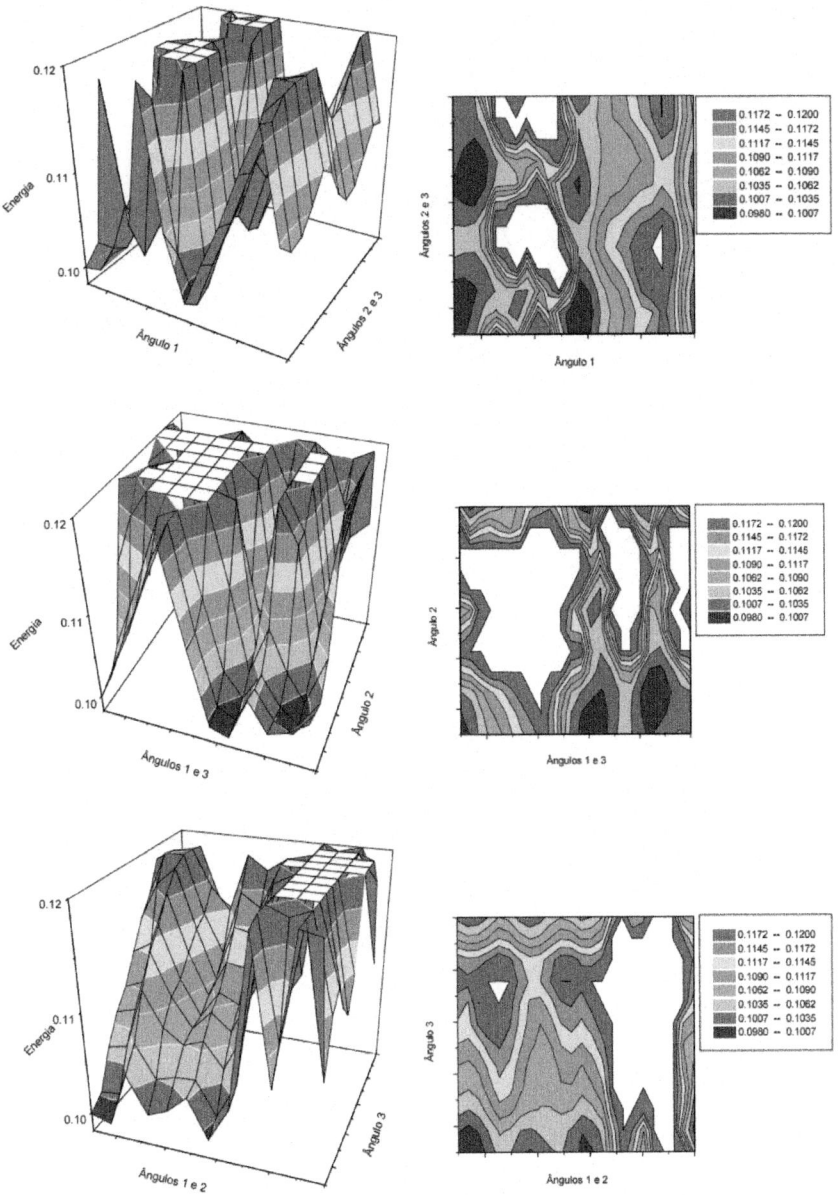

Figure 10: PES obtained from PDA-SA method for structure from Fig. 8, with a 0,12 hartress cutoff.

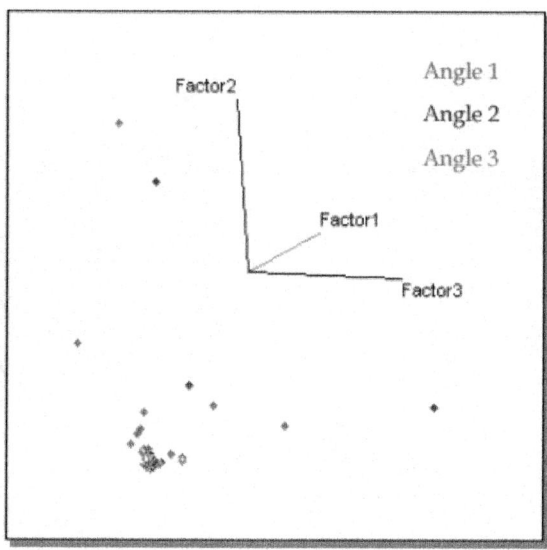

Figure 11: PCA for data from original PES (Fig.9).

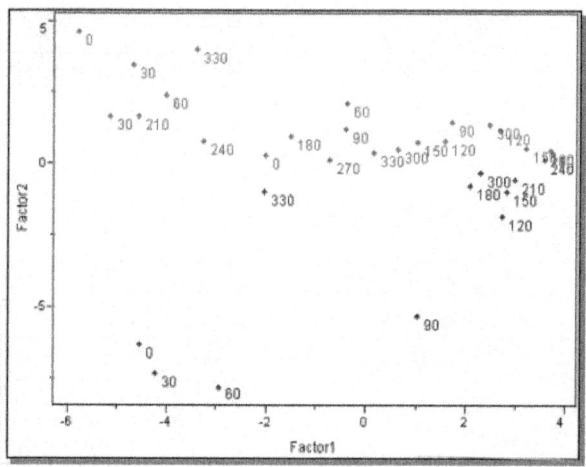

Figure 12: Principal Component Analysis forPES from Fig. 10, with a 0,12 cutoff.

Factor 1 accounts to the minimum region in each case and Factor 2 accounts for the energy range for the different combinations. Table 1 shows the selected minimum energy for each angle. The first column shows that two different regions were chosen for Angle 1 and only one region for Angles 2 and 3. Second column shows the rotation over the initial angle value (third column) resulting in the fourth column.

Table 1: Regions separated by PCA

Angle	Rotation	Initial Value	Value obtained by PCA
1 (a)	0° - 60°	48,48°	48,48° - 108,48°
1 (b)	180° - 240°	48,48°	228,48° - 288,48°
2	0° - 60°	209,79°	209,79° - 269,79°
3	330° - 30°	289,09°	259,09° - 319,09°

Once minima energy regions were defined, a small angle increment (5°) was used on them. Results for PCA are in Figure 13. In all cases a parabolic behavior was observed. When data variation decreases, curves are more easily observed and the minimum point is detectable. The amount of information accounted for both first and second PC's (Factors) is around 90%. Table 2 shows the final values for each angle. When these values are combined, two different geometries were obtained with similar energy values (Table 3). These conformations are shown in Figure 14.

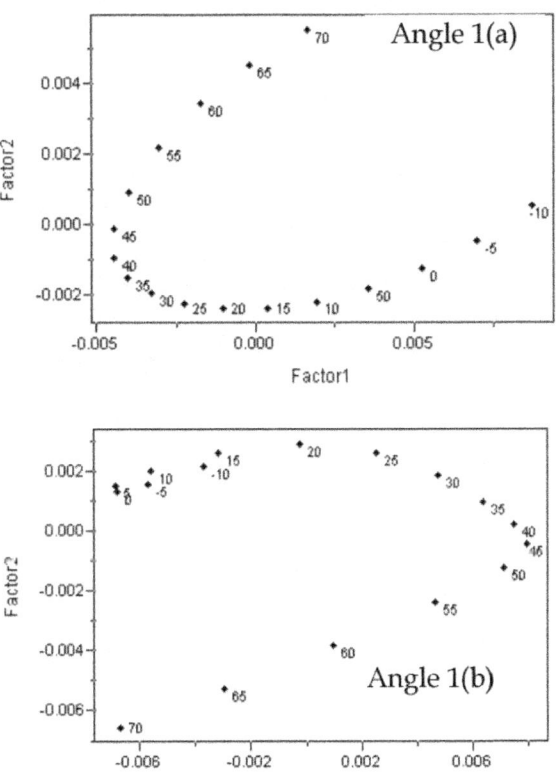

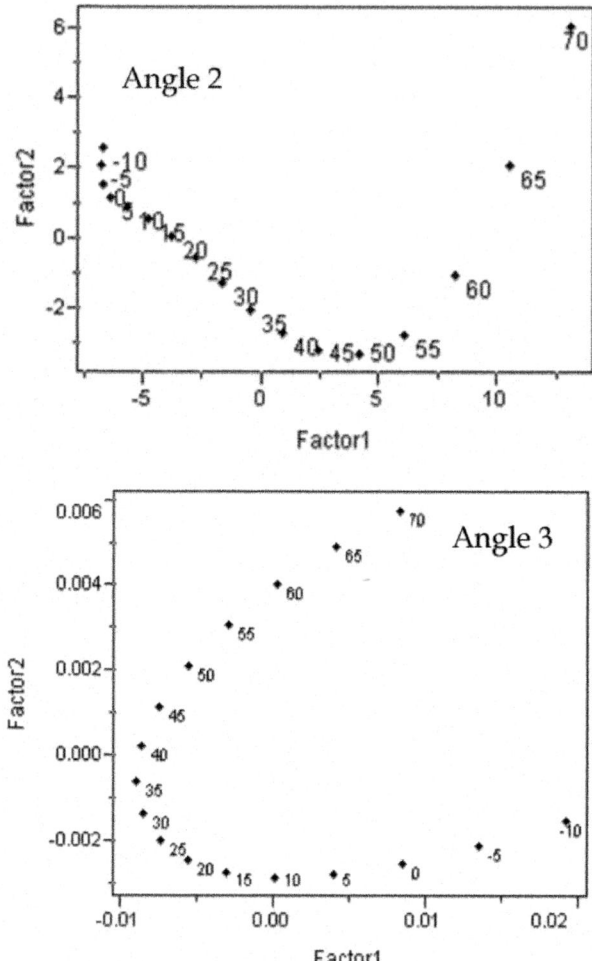

Figure 13: PCA results for 5° angle increment refinement.

Table 2: Regions obtained through PCA

Angle	Rotation	Initial Value	Value obtained by PCA
1 (a)	45°	48,48° - 108,48°	93,48°
1 (b)	45°	228,48° - 288,48°	273,48°
2	45°	209,79° - 269,79°	254,79°
3	35°	259,09° - 319,09°	294,09°

Table 3: Minimum conformation characteristics (basic structure)

Conformation	Angle	Obtained value	$\Delta H_{f\,(PM3)}$/ kcal mol^{-1}	$E_{e(6\text{-}31G^{**})}$/ hartree
A	1	92.29	54.96	-1134.33
	2	107.10		
	3	298.32		
B	1	266.45	54.71	-1134.35
	2	175.70		
	3	296.45		

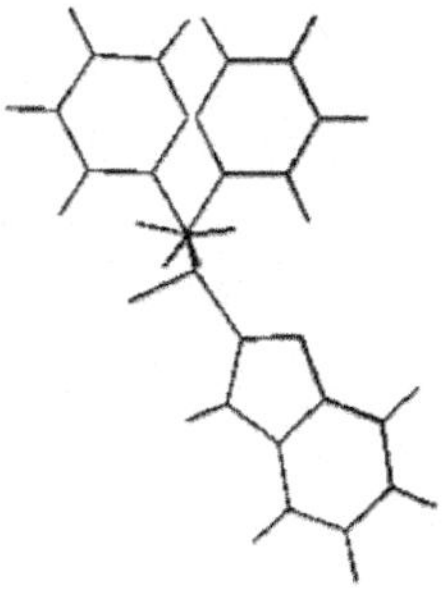

Figure 14: Optimized superposed conformations for structure from Fig. 8.

In the second approach, the conformational analysis was made according to Equation 1 and took into account all possible conformations. PCA was performed on data matrix and minimum energy regions were selected. The next step a lower dihedral increment of 5° was used to refine those selected regions. PCA was performed again, and the same structures and energy, shown in Table 3, were obtained. This indicates that the two approaches are equivalent. The details of this complete systematic search can be found in (Bruni et al., 2002).

IAN Preliminary Studies

IAN (isobutyryl-Ala3-NH-methyl) tetrapeptide has also been studied to validate PDA-SA methodology. IAN has 11 consecutives dihedrals and its main characteristic is to be the shorter peptide able to make a complete helix turn. Figure 15 shows the IAN 2D structure (Becker, 1998). Red arrows indicate the ψ, Φ e ω dihedrals. The dihedral angles ψ, ω and Φ are related to the rotations of single bonds between atoms in the main chain C (i)-C, OC-NH and N-C(i+1), respectively, where C (i) is the ith alpha carbon of the polypeptide chain. Angles $\top$ and Φ are connected to two arrays of functional protein chain: alpha-helix or beta-sheet.

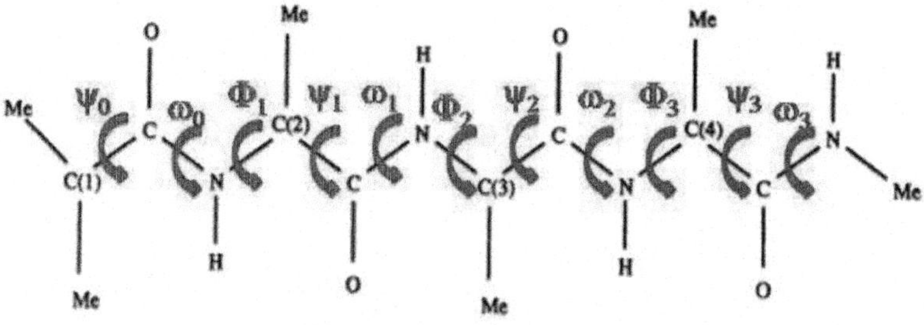

Figure 15: 2D IAN peptide structure.

Ten random different starting conformations were studied. Table 4 shows the angles and energy values corresponding to these initial conformations. The red values indicate dihedrals that were changed in comparison to initial conformation number 1. The starting conformation 2 is close to an alpha-helix. Energy values correspond to single point AM1 semi-empirical calculation, in kcal mol^{-1}.

Table 4: Energy(kcal mol^{-1}) and dihedrals values (degrees) for each starting IAN structure

Number	Energy	ψ_0	ω_0	ϕ_0	ψ_1	ω_1	ϕ_1	ψ_2	ω_2	ϕ_2	ω_3	ϕ_3
1	-179.66	79.45	-169.01	-64.73	-44.75	171.78	-84.62	44.47	179.30	-144.22	-59.37	178.07
2	-122.24	-60.55	179.00	-64.73	-64.75	-180.00	-64.62	-65.55	180.00	-64.22	-59.37	178.07
3	-156.02	79.45	170.99	-64.73	-34.75	171.78	-84.62	74.47	179.30	-124.22	-59.37	178.07
4	195.46	79.45	-169.01	-14.73	-44.75	171.78	-134.62	44.47	179.30	-144.22	-39.37	178.07
5	-80.48	49.45	-169.01	-64.73	-44.75	151.78	-84.62	44.47	179.30	-144.22	-39.37	178.07
6	-169.24	79.45	-149.01	-64.73	-44.75	171.78	-84.62	44.47	159.30	-144.22	-79.37	178.07
7	-149.11	59.45	-169.01	-84.73	-44.75	-168.22	-84.62	74.47	179.30	-144.22	-59.37	178.07
8	-147.67	79.45	-169.01	-64.73	-24.75	171.78	-104.62	44.47	-160.70	-144.22	-39.37	178.07
9	-29.05	109.45	-169.01	-64.73	-44.75	171.78	-84.62	24.47	179.30	-174.22	-59.37	178.07
10	-167.76	79.45	-169.01	-44.73	-44.75	-178.22	-84.62	44.47	179.30	-144.22	-59.37	178.07

IAN was analyzed using the PDA-SA procedure. The eleven dihedral angles provide 55 different conformations according to all possible combinations. Conformational analysis was performed with a 20° increment. PCA was carried out and Figure 16 shows that all points converge to specific regions of the phase space. Each selected region for each angle was refined with a 5° angle increment. PCA was performed again and the final structures characteristics are shown in Table 5.

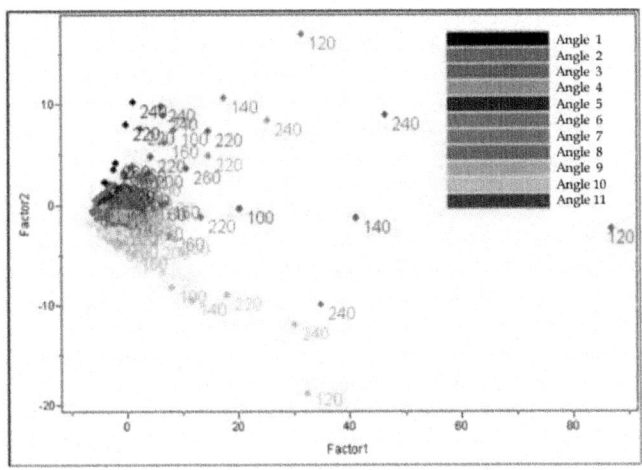

Figure 16: PCA results for IAN peptide.

Table 5 shows the obtained energy values for the final structures and they indicate that some correspond to identical conformations. Three different groups were identified. Figure 17 shows the group that corresponds to structures 1, 5, 6, 9 and 10 superposed (blue ones in Table 5). Structure 9 shows a slightly different value on ψ0 but it does not change the energy value. These five structures have two stabilizing hydrogen bonds which are indicated by the red circle and the resulting conformations for them resemble a beta-sheet.

Table 5: Energy (kcal mol-1) and dihedrals values (degrees) for each obtained IAN structure

Number	Energy	ψ0	ω0	φ0	ψ1	ω1	φ1	ψ2	ω2	φ2	ω3	φ3
1	- 181.498	70.50	176.87	85.57	66.93	-177.14	-85.01	67.39	179.60	-112.24	-46.78	177.92
2	- 180.70	-61.83	-172.66	-78.26	25.05	157.24	-90.68	-29.19	177.15	-110.70	-45.10	178.62
3	- 179.661	79.45	-168.99	-64.80	-44.71	171.76	-84.61	44.32	179.26	143.99	-59.33	178.07
4	-179.276	72.93	-174.15	-104.24	-32.54	-179.9	-82.66	70.38	-179.69	-115.57	-51.50	179.42
5	-181.498	70.49	-176.89	-85.56	66.89	-177.17	-84.96	67.52	179.62	-112.34	-46.80	177.93
6	-181.498	70.49	-176.88	-85.57	66.93	-177.15	-85.01	67.40	179.61	-112.26	-46.78	177.92
7	-179.276	72.93	-174.15	-104.24	-32.52	-179.9	-82.65	70.39	-179.69	-115.58	-51.50	179.44
8	-179.276	72.94	-174.14	-104.27	-32.45	-179.89	-82.66	70.40	-179.67	-115.55	-51.48	179.36
9	-182.385	-62.43	-177.31	-84.72	67.46	-177.22	-84.95	67.38	179.58	-112.28	-46.85	177.94
10	-181.498	70.49	-176.89	-85.56	66.93	-177.15	-85.00	67.43	179.62	-112.32	-46.82	177.91

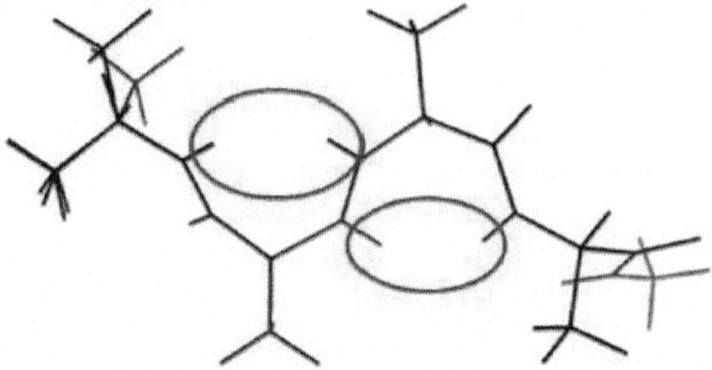

Figure 17: Superposed final conformations for 1, 5, 6, 9 and 10 structures (blue ones in Table 5).

The second group is composed by final conformations 4, 7 and 8 (green ones in Table 5). The superposed conformations can be observed in Figure 18. These conformations are more open and have only one hydrogen-bond (red circle in Fig.18). The last group, the black ones in Table 5, are superposed in Figure 19. The resulting structures show an alpha-helix like behavior, with two stabilizing hydrogen bond (red circles, Fig. 19).

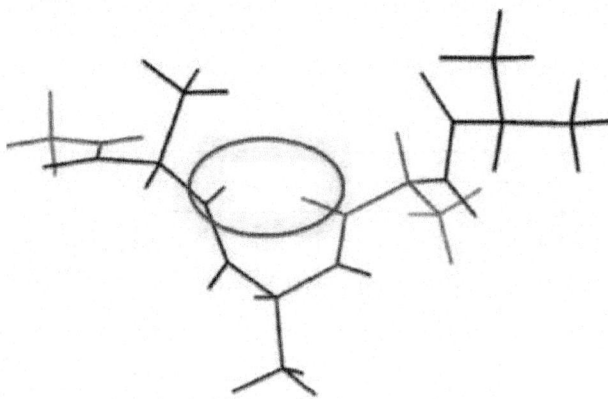

Figure 18: Superposed final conformations for 4,7and 8 structures (green ones in Table 5).

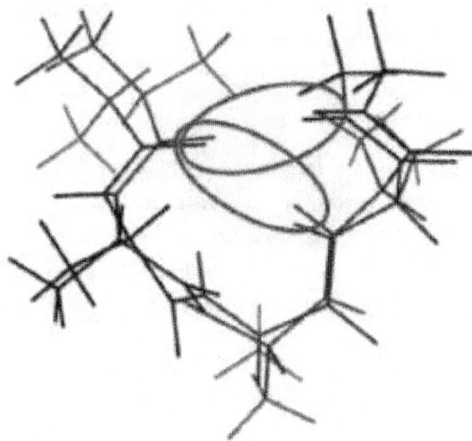

Figure 19: Superposed final conformations for structures 2 and 3 (black ones in Table 5).

Results presented for IAN peptide are partial and were only performed for one minimum region for each starting structure. Other minimum energy regions of this system are being investigated. A gradual increase in the size of the chain is also been explored.

CONCLUSION

The arrangement of atoms in a molecule or its structure determination has intrigued scientists through history. However only with recent experimental and computational advances the discussions on this theme became more effective and elucidative. The nature of PES is intrinsically multidimensional, usually has a very complex landscape. The global minima search, like the one encountered in the protein folding problem, is a NP-hard problem. This means that this task belongs to a large set of computational problems, assumed to be very hard ("conditionally intractable") (Fraenkel, 1993). The search for its relevant minima in molecular modeling has motivated the development of methods with very specific applications, as discussed in this chapter. For each particular problem one finds a variety of methods that allows feasible solutions, and most likely a combination of methods provides the optimum solution. In this chapter, we discussed some aspects of conformational search that controls the combinatorial explosion. In particular, Principal Component Analysis was associated with a systematic search method to find structures with low energy in PES. The methodology can be useful to handle small- and medium-size molecules. The maximum size which the method can efficiently handle is being investigated (Nascimento et al., 2009). Due to the PCA dimension reduction, the method's efficiency is highly increased, allowing it to be of practical use in the study of more complex molecules.

ACKNOWLEDGMENT

We thank Prof. Márcia M.C. Ferreira (Unicamp) for the helpful discussions. We were supported by Fundação de Amparo à Pesquisa do Estado de São Paulo (FAPESP) and Conselho Nacional de Desenvolvimento Científico e Tecnológico (CNPq), Brazil. Computational resources were provided by Centro Nacional de Processamento e Alto Desempenho em São Paulo (CENAPAD-SP), Brazil.

Apendix A

Matrix with the discrete values for each rotation angle and its corresponding energy value for the first rotation for basic structure in Figure 8 Labels in bold were not used in PCA analysis, they are shown to help the matrix notation and visualization.

Angle 1

Rotation	0	30	60	90	120	150	180	210	240	270	300	330	0	30	60	90	120	150	180	210	240	270	300	330
0	0.0998	0.100	0.103	0.111	0.119	0.113	0.106	0.118	0.146	0.133	0.106	0.099	0.099	0.100	0.104	0.107	0.108	0.111	0.114	0.115	0.118	0.116	0.106	0.102
30	0.0984	0.099	0.101	0.105	0.106	0.105	0.102	0.106	0.117	0.111	0.103	0.099	0.098	0.099	0.105	0.106	0.108	0.110	0.113	0.114	0.115	0.115	0.107	0.101
60	0.0992	0.099	0.100	0.102	0.103	0.104	0.103	0.105	0.107	0.115	0.125	0.108	0.099	0.101	0.104	0.107	0.109	0.112	0.117	0.125	0.119	0.115	0.107	0.101
90	0.1020	0.101	0.101	0.103	0.107	0.115	0.126	0.108	0.132	0.201	0.310	0.151	0.102	0.103	0.107	0.110	0.112	0.116	0.120	0.119	0.120	0.117	0.109	0.103
120	0.1046	0.104	0.104	0.107	0.124	0.196	0.174	0.345	0.322	0.634	0.383	0.116	0.104	0.106	0.110	0.113	0.115	0.117	0.119	0.120	0.122	0.119	0.111	0.106
150	0.1057	0.103	0.104	0.109	0.173	0.286	0.136	0.282	0.516	0.439	0.237	0.105	0.105	0.109	0.110	0.112	0.115	0.117	0.318	0.120	0.118	0.110	0.105	0.102
180	0.1010	0.100	0.100	0.103	0.156	0.116	0.121	0.166	0.263	0.145	0.107	0.100	0.101	0.103	0.107	0.110	0.112	0.113	0.114	0.116	0.112	0.106	0.102
210	0.1003	0.099	0.098	0.100	0.104	0.106	0.104	0.112	0.119	0.111	0.104	0.101	0.100	0.103	0.106	0.107	0.108	0.109	0.110	0.111	0.114	0.112	0.106	0.101
240	0.1019	0.099	0.098	0.099	0.102	0.105	0.107	0.107	0.109	0.114	0.118	0.108	0.101	0.104	0.106	0.106	0.107	0.108	0.109	0.110	0.114	0.115	0.108	0.102
270	0.1033	0.101	0.100	0.101	0.107	0.118	0.121	0.114	0.120	0.150	0.146	0.117	0.103	0.104	0.106	0.107	0.108	0.110	0.113	0.117	0.115	0.115	0.109	0.104
300	0.1036	0.102	0.102	0.106	0.121	0.132	0.127	0.121	0.152	0.249	0.163	0.111	0.103	0.104	0.106	0.108	0.109	0.111	0.114	0.117	0.120	0.118	0.111	0.106
330	0.1023	0.102	0.103	0.112	0.128	0.128	0.137	0.126	0.180	0.216	0.126	0.103	0.102	0.103	0.106	0.108	0.110	0.112	0.115	0.117	0.120	0.118	0.111	0.105

Angle 2

Rotation	0	30	60	90	120	150	180	210	240	270	300	330	0	30	60	90	120	150	180	210	240	270	300	330
0	0.0998	0.100	0.104	0.107	0.108	0.111	0.114	0.115	0.118	0.116	0.108	0.102	0.099	0.098	0.099	0.102	0.104	0.105	0.101	0.100	0.101	0.103	0.103	0.102
30	0.1006	0.101	0.105	0.108	0.110	0.113	0.115	0.118	0.123	0.115	0.108	0.103	0.100	0.099	0.099	0.101	0.104	0.103	0.100	0.099	0.099	0.101	0.102	0.102
60	0.1032	0.104	0.108	0.111	0.113	0.116	0.118	0.120	0.120	0.117	0.111	0.105	0.101	0.101	0.101	0.103	0.104	0.100	0.098	0.098	0.100	0.102	0.103	0.103
90	0.1115	0.113	0.117	0.120	0.122	0.125	0.127	0.126	0.129	0.128	0.120	0.113	0.111	0.105	0.102	0.103	0.107	0.109	0.105	0.100	0.099	0.101	0.106	0.112
120	0.1194	0.121	0.126	0.129	0.132	0.134	0.136	0.138	0.141	0.138	0.127	0.121	0.119	0.106	0.103	0.107	0.124	0.173	0.116	0.104	0.102	0.107	0.121	0.128
150	0.1136	0.116	0.122	0.127	0.130	0.131	0.132	0.134	0.139	0.129	0.120	0.114	0.113	0.103	0.104	0.115	0.196	0.186	0.116	0.104	0.105	0.118	0.132	0.128
180	0.1062	0.109	0.119	0.134	0.141	0.135	0.129	0.123	0.127	0.150	0.171	0.116	0.106	0.102	0.105	0.126	0.174	0.136	0.121	0.104	0.107	0.121	0.127	0.117
210	0.1182	0.133	0.172	0.238	0.189	0.147	0.133	0.147	0.319	0.241	0.376	0.140	0.118	0.106	0.103	0.108	0.148	0.252	0.166	0.112	0.107	0.114	0.121	0.126
240	0.1467	0.225	0.339	0.279	0.174	0.146	0.167	0.434	2.168	0.552	0.181	0.135	0.146	0.117	0.107	0.132	0.322	0.916	0.243	0.119	0.109	0.120	0.152	0.180
270	0.1467	0.133	0.166	0.167	0.140	0.121	0.126	0.178	0.333	0.368	0.167	0.120	0.133	0.117	0.115	0.201	0.634	0.459	0.145	0.111	0.114	0.150	0.249	0.216
300	0.1196	0.106	0.111	0.113	0.110	0.108	0.111	0.123	0.146	0.117	0.108	0.126	0.103	0.125	0.310	0.365	0.137	0.107	0.104	0.115	0.146	0.103	0.126
330	0.1047	0.099	0.101	0.104	0.105	0.107	0.109	0.133	0.115	0.116	0.112	0.107	0.099	0.099	0.108	0.191	0.116	0.105	0.100	0.101	0.108	0.117	0.111	0.103

Angle 3

Rotation	0	30	60	90	120	150	180	210	240	270	300	330	0	30	60	90	120	150	180	210	240	270	300	330
0	0.0998	0.098	0.099	0.102	0.104	0.103	0.101	0.101	0.103	0.103	0.102	0.099	0.100	0.103	0.111	0.119	0.113	0.106	0.100	0.101	0.103	0.146	0.119	0.104
30	0.1009	0.099	0.101	0.103	0.106	0.105	0.103	0.103	0.104	0.104	0.103	0.100	0.101	0.104	0.113	0.121	0.116	0.109	0.135	0.225	0.133	0.106	0.099
60	0.1045	0.103	0.104	0.107	0.110	0.107	0.106	0.106	0.106	0.106	0.106	0.103	0.105	0.108	0.117	0.126	0.122	0.119	0.172	0.239	0.166	0.111	0.101
90	0.1071	0.106	0.107	0.110	0.113	0.112	0.109	0.107	0.106	0.107	0.106	0.105	0.107	0.108	0.111	0.120	0.129	0.127	0.134	0.278	0.279	0.167	0.115	0.104
120	0.1068	0.108	0.109	0.112	0.115	0.113	0.110	0.108	0.107	0.107	0.109	0.110	0.108	0.110	0.113	0.122	0.132	0.130	0.141	0.189	0.174	0.140	0.110	0.105
150	0.1115	0.110	0.112	0.116	0.117	0.115	0.112	0.109	0.108	0.111	0.111	0.113	0.113	0.115	0.125	0.134	0.131	0.146	0.196	0.174	0.146	0.121	0.108	0.107
180	0.1140	0.113	0.117	0.120	0.119	0.117	0.113	0.110	0.109	0.110	0.114	0.115	0.114	0.115	0.118	0.127	0.136	0.132	0.125	0.135	0.167	0.126	0.111	0.106
210	0.1157	0.114	0.123	0.119	0.120	0.118	0.114	0.111	0.110	0.113	0.117	0.117	0.115	0.118	0.120	0.128	0.138	0.134	0.125	0.147	0.434	0.178	0.113	0.113
240	0.1187	0.118	0.119	0.120	0.122	0.126	0.114	0.114	0.117	0.120	0.118	0.123	0.120	0.129	0.141	0.159	0.127	0.319	2.168	0.553	0.146	0.115
270	0.1166	0.115	0.115	0.117	0.119	0.118	0.114	0.112	0.113	0.115	0.118	0.118	0.116	0.115	0.117	0.125	0.138	0.129	0.130	0.241	0.582	0.368	0.140	0.116
300	0.1086	0.107	0.107	0.109	0.111	0.110	0.107	0.106	0.106	0.109	0.111	0.111	0.108	0.108	0.111	0.120	0.127	0.120	0.171	0.376	0.181	0.367	0.117	0.112
330	0.1025	0.101	0.101	0.103	0.106	0.105	0.102	0.101	0.102	0.104	0.106	0.105	0.102	0.103	0.105	0.113	0.121	0.114	0.116	0.140	0.135	0.120	0.108	0.107

REFERENCES

1. Allen, F. H., Galek, P. T. a, & Wood, P. a. (2010). Energy matters! Crystallography Reviews, 16(3), 169-195. doi:10.1080/08893110903476919

2. Araujo-Andrade, C., Lopes, S., Fausto, R., & Gómez-Zavaglia, A. (2010). Conformational study of arbutin by quantum chemical calculations and multivariate analysis. Journal of Molecular Structure, 975(1-3), 100-109. doi:10.1016/j.molstruc.2010.04.002

3.   Becker, O M. (1998). Principal coordinate maps of molecular potential energy surfaces. Journal of Computational Chemistry, 19(11), 1255-1267. 605 Third Ave, New York, NY 10158-0012 Usa: John Wiley & Sons Inc. doi:10.1002/(SICI)1096- 987X(199808)19:11<1255::AID-JCC5>3.3.CO;2-H

4.   Becker, Oren M., & Karplus, M. (1997). The topology of multidimensional potential energy surfaces: Theory and application to peptide structure and kinetics. The Journal of Chemical Physics, 106(4), 1495. doi:10.1063/1.473299

5.   Beckers, M. L. M., Derks, E. P. P. A., Melssen, W. J., & Buydens, L. M. C. (1996). Pergamon oo!n-8485(%~6-0. Science, 20(4), 449-457.

6.   Beckers, M. L., Buydens, L. M., Pikkemaat, J. a, & Altona, C. (1997). Application of a genetic algorithm in the conformational analysis of methylene-acetal-linked thymine dimers in DNA: comparison with distance geometry calculations. Journal of biomolecular NMR, 9(1), 25-34. Retrieved from http://www.ncbi.nlm.nih.gov/pubmed/9081542

7.   Beebe, K. R. (1998). Chemometrics: A Practical Guide (p. 360). Wiley-Blackwell. Retrieved from erscienceLaboratoryAutomation/dp/0471124516/ref=sr_1_1?ie=UTF8&qid=1320084260&sr=8-1

8.   Beusen, D. D., Shands, E. F. B., Karasek, S. F., Marshall, G. R., & Dammkoehler, R. A. (1996). Systematic search in conformational analysis. Theochem-Journal of Molecular Structure, 370(2-3), 157-171. Po Box 211, 1000 Ae Amsterdam, Netherlands: Elsevier Science Bv.

9.   Biarnés, X., Bongarzone, S., Vargiu, A. V., Carloni, P., & Ruggerone, P. (2011). Molecular motions in drug design: the coming age of the metadynamics method. Journal of computer-aided molecular design, 25(5), 395-402. doi:10.1007/s10822-011-9415-3

10. Brodmeier, T., & Pretsch, E. (1994). Application of genetic algorithms in molecular modeling. Journal of Computational Chemistry, 15(6), 588-595. doi:10.1002/jcc.540150604

11. Bruni, A. T., Leite, V. B. P., & Ferreira, M. M. C. (2002). Conformational analysis: A new approach by means of chemometrics. Journal Of Computational Chemistry, 23(2), 222- 236. Commerce Place, 350 Main St, Malden 02148, MA USA: Wiley-Blackwell. doi:10.1002/jcc.10004

12. Bruni, A. T., & Ferreira, M. M. C. (2008). Theoretical study of omeprazole behavior: Racemizatin barrier and decomposition reaction. International Journal of Quantum Chemistry, 108(6), 1097-1106. doi:10.1002/qua.21597

13. Bruni, A. T., & Ferreira, M. M. C. (2002). Omeprazole and analogue

compounds: a QSAR study of activity againstHelicobacter pylori using theoretical descriptors. Journal of Chemometrics, 16(8-10), 510-520. doi:10.1002/cem.737

14. Brush, S. G. (1999). Dynamics of Theory Change in Chemistry : Part 1 . The Benzene Problem 1865 – 1945. Science, 30(1), 21-79.

15. Cintas, P. (2007). Tracing the origins and evolution of chirality and handedness in chemical language. Angewandte Chemie (International ed. in English), 46(22), 4016-24. doi:10.1002/anie.200603714

16. Das, G., Gentile, F., Coluccio, M. L., Perri, a M., Nicastri, a, Mecarini, F., Cojoc, G., et al. (2011). Principal component analysis based methodology to distinguish protein SERS spectra. Journal of Molecular Structure, 993(1-3), 500-505. Elsevier B.V. doi:10.1016/j.molstruc.2010.12.044

17. Derks, E. P. P. A., & Buydens, L. M. C. (1996). E. P. P. A DERKS,* M. L. M. BECKER& W. J. MELSSEN and L. M. C. BUYDENS, 20(4), 439-448.

18. Drayer, D. (1993). The Early History of Stereochemistry: From the Discovery of Molecular Asymmetry and the First Resolution of a Racemate by Pasteur to the Asymmetrical Chiral Carbon of van't Hoff and Le Bel. Clinical Pharmacology-New York-Marcel Dekker Incorporated-, 18(3), 1–1. Marcel Dekker Ag. Retrieved from http://scholar.google.com/schol ar?hl=en&btnG=Search&q=intitle:The+Early+His tory+of+Stereochem istry+:+From+the+Discovery+of+Molecular+Asymmetry+and +the+Fi rst+Resolution+of+A+Racemate+By+Pasteur+To+The+Asymmetrical+ Chir al+Carbon+Of+Van+?+T+Hoff+And+Le+Bel+*#0

19. Fraenkel, a S. (1993). Complexity of protein folding. Bulletin of mathematical biology, 55(6), 1199-210. Retrieved from http://www. pubmedcentral.nih.gov/articlerender.fcgi?artid=3042729&tool=pmce ntrez&rendertype=abstract

20. Gal, J. (2007). Review Article Carl Friedrich Naumann and the Introduction of Enantio Terminology : A Review and Analysis on the 150th Anniversary. Chirality, 98(May 2006), 89-98. doi:10.1002/chir

21. Gal, J. (2011). Review Article Louis Pasteur , Language , and Molecular Chirality . I . Background and Dissymmetry. Clinical Laboratory, 16(March 2010), 1-16. doi:10.1002/chir

22. Geladi, P. (2003). Chemometrics in spectroscopy. Part 1. Classical chemometrics. Spectrochimica Acta Part B Atomic Spectroscopy, 58(5), 767-782. doi:10.1016/S0584- 8547(03)00037-5

23. Golub, G. H., & Loan, C. F. van V. (1996). Matrix Computations (Johns Hopkins Studies in Mathematical Sciences)(3rd Edition) (p. 728). The

Johns Hopkins University Press. Retrieved from http://www.amazon.com/Computations-Hopkins-StudiesMathematical- ciences/dp/0801854148

24. Grouleff, J., & Jensen, F. (2011). Searching Peptide Conformational Space. Journal of Chemical Theory and Computation, 1783-1790.

25. Hanselman, D., Littlefield, B., Inc., M., & Mathworks. (1997). The Student Edition of Matlab Version 5 User's Guide (p. 429). Prentice Hall College Div. Retrieved from http://www.amazon.com/Student-Matlab-Version-Users-Guide/dp/0132725509

26. Havel, T. F., Crippen, G. M., Kuntz, I. D., & Blaney, J. M. (1983). The combinatorial distance geometry method for the calculation of molecular conformation. II. Sample problems and computational statistics. Journal of theoretical biology, 104(3), 383-400. Retrieved from http://www.ncbi.nlm.nih.gov/pubmed/6197591

27. Havel, T. F., Kuntz, I. D., & Crippen, G. M. (1983). The combinatorial distance geometry method for the calculation of molecular conformation. I. A new approach to an old problem. Journal of theoretical biology, 104(3), 359-81. Retrieved from

28. Hunger, J., & Huttner, G. (1999). Optimization and analysis of force field parameters by combination of genetic algorithms and neural networks. Journal of Computational Chemistry, 20(4), 455-471. doi:10.1002/(SICI)1096-987X(199903)20:4<455::AIDJCC6>3.0.CO;2-1

29. Izgorodina, E. I., Lin, C. Y., & Coote, M. L. (2007). Energy-directed tree search: an efficientsystematic algorithm for finding the lowest energy conformation of molecules.

30. Physical chemistry chemical physics : PCCP, 9(20), 2507-16. doi:10.1039/b700938k Jordan, S. N., Leach, A. R., & Bradshaw, J. (1995). The Application of Neural Networks in Conformational Analysis. 1. Prediction of Minimum and Maximum Interatomic Distances. Journal of Chemical Information and Modeling, 35(3), 640-650. doi:10.1021/ci00025a035

31. Jorgensen, W. L., & TiradoRives, J. (1996). Monte Carlo vs molecular dynamics for conformational sampling. Journal of Physical Chemistry, 100(34), 14508-14513. 1155 16th St, Nw, Washington, Dc 20036: Amer Chemical Soc. doi:10.1021/jp960880x

32. Kiralj, R., Ferreira, M. C., Donate, P. M., & Silva, R. (2007). Combined Computational, Database Mining, NMR, and Chemometric Approaches. Analysis, 6316-6333.

33. Lucasius, C. B., & Kateman, G. (1994). Understanding and Using Genetic Algorithms.2. Representation, Configuration and Hybridization.

Chemometrics and Intelligent Laboratory Systems, 25(2), 99-145. Po Box 211, 1000 Ae Amsterdam, Netherlands:

34. Elsevier Science Bv. doi:10.1016/0169-7439(94)85038-0 Leach, A. (2001). Molecular Modelling: Principles and Applications (2nd Edition). Prentice Hall. Retrieved from http://www.amazon.ca/exec/obidos/redirect?tag=citeulike09- 20&path=ASIN/0582382106

35. Leach, A. R., & Smellie, A. S. (1992). A combined model-building and distance-geometry approach to automated conformational analysis and search. Journal of Chemical Information and Modeling, 32(4), 379-385. doi:10.1021/ci00008a019

36. Lucasius, C. (1993). Understanding and using genetic algorithms Part 1. Concepts, properties and context. Chemometrics and Intelligent Laboratory Systems, 19(1), 1-33. doi:10.1016/0169-7439(93)80079-W

37. Malinowski, E. R. (2003). Factor Analysis in Chemistry. Technometrics (Vol. 45, pp. 180-181). Wiley. doi:10.1198/tech.2003.s145

38. Li Manni, G., Barone, G., Duca, D., & Murzin, D. Y. (2009). Systematic conformational search analysis of the SRR and RRR epimers of 7-hydroxymatairesinol. Journal of Physical Organic Chemistry, (June 2009), n/a-n/a. doi:10.1002/poc.1595

39. Massart, D. L., Heyden, Y. V., & Brussel, V. U. (2004). What Can Chemometrics Do for Separation Science ? Europe, 17(9).

40. Mo, Y., & Gao, J. (2007). Theoretical analysis of the rotational barrier of ethane. Accounts of chemical research, 40(2), 113-9. doi:10.1021/ar068073w

41. Nair, N., & Goodman, J. M. (1998). Genetic Algorithms in Conformational Analysis. Journal of Chemical Information and Modeling, 38(2), 317-320. doi:10.1021/ci970433u

42. Nascimento, R. R., Bruni, A. T. , & Leite, V. B. P. (2009). Estudo conformacional do peptide IAN e seus fragmentos pelo método de análise sistemática reduzida. 07/10/09. Retrieved November 1, 2011, from http://www.athena.biblioteca.unesp.br/exlibris/bd/brp/33004153068P9/2009/na scimento_rr_me_sjrp_parcial.pdf Oblinsky, D. G., Vanschouwen, B. M. B., Gordon, H. L., & Rothstein, S. M. (2009).

43. Procrustean rotation in concert with principal component analysis of molecular dynamics trajectories: Quantifying global and local differences between conformational samples. The Journal of chemical physics, 131(22), 225102. doi:10.1063/1.3268625

44. Pietropaolo, A., Branduardi, D., Bonomi, M., & Parrinello, M. (2011). A

Chirality-Based Metrics for Free-Energy Calculations in Biomolecular Systems. Journal of Computational Chemistry. doi:10.1002/jcc

45. Ramberg, P. J., & Somsen, G. J. (2001). Annals of Science The Young J . H . van ' t Hoff : The Background to the Publication of his 1874 Pamphlet on the Tetrahedral Carbon Atom , Together with a New English Translation. Annals of Science, (September 2011), 51-74.

46. Ramos, L. S., Beebe, K. R., Carey, W. P., M, E. S., Erickson, B. C., Wilson, B. E., Wangen, L. E., et al. (1986). L. Scott Ramos, Kenneth R. Beebe, W. Patrick Carey, Eugenio Sfinchez M., Brice C. Erickson, Bruce E. Wilson, Lawrence E. Wangen,' and Bruce R. Kowalski* Laboratory for Chemometrics, Department. Education, (300), 31-49.

47. Seabra, G. D. M., Walker, R. C., & Roitberg, A. E. (2009). Are current semiempirical methods better than force fields? A study from the thermodynamics perspective. The journal of physical chemistry. A, 113(43), 11938-48. doi:10.1021/jp903474v

48. Silva, D.-A., Domínguez-Ramírez, L., Rojo-Domínguez, A., & Sosa-Peinado, A. (2011). Conformational dynamics of L-lysine, L-arginine, L-ornithine binding protein reveals ligand-dependent plasticity. Proteins, 79(7), 2097-108. doi:10.1002/prot.23030

49. Smellie, A., Stanton, R., Henne, R., & Teig, S. (2003). Conformational analysis by intersection: Conan. Journal of Computational Chemistry, 24(1), 10-20. 111 River St, Hoboken, Nj 07030 Usa: John Wiley & Sons Inc. doi:10.1002/jcc.10175

50. Spellmeyer, D. C., Wong, a K., Bower, M. J., & Blaney, J. M. (1997). Conformational analysis using distance geometry methods. Journal of molecular graphics & modelling, 15(1), 18-36. Retrieved from http://www.ncbi.nlm.nih.gov/pubmed/9346820

51. Wold, S., Esbensen, K., & Geladi, P. (1987). Principal Component Analysis. Chemometrics And Intelligent Laboratory Systems, 2(1-3), 37-52. Po Box 211, 1000 Ac Amsterdam, Netherlands: Elsevier Science Bv. doi:10.1016/0169-7439(87)80084-9

52. Wold, S. (1995). Chemometrics; what do we mean with it, and what do we want from it? Chemometrics and Intelligent Laboratory Systems, 30(1), 109-115. doi:10.1016/0169- 7439(95)00042-9

53. Yanmaz, E., Sarıpınar, E., Şahin, K., Geçen, N., & Çopur, F. (2011). 4D-QSAR analysis and pharmacophore modeling: electron conformational-genetic algorithm approach for penicillins. Bioorganic & medicinal chemistry, 19(7), 2199-210. doi:10.1016/j.bmc.2011.02.035

Chapter 6

CHARGE CARRIER MOBILITY IN PHTHALOCY-ANINES: EXPERIMENT AND QUANTUM CHEMICAL CALCULATIONS

Irena Kratochvilova

Institute of Physics, Academy of Sciences of the Czech Republic, Prague, Czech Republic

INTRODUCTION

The main goal of this chapter is to show how and why quantum chemistry modeling can /should be applied on class of organic materials with relatively high carrier mobility – phthalocyanines (H2Pc, NiPc and $NiPc(SO_3Na)_x$) . It will be shown how Density Functional Theory (DFT) can be used to calculate/ model main parameters that influence the group of material properties that are crucial from practical point of view. With the ongoing miniaturization of microelectronics, functional elements in electronic circuits may soon consist of only a couple of electrons or molecules. We therefore address the question how the physical laws which hold for macroscopic solids become modified when one deals with very small structures. It turns out that down-scaling of electronic properties from the macro world to the atomic or molecular level does not work at all. For example, the famous Ohm's law does not hold anymore, because the resistance does not scale with the length of a ''quantum wire'' [1-15].

Once having realized this fundamental issue, one immediately conceives this as a chance to develop new concepts. Instead of continuously scaling down, as in industrial chip designs, one considers building up electronic circuits with tailored properties ''bottom up''. The first questions which have to be answered now are: ''Which atoms or molecules have to be combined in which way to achieve the desired properties?'' and: ''Which physical and chemical properties determine the electrical conductance of atomic-size or molecular-size circuits?''. In most areas of science, there are free major steps that need to be taken for understanding to be gleaned: the first is synthesis and preparation, the second is measurements and characterization and third is theory and modelling [12-20].

The great advance in measurement and characterization was clearly the advent of scanning probe microscopy which permitted measurements both of structure and of transport at the level of one to a few molecules. The critical advances in theory and modelling came with adaption of the coherent tunnelling models originally developed by Landauer to the study of transport in molecular tunnel junctions [20-31].

Current research focuses, among others, on the design and implementation of nanometer scale electronic systems which exhibit new classical and quantum mechanical effects. The motivation for creating such elements has been two-fold: first, to create nanoscale laboratories to explore physics in a new way, and second, to develop novel devices with significant applications. The architecture of molecular-scale electronic devices can be designed starting from molecular segments whose properties have been known from experiment and/or suitable theoretical models. The field of ''Molecular Electronics'' has been opened by the seminal proposition of A. Aviram and M. Ratner in 1974 to build a diode from a single molecule

[1-2]. Since then, it took almost 20 years before the first molecular diode was experimentally realized. By now, molecular electronics is a broad field of research world-wide [3-21].

Some examples of molecular nanostructures that might be used as switching units, memories, logic elements and devices embodying a negative resistance have recently been demonstrated (e.g., [1-5]). A pioneering construction of a molecular switch was based on the electron tunneling principle [1]. An electron travels along a 'molecular wire' (e.g., a conjugated polymer chain) containing a finite series of periodic potential walls. The tunnel switch is 'on' if the transmission coefficient of the electron is close to unity, i.e., if the electron energy matches pseudostationary energy levels of the walls, and can be turned off by either changing a barrier height or the depth of a potential well, which can be controlled by the dipole moment of polymer side groups. As we have already mentioned the charge transport in molecular electronic materials is a very complex process which can be affected by many physical and chemical parameters; the knowledge of its nature is crucial for the development and optimization of molecular-based devices. Each step forward in understanding and controlling the charge transport is extremely important for the practical applications of molecular systems in electronics [1-16].

One of the reasons for the difficulties in charge-transport phenomena/ conditions description is the lack of a well-understood mechanism of charge transport in organic materials. The charge-carrier transport in molecular systems is a very complicated and comprehensive event affected by many parameters. In the early days, band theory was applied to predict the charge-carrier mobility in organic materials. However, it was repeatedly pointed out by several authors 18-21. that this approach is not suitable. Comparing organic and inorganic semiconductors, the materials from the latter class are usually less disordered and have molecular sites closer to one another. For disordered materials, the hopping mechanism seems to be more appropriate for the description of the charge-carrier behaviour. In this approach, the charge carrier is localised on the molecular site and jumps to the other site by overcoming some energy barrier. However, for materials possessing a preferred direction of charge-carrier transport like phtalocyanines we need a combination of different approaches for the description of the charge-carrier motion along the preferred paths and for hopping among them.

Molecular materials are generally not very good conductors. Main limitation follows from low charge carrier "on-chain" mobility which mostly using the microwave photoconductivity was found to be 10^{-5} m^2 V^{-1} s^{-1} and 10^{-4} m^2 V^{-1} s^{-1}, for s- and p-conjugated molecular materials, respectively. The mobility is limited by polaron formation and by the dispersion of transfer integrals among the monomer units the wire. Electrical current through a single molecule is influenced by charge tunnelling – the Fowler-Nordheim model seems to be a good approximation for the description of charge transport. The presence of dipolar species results in the mobility decrease due to the increase of the transfer integral dispersion. Polar group chemically attached to the molecular wire, can cause the orbital localization. The charge transport in 3D samples can be described by the theory of disordered polarons which postulates that the activation energy of the charge carrier mobility is composed of contribution both from the dynamic disorder, i.e. the polaronic barrier, and from the static disorder, i.e. the variation of the energy of transport states as a result of the environment. The main contribution to the polaron binding energy results from molecular deformation; electron-phonon term makes for 20 % only. Dipolar additives make the distribution of hopping states broadened and new localized states for charge carriers are formed; it results in the reduction of charge mobility [17-19].

Experimentally measured conductivity is a macroscopic phenomenon and reflects on-stack mobility plus intra-stack hopping mobility. Intra-stack transport mainly contributes to the whole resistivity – i.e. the main charge-transport obstruction is between stacks. From the charge carrier mobility point of view phtalocyanines seem to be promising materials - in devices with vacuum evaporated phthalocyanine thin films, the values reached from 10^{-5}

to 10^{-4} cm^2 V^{-1} s^{-1}, when phthalocyanine was evaporated on hot substrates the value increased up to 10^{-2} cm^2 V^{-1} s^{-1} [15-22].

Preliminary results obtained on sulphonated phthalocyanines, i.e., materials containing both electronic and electrolytic segments, seem to be very promising. The charge carrier mobility determined from the dependences of the source-drain current vs. the source-drain voltage of the OFET was surprisingly high. In the case of NiPc(SO$_3$Na)$_4$, the field-effect mobility was 0.02 cm^2 V^{-1} s^{-1} [12].

Phthalocyanines can be organized in columns at a supramolecular level, giving rise to conducting properties. The cofacial stacking of metallophthalocyanines enables electron delocalization along the main axis of the column through π-π orbital overlapping. Metallophthalocyanines generally crystallize in an inclined stacked insulating arrangements called α or β -modifications that do not allow an appropriate overlap of π-orbitals and hence no formation of a conduction band. Only in few cases stacked arrangements are found, being the most representative the nonplanar cone-shaped phthalocyaninatolead (II) (PbPc) in its monoclinic modification. In the last few years, phthalocyanines (Fig. 1) are being intensively studied as targets for optical switching and limiting devices, organic field effect transistors, sensors, light-emitting devices, low band gap molecular solar cells, optical information recording media, photosensitizers for photodynamic therapy, and nonlinear optical materials, among others. Phthalocyanines will burst also in a very near future into the nanotechnology field. Phthalocyanine is an intensely blue-green coloured macrocyclic compound. Phthalocyanines form coordination complexes with most elements of the periodic table. Phthalocyanines are structurally related to other macrocyclic pigments, especially the porphyrins. Four pyrrolelike subunits are linked to form a 16-membered ring. The pyrrole-like rings within H2Pc are closely related to isoindole. Both porphyrins and phthalocyanines function as planar tetradentate dianionic ligands that bind metals through four inwardly projecting nitrogen centers. Such complexes are formally derivatives of Pc^{2-}, the conjugate base of H$_2$Pc.

Many derivatives of the parent phthalocyanine are known, where either carbon atoms of the macrocycle are exchanged for nitrogen atoms or where the hydrogen atoms of the ring are substituted by functional groups like halogens, hydroxy, amino, alkyl, aryl, thiol, alkoxy, nitro, etc. Due to phthalocyanine stack formation and the presence of ionic groups charge transport consists of an electronic feature through the stack, charge hopping among the stacks and ionic type of the transport.

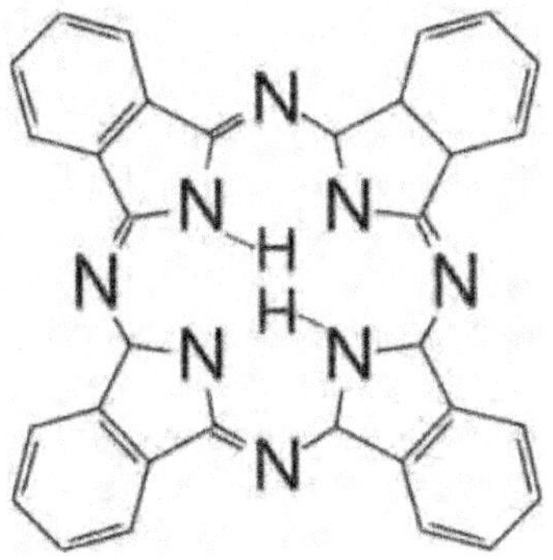

Figure 1: Chemical structure of H2 phthalocyanine (H2Pc).

In order to model the physical properties of H_2Pc layer, quantum chemical calculations on H_2Pc dimers and tetramers were performed at DFT level. The a and b types of polymorph of H_2Pc dimer structures were optimized using several density functionals. The optimizations lead to geometry with almost parallel orientation of H_2Pc planes. The optimised conformer structures are depicted in Fig. 2, the structural parameters calculated for different functional and basis sets are listed in Table 1.

a)   b)   c)

Figure 2: DFT calculated optimized structures of H2Pc dimers and tetramer. In the figure, a) a- modification of H2Pc dimer b) b-modification of H2Pc dimer c) a-modification of H2Pc tetramer, respectively.

Calculated geometry of H_2Pc dimer reasonably well represents the local part of the experimental structure of phthalocyanine films [15-19]. The best performance was found for MPW1B95/6-31+G* calculations, where calculated structures fit the phthalocyanine stacking in the crystal. For a-modification, the calculated structural parameter b = 3.77 Å, derived as averaged distances of corresponding atoms at individual monomers well reproduces the experimental value of 3.81 Å . The calculated approximate interplane distance (3.48 Å) is close to experimental value of 3.4 Å [15] as well. For b-modification, the calculations give the interplane distance 3.46 Å (experiment 3.4 Å) and parameter b = 4.87 Å (experiment 4.72). The inclusion of diffusion functions (6-31+G* basis) does not substantially change the structural parameters. As shown in Table 1, for a-modification the functional MPW1B95 slightly better describes the distance parameter b than MPWB1K. For b-modification, the performance of both functional is comparable. Standard B3LYP functional strongly overestimates the separation of molecular planes and fails to predict the real structure.

Table 1: Calculated structural parameters (in Å) and stabilization energies (kcal/mol) of phthalocyanine dimers and tetramers

DFT Functional	MPW1B95		MPWB1K		Experiment
Basis Set	6-31G*	6-31+G* (Å)	6-31G*	6-31+G* (Å)	(Å)
H_2Pc dimer polymorph α					
neutral					
a[1]	3.463	3.478	3.442	3.466	3.4
b[2]	3.744	3.768	3.721	3.752	3.81
ΔE[3]	5.1	4.6			
cation					
a[1]	3.440				
b[2]	3.665				
anion					
a[1]	3.450				
b[2]	3.804				
H_2Pc dimer polymorph β					
neutral form					
a[1]	3.428	3.456	3.428	3.432	3.4
b[2]	4.849	4.871	4.840	4.841	4.72
ΔE[3]	3.6	3.0			
cation					
a[1]	3.289				
b[2]	5.255				
anion					
a[1]	3.325				
b[2]	4.678				
H_2Pc tetramer polymorph α					
a[1]	3.388				3.4
b[2]	3.715				3.81
H_2Pc tetramer polymorph β					
a[1]	3.399				3.4
b[2]	4.833				4.72

[1] interplanar distance
[2] the average distance of corresponding atoms
[3] stabilization energy of neutral dimers in kcal/mol

Calculated stabilization energies listed in Table 1 are typical for p – p stacking interactions and indicate that the polymorph a is more stable than the b-modification; the calculated difference of stabilization energies is 1.5 kcal/mol for the functional MPW1B95. Table 1 shows that the withdrawal of electron from b polymorph leads to geometry with more closely lying interplanar arrangements of monomeric subunits characterized by the distances a = 3.44

Å and b = 3.67 Å. Optimized geometry of anionic form is characterized by similar interplane distance 3.45 Å and larger parameter b = 3.80 Å. Cationic form of b polymorph is characterized by the distances a = 3.29 Å and b = 5.26 Å, anionic form of this polymorph by distances a = 3.33 Å and b = 3.68 Å.

All following single point calculations were done at optimized geometries. On the basis of the dimer calculations, the geometry optimization of H2Pc tetramer was done by MPW1B95/6-31G* calculations on both polymorphs. The middle part of the optimized tetramer structure interprets well the H2Pc crystal structure (the calculated interplane distance is 3.39 Å 15.). The optical spectroscopy was used for the characterization of the phthalocyanine layer. Figure 3 shows the experimental spectrum in the region of the phtalocyanine Q band. The comparison of characteristic features of this spectrum with the previously measured spectra 36. of different polymorphs of H2Pc layers indicates the presence of the a polymorph.

TD DFT calculated transitions on this polymorph reasonably well reproduce the strong features of characteristic Q band and the effect of aggregation. The experimental transition energies measured at 1.77 and 1.93 eV are slightly overestimated by TD DFT calculations (obtained at 2.15 and 2.21 eV) with oscillator strengths 0.52 and 0.58, respectively. Transitions with low oscillator strength calculated at 1.65 and 1.71 eV describe the appearance of the week features in the long wave region. The TD DFT calculation on the b polymorph describes correctly the shift of the intense transitions to the longer wavelengths.

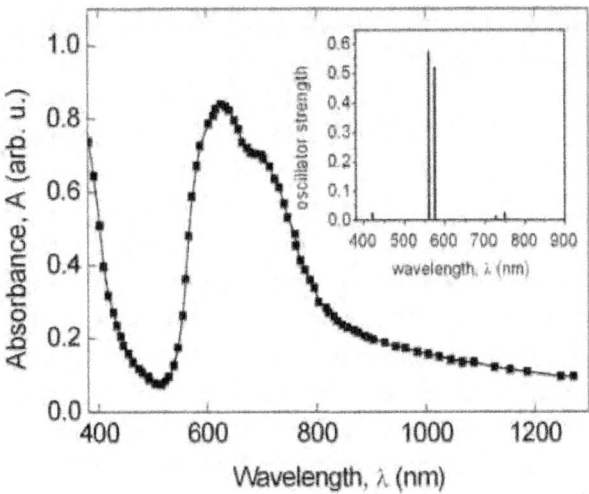

Figure 3: The experimental spectrum of H_2Pc layer in the region of the Q band. Inset shows calculated transitions for H_2Pc dimer.

At room temperature, the charge carrier transport can be described by a hopping within the Marcus theory. Subsequently, the rate expression of a self-exchange process is provided by the expression [19].

$$k_{et} = \frac{2\pi}{\hbar} \frac{1}{\sqrt{4\pi\lambda^+ kT}} (t^+)^2 \exp\left(-\frac{\lambda^+}{4kT}\right),$$

(1)

where $\hbar$ is the reduced Planck constant, k is the Boltzmann constant, T is the temperature an λ^+ is the reorganisation energy and t $^+$ is the electronic coupling matrix element (the measure of charge transport probability – electron transfer integral) for the charge transfer. In order to estimate the electronic coupling between individual monomeric units, electron transfer integrals were calculated for α and β crystallographic modifications of the phthalocyanine dimer in HF approximation.

For the configuration with doubly occupied HOMO on the first monomer and singly occupied HOMO of the second one, calculation gives electron transfer integral (for otimized geometry MPW1B95/6-31G*) values for non-relaxed cation radical 11.7 meV and 5.7 meV (hole transfer) for the α and β modification, respectively. It is supposed that in the course of electron or hole hopping, the atoms can relax in order to reach the minimum of energy at the potential surface. Transfer integral for optimized cation radical α crystal form with relaxed atoms (UMPW1B95/6-31G*) is 18.7 meV. The calculations indicate that electron mobility is smaller than hole mobility because optimized geometry of anionic form is characterized by larger interplane parameter b (see Tab. 1) - for anionic form b=3.804 Å, for cationic form b=3.665 Å.

The transfer integral t $^+$ was calculated according to

$$t^+ = \frac{H_{RP} - S_{RP}(H_{RR} + H_{PP})/2}{1 - S_{RP}^2},$$

(2)

H_{RP} is the interaction energy between reactant and product states, S_{RP} is the overlap between the reactant and product states and H_{RR} is electronic energy of the reaktant state and H_{PP} is electronic energy of the product state. All these terms were obtained via the direct coupling of localised monomer orbitals. On the molecular level, three factors, the electronic coupling (transfer integral t $^+$) between the individual parts of the molecule, the reorganisation energy λ^+ during charge transport and the effective length of hole transfer L, are usually considered to be important for charge transport in organic materials. The reorganization energy λ^+ consists of the sum of λ_1^+ and λ_2^+, $\lambda^+ = \lambda_1^+ + \lambda_2^+$, where the deformation energy of the system was calculated as the difference between the vertical and cationic state:

$\lambda_1^+ = E_+(Q_N) - E_+(Q_+)$ and $\lambda_2^+ = E_+(Q_+) - E_+(Q_N)$. Here, $E_+(Q_N)$ is the total electronic energy of the cationic state in the neutral geometry, $E_+(Q_+)$ is the total energy of the cationic state in the cationic state geometry, $E_N(Q_+)$ is the total energy of the neutral state in the cationic state geometry and $E_N(Q_N)$ is the total energy of the neutral state in the neutral geometry. λ_1^+ is frequently called as deformation energy of the system.

The diffusion coefficient D of charge carriers can be expressed using the EinsteinSmoluchowski equation

$$D = \frac{L^2 k_{et}}{2}.$$

(3)

This makes it possible to evaluate the drift mobility of the charge carriers using the Einstein relation $\mu = eD/kT$. It should also be noted that the calculated charge mobility mentioned here represents the zero electric field approximation value.

To explain the differences between the charge carrier mobility in sulphonated and nonsulphonated Ni phthalocyanines (NiPc), quantum chemical calculations were performed on the DFT (density functional theory) level for H_2Pc, NiPc and $NiPc(SO_3Na)_x$ (x = 1, 2) dimers and their cationic and anionic forms. Quantum chemical modeling was found to be very useful – we were able to see states of various molecular systems from new and comprehensive perspectives. The optimized structure of the $Ni(Pc(SO_3Na)_2)_{\cdot 2}$ is depicted in Fig.4. The calculated geometry of the NiPc dimer and tetramer reasonably well represents the local part of the NiPc crystal structure [34]. The best performance for the NiPc dimer was found by the MPW1B95/6-31+G* method. The calculated approximate interplane distance (3.383 Å) is close to the experimental value of 3.4 Å [15,19].

The calculated stabilization energies are typical for π–π stacking interactions. The calculated stabilization energies with the BSSE correction are 4.1, 5.4, 7.4 and 11.3 kcal/mol for the $H_2Pc._2$, $NiPc._2$, $NiPc(SO_3Na)._2$ and $NiPc(SO_3Na)_{2\cdot2}$ dimers, respectively. The MPWB1K/6- 31G* calculations yield similar results, whereas the calculated stabilization energies are slightly higher. The increasing stabilization energies in the series of dimers going from $H_2Pc._2$ to $NiPc(SO_3Na)_{2\cdot2}$ indicate the better stability and organization of the $NiPc(SO_3Na)_2$ layers in comparison with the others.

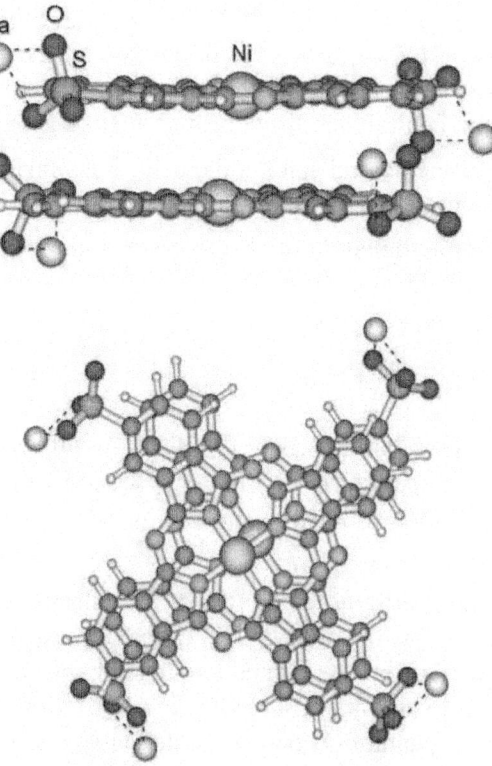

Figure 4: The DFT-calculated optimized structure of the dimer $NiPc(SO_3Na)_{2\cdot2}$. The dashed line indicates the shortest distance between the O and Na atoms.

Figure 5: The schematic representation of the HOMO (top) and HOMO-1 (bottom) of the dimer $NiPc(SO_3Na)_{2\cdot2}$.

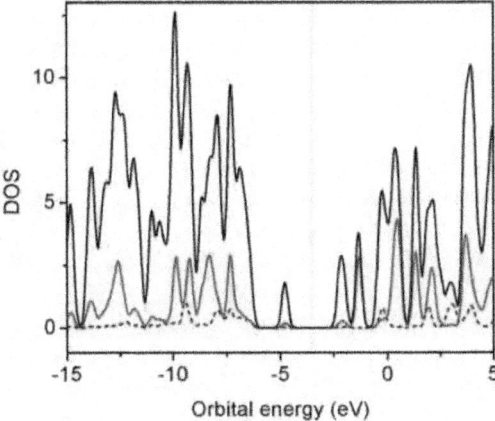

Figure 6: The density of the states (DOS) for NiPc(SO$_3$Na)$_{2'2}$. The black line indicates the total density of the states of the whole system, the blue dashed line the contributing Ni orbitals and the red one the contribution from the SO$_3$Na group. The vertical dashed line indicates an approximate midpoint of the HOMO–LUMO levels.

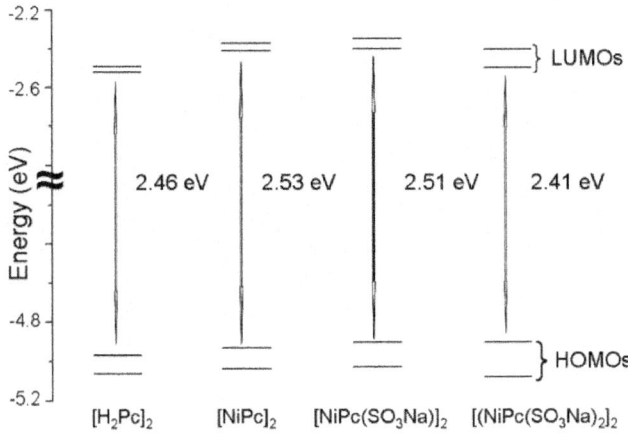

Figure 7: The energies of the frontier orbitals of the studied phthalocyanine dimers.

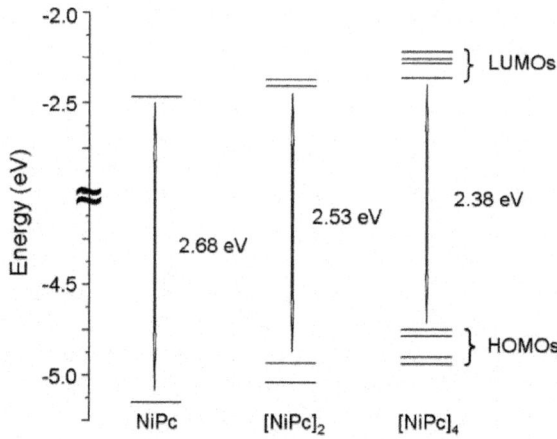

Figure 8: The energies of the frontier orbitals of the NiPc monomer, dimer and tetramer.

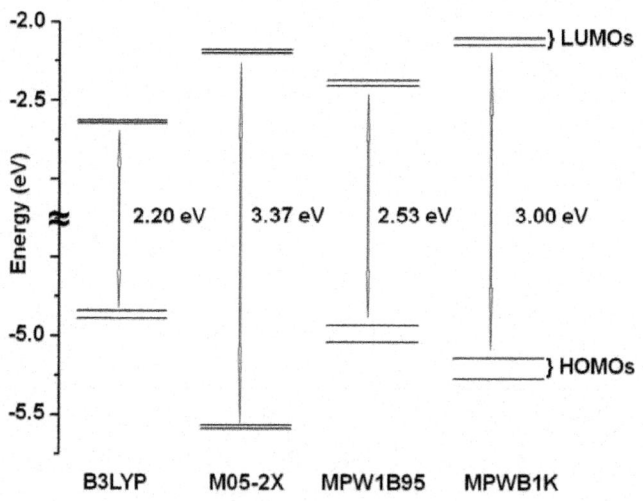

Figure 9: The energies of the frontier orbitals of the NiPc dimer, calculated with different density functional for geometry optimized with corresponding functional.

Figure 5 depicts two highest occupied molecular orbitals (HOMO, HOMO-1) of $NiPc(SO_3Na)_{2'2}$, the character of which can influence the hole conductivity. The plot of the density of the states in Fig. 6 shows the distribution of one-electron molecular orbitals and indicates that the Ni and SO_3Na orbitals do not substantially contribute to the frontier molecular orbitals of the dimer $NiPc(SO_3Na)_{2'2}$. These orbitals are formed by Pc p orbitals. Owing to the mutual interaction of the p monomer orbitals, the frontier orbitals of the supersystem form groups of two closely lying p molecular orbitals. The contribution of the 3d metal orbitals is negligible as in the case of the monomeric NiPc. The second pair of lower-lying occupied orbitals lies about 1.57 eV lower. Thus, it can be supposed that these orbitals do not strongly influence the charge carrier transport. The calculated HOMO– HOMO-1 separations are visualized in Figs. 7 and 8 while HOMO–LUMO gaps are listed in Tables 4 and 5. Figure 7 depicts the splitting of the frontier orbitals and their mutual position in the series of the systems studied; Fig. 8 compares orbital splitting in the series of the NiPc monomer, dimer and tetramer. The splitting of the frontier orbitals reflects the mutual interaction and thus the probability of charge transfer between individual monomeric subsystems. It should be mentioned that calculated HOMO-LUMO gap (Fig. 8) is diminishing in the series NiPc, $NiPc._2$ and $NiPc._4$ and approaching to the experimental value of 1.8 eV. Fig. 9 shows how the variation of functional influences the energies of frontier orbitals.

The rate constant k_{ET} of the electron transfer between molecular orbitals was expressed – see (1). The electronic coupling between the individual monomeric units V_{RP}, the electrontransfer integral, can be calculated by several procedures 8-13. Koopman's approximation, where the electronic coupling is estimated as half of the corresponding orbital energies, was used for the matrix-element estimation 14-18. This method has already been used in the case of interacting metal-containing dimmers 22.

The values of the electron-transfer integrals for the closed-shell systems were estimated from the separation of HOMO and HOMO-1 for the optimized structures. The values of the electron-transfer integrals based on the MPW1B95/6-31G* functional model are 47.5 meV, 54.3 meV, 63.1 meV and 88.6 meV for $H_2Pc._2$, $NiPc._2$, $NiPc(SO_3Na)._2$ and $NiPc(SO_3Na)_{2}._2$, respectively (see Table 2).

From these values, it follows that the electronic contribution to the rate constant of the electron transfer (proportional to the power of the electronic coupling – V^2_{RP}) for $NiPc(SO_3Na)_{2}._2$ should be 2.7 times larger than for $NiPc._2$ and 3.5 times larger than for $H_2Pc._2$, whereas the electronic contribution for $NiPc(SO_3Na)._2$ is only about 1.4 times larger than for $NiPc._2$. The dependence of the electron-transfer integrals on the Ni-Ni separation for $NiPc._2$ and $NiPc(SO_3Na)_{2}._2$ dimers is depicted in Fig. 10. The electron-transfer integral value varies strongly with the intermolecular separation, nevertheless the ratio between V_{RP} calculated for SO_3Na substituted and unsubstituted systems is close to this ratio obtained from equilibrium intermolecular separation.

Table 2: The calculated stabilization energies – ΔE, (kcal/mol), electron-transfer integrals – ET coupling (meV) and magnitude of the gap between the HOMO a LUMO orbitals of the various types of phthalocyanine dimers

| Dimer | MPW1B95/6-31G* | | | |
| | ΔE | ΔE(BSSE) | ET coupling | Gap |
	kcal mol^{-1}		meV	eV
$[H_2Pc]_2$	14.8	4.1	47.5	2.46
$[NiPc]_2$	19.4	5.4	54.3	2.53
$[NiPc(SO_3Na)]_2$	22.1	7.4	63.1	2.51
$[NiPc(SO_3Na)_2]_2$	27.6	11.3	88.6	2.41
$[(NiPc(SO_3Na)_2)(NiPc(SO_3Na)SO_3)]^-$	31.2[a]	13.2[a]		2.34[a]
$[NiPc(SO_3Na)(SO_3)]_2^{2-}$	6.3[a]	-13.1[a]		2.41[a]
$[NiPc(SO_3)_2]_2^{4-}$	-79.4	-100.5		2.44

[a] an average of the calculated values for the two different Na ions locations

Both water molecules surrounding or the application of an electric field can cause the dissociation of the ionic Na-SO$_3$ bond. In order to explain the effect of the Na$^+$ dissociation, calculations were performed on the model dimeric systems, where Na$^+$ ions were stepwise removed from the NiPc(SO$_3$Na)$_2$ dimer . The stabilization energy of the system with one Na$^+$ removed (13.2 kcal mol^{-1}) is even larger than that of NiPc(SO$_3$Na)$_{2\cdot2}$. From Table 2, it can be seen that Ni$_2$Pc$_2$(SO$_3$Na)$_3$(SO$_3$).- is the most stable and thus the best organized of all the structures investigated. The density of the states plot depicted in Fig. 11 shows how the withdrawal of one Na$^+$ influences the electronic structure. Projected Densities of States reported in this chapter were broadened using discrete molecular orbital levels – the broadening parameter used in our PDOS calculations (the full width at half maximum) was 0.3 eV.

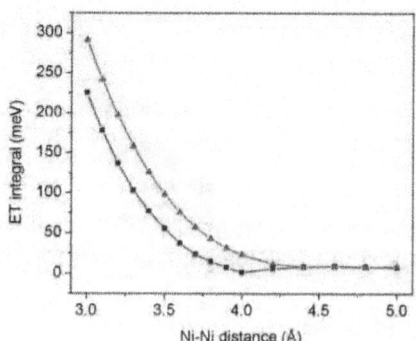

Figure 10: The dependence of electron-transfer integrals on the Ni-Ni separation. NiPc.$_2$ - blue squares,

NiPc(SO$_3$Na)$_{2·2}$ – red triangles.

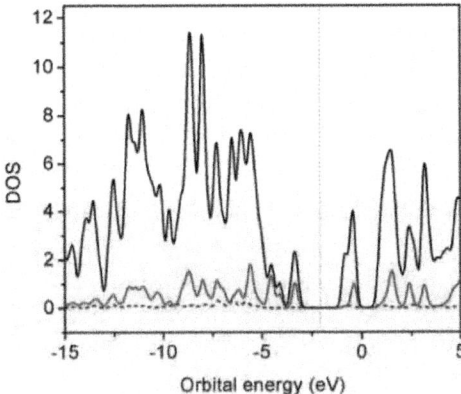

Figure 11: The density of the states (DOS) for the system Ni$_2$Pc$_2$(SO$_3$Na)$_3$(SO$_3$).-. the black line indicates the total density of the states of the whole system, the blue dashed line the contributing Ni orbitals and the red one the total contribution from the SO$_3$Na and SO$_3$ groups; the vertical dashed line indicates an approximate midpoint of the HOMO–LUMO levels.

In comparison with the unperturbed system, the set of HOMOs is no longer separated from the lower-lying occupied orbitals and contains contributions from the SO$_3^-$ group. It can be supposed that after Na$^+$ ion dissociation the whole layer contains mobile Na$^+$ and fixed NiPc(SO$_3$Na)(SO$_3$).- ions surrounded by water molecules. In this case, the highest molecular orbitals, which contain contributions from the SO$_3^-$ group and are close to the lower-lying occupied orbitals, allow the whole system to contain electronic states that are more extended and therefore more conductive in character.

Due to the withdrawal of the remaining Na+ ions, the stabilization energies decrease. The least stable dimer in the series NiPc(SO$_3$Na)$_{2·2}$, NiPc(SO$_3$Na)(SO$_3$).$_2$ $^{2-}$ and NiPc(SO$_3$)$_{2·2}$ $^{4-}$ is the last, completely dissociated dimer, i.e. NiPc(SO$_3$)$_{2·2}$ $^{4-}$. This fact can induce reduced layer organization and thus lower the conductivity of the systems with more than one Na$^+$ ion removed.

How can the selected chemical modifications of Ni phthalocyanine affect on the molecular level the charge carrier mobility? At zero-gate bias, the HOMO band is filled. As the gate voltage is swept more negative, the positions of the LUMO and HOMO bands rise in energy, allowing mobile holes to sustain a source-drain current via the HOMO band. Materials based on NiPc and NiPc(SO$_3$Na)$_x$. are hole semiconductors, which means that the source of the charge carriers in the structure electrode-semiconductor – electrode is not the injecting metal electrode but the semiconductor. In the case of NiPc(SO$_3$Na)

(SO_3).⁻ , the contribution from the SO_3 anions to the whole HOMO orbital system enables the mobile holes to be more delocalized; thus the change charge transport (hopping) mechanism becomes more ballistic (less scattered and more effective). The stability of dimers (the calculated stabilization energy), which induces better layer organization and consequently better charge mobility[37] grows in the series H_2Pc.⁻$_2$, $NiPc$.⁻$_2$, $NiPc(SO_3Na)$.⁻$_2$, $NiPc(SO_3Na)_2$.⁻$_2$ and $(NiPc(SO_3Na)_2)(NiPc(SO_3Na)SO_3)$.⁻ The withdrawal of more than one Na cation leads to destabilization and consequently to a decrease in layer organization. The layer of $Ni_2Pc_2(SO_3Na)_3(SO_3)$.⁻ should be the most regular (is supposed to be the most conductive) of all the structures examined. In the case of $NiPc(SO_3Na)_{3.3}$ the charge carrier mobility (0.02 cm² V⁻¹ s⁻¹) is lower than for $NiPc(SO_3Na)1.5$ (1.08 cm² V⁻¹ s⁻¹) ⁻ see Tab. 3.

Table 3: The mobilities for the various types of nickel phthalocyanines

Phthalocyanine	Film preparation	Mobility (cm² V⁻¹ s⁻¹)
NiPc	evaporation	10⁻⁵
NiPc(SO₃Na)₃.₃	spin-coating	0.02
iPc(SO₃Na)₁.₅	spin-coating	1.08

From calculating spatial models it was figured out that due to the sterical communication hindrance caused by third and fourth SO_3Na groups the $NiPc(SO_3Na)_{3.3}$ dimers are less stable. This fact results in reduced organized layer structure with reduced charge mobility. Using DFT calculations, several aspects that influence the charge-carrier mobility in the case of the materials were compared:

1. Charge transfer probability grows in the series H_2Pc.⁻$_2$, $NiPc$.⁻$_2$, $NiPc(SO_3Na)$.⁻$_2$ and $NiPc(SO_3Na)_2$.⁻$_2$ because of the increasing electronic coupling. The presence of (SO_3Na) groups increases the probability of electron transfer between the composing units.

2. At the same time, the stabilization energies grow in the series H_2Pc.⁻$_2$, $NiPc$.⁻$_2$ and $NiPc(SO_3Na)_2$.⁻$_2$. Due to larger stabilization energies of the systems with (SO_3Na) groups, it can be assumed that layers with (SO_3Na) groups are organized more regularly – resulting in better p–p interaction and consequently better charge carrier mobility than in the case of nonsubstituted ones. The withdrawal of one Na^+ ion leads to a further stabilization of the structure and thus a higher probability of charge-carrier transport. The dissociations of more than one Na^+ ion strongly destabilize the layer structure. The probability of charge transfer through sulphonated Ni-phthalocyanine dimers with totally removed Na^+ ions should, therefore, be lower than in the case of standard sulphonated Ni phthalocyanines dimers.

3. The system is strongly affected by an external electric field and water molecules in the vicinity, which can cause the dissociation of the ionic Na–SO$_3$ bond. After the Na–SO$_3$ bond's dissociation, the whole layer contains mobile Na$^+$ and fixed Ni$_2$Pc$_2$(SO$_3$Na)$_3$(SO$_3$).$^-$ anions which allows the mobile holes to delocalize and increase the probability of hole transfer through the layer structure. We found that the highest molecular orbitals in Ni$_2$Pc$_2$(SO$_3$Na)$_3$(SO$_3$).$^-$ contain new electronic contributions to the density of the states from the SO$_3$ - group, which should lead to an increase in the hole mobility and thus conductivity.

Other aspects which can increase charge carrier mobility in of the OFET structures based on sulphonated phthalocyanine derivatives should be mentioned: 1. Better resistivity to oxygen and thus the stability of the system in atmospheric air. 2. The interface SiO$_2$/NiPc(SO$_3$Na)$_{2-x}$ is between two hydrophilic groups and thus adheres materials enabling the formation of a molecular layer on the SiO$_2$ surface with proper and regular molecular orientation which increases the conductivity .

The objective of this chapter was to present results of DFT studies of class of organic materials with relatively high carrier mobility – phthalocyanines. It was shown how DFT can be used to calculate/model some parameters that influence charge carrier mobility on essentially level 18.

The calculations indicate that electron mobility in H$_2$Pc is smaller than hole mobility because optimized geometry of anionic form is characterized by larger interplane parameter. In the case of NiPc(SO$_3$Na)(SO$_3$).$^-$, the contribution from the SO$_3$ anions to the whole HOMO orbital system enables the mobile holes to be more delocalized; thus the change charge transport (hopping) mechanism becomes more ballistic (less scattered and more effective). The withdrawal of one Na$^+$ ion leads to a further stabilization of the structure and thus a higher probability of charge-carrier transport.

- Larger stabilization energies of the systems with (SO$_3$Na) groups - layers with (SO$_3$Na) groups are organized more regularly – resulting in better π–π interaction and consequently better charge carrier mobility than in the case of nonsubstituted ones.

- The withdrawal of one Na$^+$ ion leads to a further stabilization of the structure and thus a higher probability of charge-carrier transport.

- The dissociations of more than one Na$^+$ ion strongly destabilize the layer structure. The probability of charge transfer through sulphonated Ni-phthalocyanine dimers with totally removed Na$^+$ ions is lower than in the case of standard sulphonated Ni phthalocyanines dimers.

REFERENCES

1. M. A. Reed, Proc. IEEE , 1999 , 87, 652.

2. D. Braga, G. Horowitz, Adv. Mater., 2009, 21, 1.

3. J. Health, M. Read, Molecular Electronics, Physics Today, 2003, 43.

4. C.P. Collier et al., Science, 2000, 289, 1172

5. A.J. Heinrich et al. Science, 2002, 298, 1381

6. J. Park et al. Nature, 2002, 417, 722

7. M. Taniguchi and T. Kawai, Physica E 2006, 33 1-12.

8. S. Datta, W. Tian, S. Hong, R. Reifenberger, I. Henderson, and C. Kubiak, Phys.Rev.Lett. 1997, 79 2530-2533.

9. F. Zahid, M. Paulsson and S. Datta, Chapter published in "Advanced Semiconductors and Organic nano-Techniques", edited by H. Morkoc, Academic Press 2003. See also arXiv: Cond-Mat/0208183.

10. K. Müllen, G. Wegner (Eds.), Electronic Materials: The Oligomeric Approach (Wiley-VCH, Weinheim, 1998).

11. J. Simon, P. Bassoul, Design of Molecular Materials. Supramolecular Engineering (Wiley, Chichester, 2000).

12. R.G. Enders, D. L. Cox, and R.R.P. Singh, Colloquium: Rev. Modern Phys.2004, 76 195-217.

13. L. Cai, H. Tabata, and T. Kawai, Self-assembled DNA networks and their electrical conductivity, Appl. Phys. Let. 2000, 77 3105-3106.

14. I. Kratochvílová, S. Nešpůrek, J. Šebera, S. Záliš, M. Pavelka, G. Wang, J. Sworakowski, Eur. Phys. J. E, 2008, 25, 299.

15. I. Kratochvílová, K. Král, M. Bunček, A. Víšková, S. Nešpůrek, A. Kochalska, T. Todorciuc, M. Weiter, B. Schneider, Biophys. Chem., 2008, 138, 3.

16. I. Kratochvílová , K. Král, M. Bunček, S. Nešpůrek, T. Todorciuc, M. Weiter, J. Navrátil, B. Schneider and J. Pavluch, Cent. Eur. J. Phys., 2008, 6, 422.

17. S. Záliš, I. Kratochvílová, A. Zambova, J. Mbindyo, T. E. Mallouk, T. S. Mayer, Eur. Phys. J. E, 2005, 18, 201.

18. M. F Craciun, S Rogge, M – J. L. Den Boer, S Margadonna, K Prassides, Y Iwasa, A. F Morpurgo, Advanced Materials 18, (2006) 320.

19. I.I. Fishchuk, A. Kadashchuk, V.N. Poroshin, N. Volodymyr, H. Bassler, Philosophical Magezine, 90 (2010) 129.

20. J. Šebera, S. Nešpůrek, I. Kratochvílová, S. Záliš, G. Chaidogiannos, N.

Glezos, European Physical Journal B 72 (2009) 385.

21. M. C. Reese, M. Roberts, M. Ling and Z. Bao, Mater. Today, 2004, 7, 20.

22. C. D. Dimitrakopoulos, P. R. L. Malenfant, Adv. Mater., 2002, 14, 99.

23. R. Zeis, T. Siegrist and C. Kloc, Appl. Phys. Lett., 2005, 86, 022103.

24. S. Nešpůrek, G. Chaidogiannos, N. Glezos, G. Wang, S. Böhm, J. Rakušan, M. Karásková, Mol. Cryst. Liq. Cryst., 2007, 468, 355.

25. F. Yang, M. Shtein and S. R. Forrest, Nat. Mater., 2005, 4, 37.

26. Jakub Šebera, Stanislav Nešpůrek, Irena Kratochvílová, Stanislav Záliš, George Chaidogiannos, Nikos Glezos, European Physical Journal B - Condensed Matter and Complex Systems 72 (2009) 385-395.

27. S. M. Bayliss, S. Heutz, G. Rumbles, T. S. Jones, Phys. Chem. Chem. Phys. 1, 3673 (1999)

28. R.D. Gould, Coord.Chem. Rev. 156, 237 (1996)

29. M. Ashida, N. Uyeda, E. Suito, Bull. Chem. Soc. Jpn. 39, 2616 (1966)

30. A. Farazdel, M. Dupuis, E. Clementi, A. Aviram, J. Am. Chem. Soc. 112, 4206 (1990)

31. V. Coropceanu, J. Cornil, D.A. da Silva Filho, Y. Olivier, R. Silbey, J. L. Brédas, Chem. Rev. 107, 926 (2007)

32. Hynek Nemec, Irena Kratochvilova , Petr Kuzel, Jakub Sebera, Anna Kochalska, Juraj Nozar and Stanislav Nespurek , Phys. Chem. Chem. Phys., 2011, 13, 2850–2856

Chapter 7

QUANTUM CHEMISTRY STRUCTURES AND PROPERTIES OF 134 KILO MOLECULES

Raghunathan Ramakrishnan [1], Pavlo O. Dral [2,3], Matthias Rupp [1] & O. Anatole von Lilienfeld [4]

[1]Department of Chemistry, Institute of Physical Chemistry, University of Basel, Klingelbergstrasse 80, CH-4056 Basel, Switzerland

[2]Max-Planck-Institut für Kohlenforschung, Kaiser-Wilhelm-Platz 1, 45470 Mülheim an der Ruhr, Germany

[3]Computer-Chemie-Centrum, University of Erlangen-Nuremberg, Nägelsbachstr. 25, 91052 Erlangen, Germany

[4] Leadership Computing Facility, Argonne National Laboratory, 9700S. Cass Avenue, Lemont, Illinois, 60439, USA

ABSTRACT

Computational *de novo* design of new drugs and materials requires rigorous and unbiased exploration of chemical compound space. However, large uncharted territories persist due to its size scaling combinatorially with molecular size. We report computed geometric, energetic, electronic, and thermodynamic properties for 134k stable small organic molecules made up of CHONF. These molecules correspond to the subset of all 133,885 species with up to nine heavy atoms (CONF) out of the GDB-17 chemical universe of 166 billion organic molecules. We report geometries minimal in energy, corresponding harmonic frequencies, dipole moments, polarizabilities, along with energies, enthalpies, and free energies of atomization. All properties were calculated at the B3LYP/6-31G(2df,p) level of quantum chemistry. Furthermore, for the predominant stoichiometry, $C_7H_{10}O_2$, there are 6,095 constitutional isomers among the 134k molecules. We report energies, enthalpies, and free energies of atomization at the more accurate G4MP2 level of theory for all of them. As such, this data set provides quantum chemical properties for a relevant, consistent, and comprehensive chemical space of small organic molecules.

This database may serve the benchmarking of existing methods, development of new methods, such as hybrid quantum mechanics/machine learning, and systematic identification of structure-property relationships.

BACKGROUND & SUMMARY

The goal of computationally designing novel materials and molecules with desired physicochemical properties is yet to be achieved. High-throughput screening represents the most straightforward approach towards materials design[1]. However, it presupposes that all assumptions and approximations inherent to the employed modeling techniques are applicable to the entire chemical compound space, which is the space populated by all stable molecules or materials[2]. Furthermore, due to the combinatorial scaling of chemical space with molecular size, it is difficult to explore or even navigate. Conclusive insights about the domain of applicability (transferability) are lacking even for the most popular first principle quantum chemistry methods. For example, the reliability and accuracy of density functional theory is known to dramatically depend on chemical composition and atomistic configurations[3], highlighting the importance of reliable experimental[4] or high-level quantum chemistry state-of-the-art results[5,6]. Unfortunately, the systems reported are typically small, which implies the existence of severe selection bias. One can therefore question how representative they are. The problem of representative diversity has triggered the design of special purpose chemical space libraries for method validation or molecular design[7,8,9,10].

Here, we report molecular structures and properties obtained from quantum chemistry calculations for the first 134k molecules of the chemical universe GDB-17 data base[11], covering a molecular property set of unprecedented size and consistency. The data-set corresponds to the GDB-9 subset of all neutral molecules with up to nine atoms (CONF), not counting hydrogen. The molecular size distribution of all 134k molecules is shown in Fig. 1. This data set contains small amino acids, such as GLY, ALA, as well as nucleobases cytosine, uracil, and thymine. Also pharmaceutically relevant organic building blocks, such as pyruvic acid, piperazine, or hydroxy urea are included. Among the 134k molecules, there are 621 stoichiometries, among which $C_7H_{10}O_2$ dominates with 6,095 constitutional isomers for which atomization energies and radii of gyration also are on display in Fig. 1.

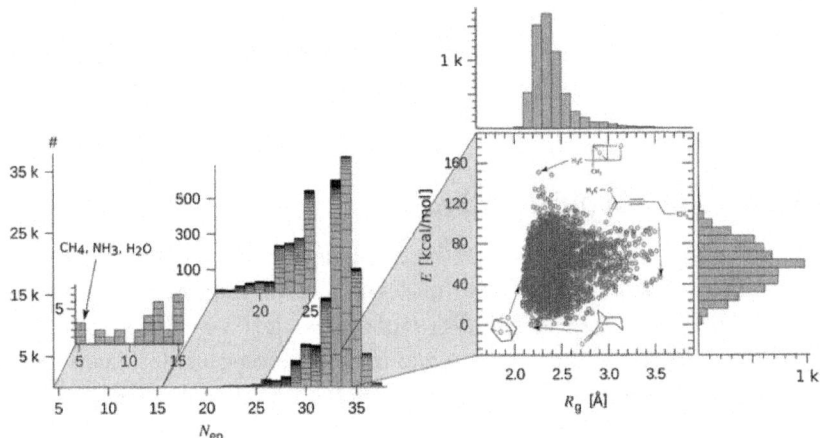

Figure 1: Illustration of the scaling of chemical space with system size.

For the smallest 134k molecules, with up to 9 heavy atoms CONF (not counting hydrogens) taken from the chemical universe GDB-17[11], the distribution of molecular size is shown as a function of number of occupied electron orbitals, i.e. number of electron pairs, $N_{ep} = N_e/2$. Each black box denotes the number of constitutional isomers for one out of the 621 stoichiometries present in the 134k molecules. The two left-hand side insets correspond to zoom-ins for smaller compounds. The right-hand side inset zooms in on the predominant stoichiometry, $C_7H_{10}O_2$, and features a scatter plot of G4MP2 relative (w.r.t. global minimum) potential energies of atomization E versus molecular radius of gyration, R_g. Joined projected distributions are shown as well.

For all 134k molecules, we have calculated equilibrium geometries, frontier orbital eigenvalues, dipole moments, harmonic frequencies, polarizabilities, and thermochemical energetics corresponding to atomization energies, enthalpies, and entropies at ambient temperature. These properties have been obtained at the B3LYP/6-31G(2df,p) level of theory which forms the basis for the more accurate state-of-the art Gn methods which are on par with experimental accuracy[12]. For the 6,095 constitutional isomers of the predominant stoichiometry, $C_7H_{10}O_2$, we report the energetics at the significantly more accurate G4MP2[12] level of theory.

This report is structured as follows. We first describe the genesis of the results. Thereafter, we discuss the validation of our DFT results by comparison to (i) G4MP2, (ii) G4, and (iii) CBS-QB3 results for 100 molecules, randomly chosen out of the 134k set. This data can serve the development, training and evaluation of inductive statistical data analysis-based machine learning (ML) models[13]. It might also assist the search and discovery of hitherto unknown trends, structure-property relationships, and molecular materials design1[,14,15].

METHODS

Generation of Atomic Coordinates

Starting with ref. 11, we use all SMILES[16] strings for molecules with up to nine heavy atoms. Cations, anions, and molecules containing S, Br, Cl, or I, have been excluded, resulting in 133,885 molecules. 1,705 zwitterions have been kept in the data due to their occurrence in small biomolecules, such as amino acids. Initial Cartesian coordinates for all molecules were generated by parsing the corresponding SMILES strings using Corina (Version 3.491 2013)[17]. We subsequently carried out geometry relaxations at the PM7 semi-empirical level of theory using MOPAC (Version 13.136L 2012)[18]. In the PM7 calculations, we invoked tight electronic and geometric convergence thresholds, using precise keyword. PM7 equilibrium geometries have subsequently been used as input for B3LYP geometry relaxations using Gaussian 09[19]. We iteratively refined the electronic and geometry thresholds. For the first iteration, Gaussian 09's default electronic and geometry thresholds have been used for all molecules. For those molecules which failed to reach SCF convergence ultrafine grids have been invoked within a second iteration for evaluating the XC energy contributions. Within a third iteration on the remaining unconverged molecules, we identified those which had relaxed to saddle points, and further tightened the SCF criteria using the keyword scf(maxcycle=200, verytight). All those molecules which still featured imaginary frequencies entered the fourth iteration using keywords, opt(calcfc, maxstep=5, maxcycles=1000). Calcfc constructs a Hessian in the first step of the geometry relaxation for eigenvector following. Within the fifth and final iteration, all molecules which still failed to reach convergence, have subsequently been converged using opt(calcall, maxstep=1, maxcycles=1000). calcall constructs a Hessian for all steps through the geometry relaxation. After all these measures taken, eleven problematic molecules still failed to converge to a minimal geometry. Out of these eleven molecules, six can be converged with low threshold using the opt(loose)-keyword. In the remaining five there were two near-linear molecules which converged to saddle points with very low imaginary frequencies ($_0 < i10 \, \text{cm}^{-1}$ for the lowest mode). In the readme. txt file of this report, all these 11 molecules are specified using their indices in the database.

In the case of the 6,095 constitutional isomers of $C_7H_{10}O_2$, all molecules converged to local minima during the B3LYP geometry relaxation. To compute atomization energies, we have also performed spin-unrestricted calculations for all atoms with spin-multiplicities 2,3,4,3,2 for the atoms H, C, N, O, F, respectively.

DATA RECORDS

Molecular structures and properties are publicly available at Figshare (Data Citation 1: Figshare http://dx.doi.org/10.6084/m9.figshare.978904) in a plain text XYZ-like format described below. Deposited files include the 133, 885 GDB-1 to GDB-9 molecules (dsgdb9nsd.xyz.tar.bz2), the 6,095 constitutional isomers of $C_7H_{10}O_2$ (dsC7O2H10nsd.xyz.tar.bz2), the 100 validation molecules (seeTable 1) enthalpies of atomization (validation.txt), and atomic reference data (atomref.txt).

Table 1: Validation of atomization enthalpies at B3LYP/6-31G(2df,p)-level.

Reference	MAE	RMSE	maxAE
G4MP2	5.0	6.1	16.0
G4	4.9	5.9	14.4
CBS-QB3	4.5	5.5	13.4

- For 100 molecules randomly drawn out of the pool of 134k molecules, mean absolute error (MAE), root mean square error (RMSE), and maximal absolute error (maxAE) with respect to more accurate reference methods are reported.
- All values are in kcal/mol.

File format

For each molecule, atomic coordinates and calculated properties are stored in a file nameddataset_index.xyz. The XYZ format (originally developed for the XMol program by the Minnesota Supercomputer Center) is a widespread plain text format for encoding Cartesian coordinates of molecules, with no formal specification. It contains a header line specifying the number of atomsn_a, a comment line, and n_a lines containing element type and atomic coordinates, one atom per line. We have extended this format as indicated in Table 2. Now, the comment line is used to store all scalar properties, Mulliken charges are added as a fifth column. Harmonic vibrational frequencies, SMILES and InChI are appended as respective additional lines.

Table 2: XYZ-like file format for molecular structure and properties.

Line	Content
1	Number of atoms n_a
2	Scalar properties (see Table 3)
$3,\ldots,n_a+2$	Element type, coordinate (x, y, z, in Å), Mulliken partial charges (in e) on atoms
n_a+3	Harmonic vibrational frequencies ($3n_a-5$ or $3n_a-6$, in cm^{-1})
n_a+4	SMILES strings from GDB-17 and from B3LYP relaxation
n_a+5	InChI strings for Corina and B3LYP geometries

n_a =number of atoms.

Properties

All molecular geometries were relaxed, and properties calculated, at the DFT/B3LYP/6-31G(2df,p) level of theory. The list of properties of the 134k molecules is summarized in Table 3. For a subset of 6,095 isomers of $C_7H_{10}O_2$, energetics (properties 12–16) were additionally calculated at the G4MP2 level of theory. For a validation set of 100 randomly drawn molecules from the 133,885 GDB-9 set, enthalpies of atomization were calculated at the DFT/B3LYP/6-31G(2df,p), G4MP2, G4 and CBS-QB3 levels of theory.

Table 3: Calculated properties

No.	Property	Unit	Description
1	tag	—	'gdb9' string to facilitate extraction
2	i	—	Consecutive, 1-based integer identifier
3	A	GHz	Rotational constant
4	B	GHz	Rotational constant
5	C	GHz	Rotational constant
6	μ	D	Dipole moment
7	α	a30	Isotropic polarizability
8	ϵ_{HOMO}	Ha	Energy of HOMO
9	ϵ_{LUMO}	Ha	Energy of LUMO
10	ϵ_{gap}	Ha	Gap ($\epsilon_{LUMO}-\epsilon_{HOMO}$)

11	$\langle R^2 \rangle$	a20	Electronic spatial extent
12	zpve	Ha	Zero point vibrational energy
13	U_0	Ha	Internal energy at 0 K
14	U	Ha	Internal energy at 298.15 K
15	H	Ha	Enthalpy at 298.15 K
16	G	Ha	Free energy at 298.15 K
17	C_v	calmolK	Heat capacity at 298.15 K

Properties are stored in the order given by the first column.

TECHNICAL VALIDATION

Validation of geometry consistency

To validate the consistency of the relaxed B3LYP geometries, we have used them to generate the corresponding InChI[20] strings with Corina and Open Babel (Version 2.3.0 2011)[21]. InChI corresponds to 'IUPAC International Chemical Identifier'. The resulting strings have been compared to the InChI strings obtained from the initial Cartesian coordinates (generated by Corina using the original GDB-17 SMILES strings). See Fig. 2 for a flow-chart of this consistency check. Out of the 134k molecules, 3,054 molecules did not pass this test. This is due to the fact that SMILES and InChI representations are not unique because transformation of Cartesian coordinates to string based chemical identifiers is prone to implementation specific artifacts. For molecules with same topology, small differences in interatomic distances, bond and dihedral angles can lead to different molecular graphs encoded by the string. To quantify this artifact, the distribution of Coulomb-matrix distances[13], D_{IJ} [Hartree], using the Manhattan or L1 metric, between Corina generated Cartesian coordinates (see Fig. 2) and B3LYP coordinates is on display in Fig. 3 for all the 3,054 molecules.

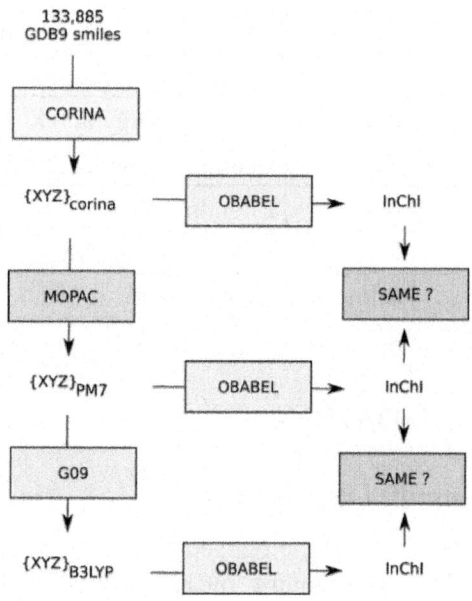

Figure 2: Schematic flow chart used for geometry consistency check.

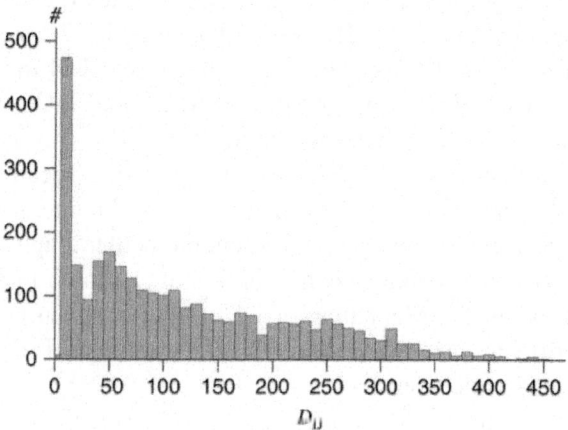

Figure 3: Histogram of Coulomb-matrix distances.

For 3,054 molecules which failed the consistency test shown in Fig. 2 Coulomb-matrix distances, D_{IJ} in Ha, between B3LYP and Corina geometries are shown.

Consider, for example, molecule indexed 58 in the 134k set, which is among the 3,054 molecules for which the consistency check failed. Its original GDB-17 SMILES corresponds to NC(=N)C#N, and Cartesian coordinates can be generated using Corina. When feeding back these coordinates to Open Babel to perform the inverse task of reproducing the initial SMILES string, [NH]C(=[NH2])C#N is obtained instead. By contrast, performing first a geometry relaxation of the Corina generated Cartesian coordinates using PM7 followed by B3LYP, and only then parsing through Open Babel, recovers the original SMILES string. In this case, D_{IJ} using the Coulomb-matrices of Corina and B3LYP geometries is rather small (~27 Ha) when compared to distances between geometries of other failed molecules, see distance distribution plot in Fig. 3. By contrast, failed molecules with large D_{IJ} between Corina and B3LYP geometries, see Fig. 3, correspond to molecules for which the B3LYP relaxation induces dramatic distortion with significant alteration and rearrangement of covalent bonding patterns. As a result, parsing back these geometries with Open Babel yields different SMILES strings. Note that all the 6,095 constitutional isomers of $C_7H_{10}O_2$ for which G4MP2 calculations have been performed, passed this geometry consistency check, shown as a flow-chart in Fig. 2.

Validation of Quantum Chemistry Results

All 134k molecules have been modeled using B3LYP/6-31G(2df,p) based DFT. Previously, B3LYP has been validated for several subsets, containing up to a few hundred small molecules. These benchmarks are of limited use since they are not necessarily sufficiently representative for gauging B3LYP's performance in general. In the case of DFT's systematic errors this issue is particularly pertinent[22]. Experimental data assembled in the NIST database is very sparse by comparison to our 134k organic molecules made up of CHONF atoms. Consequently, we have performed additional benchmark calculations for a subset of 100 randomly selected molecules using high level theories G4MP2[12], G4[23], and CBS-QB3[24,25].

The predictive power of the G4MP2 method is widely considered to be on par with experimental uncertainties. For example, comparison to the G3/05 test set[26,27] with 454 experimental energies (including enthalpies of formation, ionization potentials, electron affinities, proton affinities, and hydrogen bond energies) of small molecules yields MAE, and RMSE of 1.0, and 1.5 kcal/mol, respectively12. For the same properties and molecules, the slightly more accurate, and considerably more expensive method G4[12,23] yields errors of MAE=0.8 kcal/mol, RMSE=1.2 kcal/mol. G4MP2 has been shown to deviate only by 1.4 kcal/mol from 261 bond dissociation enthalpies computed with the highly accurate W1w composite procedure[28,29] for the BDE261 data set[28]. Consequently, we believe these calculations to be sufficiently suitable to validate the quality of the B3LYP energetics predictions. Various resulting deviations are summarized in Table 1. For the 100 molecules, the mean absolute error of B3LYP heats of atomization amounts to no more than 5 kcal/mol.

REFERENCES

1. Curtarolo, S. et al. The high-throughput highway to computational materials design. Nature Mater 12, 191–201 (2013).

2. Kirkpatrick, P. & Ellis, C. Chemical space. Nature 432, 823 (2004).

3. Koch, W. & Holthausen, M. C. A Chemist's Guide to Density Functional Theory (Wiley, 2002).

4. National institute of standards and technology. http://srdata.nist.gov (accessed 31 March 2014).

5. Jurečka, P., Šponer, J., Černý, J. & Hobza, P. Benchmark database of accurate (MP2 and CCSD(T) complete basis set limit) interaction energies of small model complexes, DNA base pairs, and amino acid pairs. Phys. Chem. Chem. Phys. 8, 1985–1993 (2006).

6. Řezáč, J., Riley, K. E. & Hobza, P. S66: a well-balanced database of benchmark interaction energies relevant to biomolecular structures. J. Chem. Theory Comput. 7, 2427–2438 (2011).

7. Lynch, B. J. & Truhlar, D. G. Small representative benchmarks for thermochemical calculations. J. Phys. Chem. A 107, 8996–8999 (2003).

8. Fink, T., Bruggesser, H. & Reymond, J.-L. Virtual exploration of the small-molecule chemical universe below 160 daltons. Angew. Chem. Int. Ed. 44, 1504–1508 (2005).

9. Martin, K. & Grimme, S. Mindless DFT benchmarking. J. Chem. Theory Comput. 5, 993–1003 (2009).

10. Virshup, A. M., Contreras-García, J., Wipf, P., Yang, W. & Beratan, D. N. Stochastic voyages into uncharted chemical space produce a representative library of all possible drug-like compounds. J. Am. Chem. Soc. 19, 7296–7303 (2013).

11. Ruddigkeit, L., van Deursen, R., Blum, L. C. & Reymond, J.-L. Enumeration of 166 billion organic small molecules in the chemical universe database GDB-17. J. Chem. Inf. Model. 52, 2864–2875 (2012).

12. Curtiss, L. A., Redfern, P. C. & Raghavachari, K. Gaussian-4 theory using reduced order perturbation theory. J. Chem. Phys. 127, 124105 (2007).

13. Rupp, M., Tkatchenko, A., Müller, K.-R. & von Lilienfeld, O. A. Fast and accurate modeling of molecular atomization energies with machine learning. Phys. Rev. Lett. 108, 058301 (2012).

14. Hachmann, J. et al. The harvard clean energy project: large-scale computational screening and design of organic photovoltaics on the world community grid. J. Phys. Chem. Lett. 2, 2241–2251 (2011).

15. Norskov, J. K., Bligaard, T., Rossmeisl, J. & Christensen, C. H. Towards the computational design of solid catalysts. Nature Chem. 1, 37–46 (2009).

16. Weininger, D. SMILES, a chemical language and information system. 1.Introduction to methodology and encoding rules. J. Chem. Inf. Comp. Sci. 28, 31–36 (1988).

17. Sadowski, J. & Gasteiger, J. From atoms and bonds to 3-dimensional atomic coordinates - automatic model builders. Chem. Rev. 93, 2567–2581 (1993).

18. Stewart, J. J. P. MOPAC2012, Version 13.136L, Stewart Computational Chemistry (Colorado Springs, 2012).

19. Frisch, M. J. et al. Gaussian 09, Revision d.01 (Gaussian, Inc., 2009).

20. Heller, S. R. & McNaught, A. D. The IUPAC international chemical identifier (InChI). Chemistry International 31, 7–9 (2009).

21. O⟩Boyle, N. M. et al. Open Babel: an open chemical toolbox. J. Chem. Inf. 3, 33 (2011).

22. Wodrich, M. D., Corminboeuf, C., Schreiner, P. R., Fokin, A. A. & Schleyer, P. v. R. How accurate are DFT treatments of organic energies? Org. Lett. 9, 1851–1854 (2007).

23. Curtiss, L. A., Redfern, P. C. & Raghavachari, K. Gaussian-4 theory. J. Chem. Phys. 126, 084108 (2007).

24. Montgomery, J. A. Jr, Frisch, M. J., Ochterski, J. W. & Petersson, G. A. A complete basis set model chemistry. VI. use of density functional geometries and frequencies. J. Chem. Phys. 110, 2282–2827 (1999).

25. Montgomery, J. A. Jr, Frisch, M. J., Ochterski, J. W. & Petersson, G. A. A complete basis set model chemistry. VII. use of the minimum population localization method. J. Chem. Phys. 112, 6532–6542 (2000).

26. Curtiss, L. A., Redfern, P. C., Raghavachari, K. & Pople, J. A. Gaussian-3X (G3X) theory: use of improved geometries, zero-point energies, and Hartree-Fock basis sets. J. Chem. Phys. 114, 108–117 (2001).

27. Curtiss, L. A., Redfern, P. C. & Raghavachari, K. Assessment of Gaussian-3 and density-functional theories on the G3/05 test set of experimental energies. J. Chem. Phys. 123, 124107 (2005).

28. Chan, B. & Radom, L. BDE261: a comprehensive set of high-level

theoretical bond dissociation enthalpies. J. Phys. Chem. A 116, 4975–4986 (2012).

29. Boese, A. D. et al. W3 theory: robust computational thermochemistry in the kJ/mol accuracy range. J. Chem. Phys. 120, 4129–4141 (2004).

Chapter 8

ADIABATIC QUANTUM SIMULATION OF QUANTUM CHEMISTRY

Ryan Babbush [1], Peter J. Love [2] & Alán Aspuru-Guzik [1]

[1]Department of Chemistry and Chemical Biology, Harvard University, Cambridge, MA 02138 USA

[2]Department of Physics, Haverford College, Haverford, PA 19041, USA

ABSTRACT

We show how to apply the quantum adiabatic algorithm directly to the quantum computation of molecular properties. We describe a procedure to map electronic structure Hamiltonians to 2-body qubit Hamiltonians with a small set of physically realizable couplings. By combining the Bravyi-Kitaev construction to map fermions to qubits with perturbative gadgets to reduce the Hamiltonian to 2-body, we obtain precision requirements on the coupling strengths and a number of ancilla qubits that scale polynomially in the problem size. Hence our mapping is efficient. The required set of controllable interactions includes only two types of interaction beyond the Ising interactions required to apply the quantum adiabatic algorithm to combinatorial optimization problems. Our mapping may also be of interest to chemists directly as it defines a dictionary from electronic structure to spin Hamiltonians with physical interactions.

INTRODUCTION

The ability to make exact quantum chemical calculations on nontrivial systems would revolutionize chemistry. While seemingly intractable for classical algorithms, quantum computers can efficiently perform such computations. There has been substantial interest in quantum algorithms for quantum chemistry involving a combination of Trotterization and phase estimation[1,2,3,4,5,6]. However, we are still technologically far from when such gate-model approaches are experimentally feasible for practical chemistry problems. Here, we propose a radically different approach based on the quantum adiabatic algorithm. In this rapidly advancing paradigm of quantum computation, there is no need for Trotterization, phase estimation or logic gates. More generally, we show the first scalable quantum simulation scheme for fermionic systems using adiabatic quantum computing.

Adiabatic quantum computing works by changing the Hamiltonian of a controllable quantum system from an initial Hamiltonian whose ground state is easy to prepare into a Hamiltonian whose ground state encodes the solution of a computationally interesting problem[7,8]. The speed of this algorithm is determined by the adiabatic theorem of quantum mechanics which states that an eigenstate remains at the same position in the eigenspectrum if a perturbation acts on the system sufficiently slowly[7,9,10]. Simply embedding a computational problem in a Hamiltonian suitable for AQC does not ensure an efficient solution. The required runtime for the adiabatic evolution depends on the energy gap between the ground state and first excited state at the smallest avoided crossing[7].

AQC has been applied to classical optimization problems that lie in the complexity class NP. For example, studies have been performed on satisfiability[11,12,13], Exact Cover[7,8], 3-regular 3-XORSAT and 3-regular Max-Cut[14], random instances of classical Ising spin glasses[15], protein folding[16,17]and machine learning[18,19]. AQC has also been applied to structured and unstructured search[20,21], search engine ranking[22] and artificial intelligence problems arising in space exploration[23]. Many of these applications follow naturally from the NP-Completeness of determining the ground state energy of classical Ising spin glasses[24]. This creates an equivalence between a large set of computational problems (the class NP) and a set of models in classical physics (classical Ising models with random coupling strengths). The advent of AQC provides a powerful motivation to study the detailed implications of this mapping. In general, we do not expect that quantum computing, including AQC, can provide efficient solutions to NP-Complete problems in the worst case[25]. However, there may exist sets of instances of some NP-Complete problems for which AQC can find the ground state efficiently, but which defy efficient classical solution by any means. If this is the case then AQC is certainly of considerable scientific interest, and likely of great industrial importance.

The potential value of a positive answer to this conjecture has motivated a commercial effort to construct an adiabatic quantum computer[26,27,28,29,30,31]. Currently, these experimental implementations of AQC are not strictly confined to the ground state at zero temperature but have considerable thermal mixing of higher lying states. Such intermediate implementations are referred to as quantum annealing devices. Quantum annealing machines with up to

509 qubits have been commercially manufactured by D-Wave Systems[32,33,34]. They are currently the subject of serious scientific investigation to determine whether their operation depends significantly on their quantum properties, and if so, whether it provides a speedup for any class of instances[15,33,35,36,37,38].

Quantum computers have been rigorously proved to provide an algorithmic advantage over the best known classical approaches for a small set of problems[39,40,41]. Adiabatic quantum computation applied to classical Ising Hamiltonians (equivalently, all problems in NP) also gives an approach to a very large class of problems where the advantage (if any) is currently unknown. The construction of medium scale (500 qubit) quantum annealing machines provides a hardware platform where the properties of AQC can be investigated experimentally. Such investigations have already been performed for many problems. At present, optimized codes on classical hardware can find the ground state of many instances in comparable time to the D-Wave device[15]. However, even if no interesting set of instances is found on which quantum annealing on the classical Ising model outperforms classical approaches, the hardware constructed to date represents an important step towards the construction of large scale quantum information technology. If quantum annealing of the classical Ising model is the first step, what is the natural next step?

Quantum simulation has provided a rich set of questions and methods in quantum computation since Feynman's suggestion that quantum devices would be best suited to computation of quantum properties[42]. This observation has been fleshed out through early work on specific systems[43,44,45,46,47,48] and through quantum algorithms for computation of eigenvalues, dynamics and other properties[49,50,51,52,53,54,55]. Recently, there have been many proposals for the simulation of quantum lattice models using trapped ions, trapped atoms and photonic systems[56,57,58,59,60]. There has been rapid experimental progress in the quantum simulation of a number of systems[61,62,63,64,65,66]. A natural target for these simulations is the phase diagram of the Fermi-Hubbard model - believed to inform our understanding of high-T_c superconductivity. For this reason many of these approaches are aimed at simulating systems of interacting fermions.

Lattice systems are a natural target for trapped ion and atom quantum simulators, with the trapping mechanism taking the place of the crystal lattice and interactions restricted to neighbors on the lattice. However, quantum chemistry applied to molecular systems is perhaps the broadest class of problems on which quantum simulation of interacting fermions could have an impact. Finding the energy of electrons interacting in the Coulomb potential of a set of fixed nuclei of an atom or molecule defines the electronic structure problem. This problem appears to be hard for classical computers because the cost of directly solving for the eigenvalues of the exact electronic Hamiltonian grows exponentially with the problem size. In spite of much progress over the last 60 years developing approximate classical algorithms for this problem, exact calculations remain out of reach for many systems of interest. Figure 1shows several of the proposals for the efficient quantum simulation of chemical Hamiltonians.

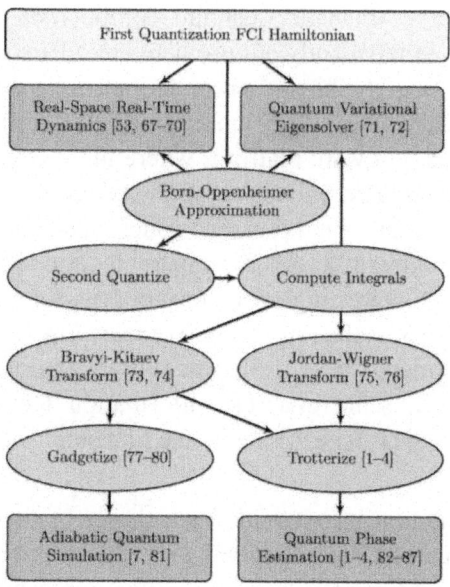

Figure 1: A diagram relating several different approaches to the quantum simulation of quantum chemistry with the procedures and approximations implicit in each approach.

Some of these approaches have been demonstrated experimentally using quantum information processors. References [68,69,70, 72] and [81,82,83,84,85,86,87] are cited in the figure above.

One may divide quantum simulation algorithms into two classes: those that address statics and compute ground state properties, and those that address dynamics, and simulate time evolution of the wavefunction. It is clear that the simulation of time evolution is exponentially more efficient on quantum computers, with significant implications for the simulation of chemically reactive scattering, in particular[67]. The computation of ground state properties naturally requires preparation of the ground state. This can be done adiabatically[1,88], or by preparation of an ansatz for the ground state[71]. Adiabatic preparation of the ground state within a gate model simulation requires time evolution of the wavefunction, which is efficient. However, the length of time for which one must evolve is determined, as for all adiabatic algorithms, by the minimum energy gap between ground and first excited states along the adiabatic path. This is unknown in general. Similarly, a successful ansatz state must have significant overlap with the true ground state, and guarantees of this are unavailable in general.

The worst case complexity of generic model chemistries (e.g. local fermionic problems studied with density functional theory) has been shown to be in the quantum mechanical equivalent of NP-Complete, QMA-Complete[89,90]. However, the subset of these generic models which correspond to stable molecules, or to unstable configurations of chemical interest such as transition states, is small and structured. Just as with adiabatic optimization, it does not matter if molecular electronic structure is QMA-Complete so long as the average instance can be solved (or even approximated) efficiently. In this case we also have considerable heuristic evidence that molecules are able to find their ground state configurations rapidly: these are the configurations in which they naturally occur. Similarly, unstable transition states of interest occur in natural processes. Given that simulation of time evolution on a quantum computer is efficient, we conjecture that simulation of the natural processes that give rise to these states will also be practical.

The proofs that Local Hamiltonian (a decision problem capturing the complexity of finding the ground state energy) is QMA-Complete relies on the construction of various specific Hamiltonians that can represent any possible instance of any problem in QMA. In general, these Hamiltonians possess couplings between more than two qubits. Hamiltonians which contain many-body interactions of order k and lower are referred to as k-local Hamiltonians; experimentally programmable couplings are 2-local. The original formulation by Kitaev was (logn)-local, he then reduced this to 5-local and that result was subsequently reduced to 3-local. To reduce 3-local Hamiltonians to 2-local Hamiltonians "perturbative gadgets" were introduced by Kempe et al.77, which can embed a k-local Hamiltonian in a subspace of a 2-local Hamiltonian using ancilla qubits. In the past decade, a growing body of work has pushed the development of different gadgets which embed various target Hamiltonians with various tradeoffs in the resources required[78,79,80,91,92,93,94].

Embedding problems in realizable Hamiltonians requires careful consideration of the availability of experimental resources. One consideration is that many-body qubit interactions cannot be directly realized experimentally. Another factor is the "control precision" of the Hamiltonian which is the dynamic range of field values which a device must be able to resolve in order to embed the intended eigenspectrum to a desired accuracy. This resource is especially important for molecular electronic structure Hamiltonians as chemists are typically interested in acquiring chemical accuracy (0.04 eV). Control precision is often the limiting factor when a Hamiltonian contains terms with coefficients which vary by several orders of magnitude. Other considerations include the number of qubits available as well as the connectivity and type of qubit couplings.

In this paper, we describe a scalable method which allows for the application of the quantum adiabatic algorithm to a programmable physical system encoding the molecular electronic Hamiltonian. Our method begins with the second quantized representation of molecular electronic structure in which the Hamiltonian is represented with fermionic creation and annihilation operators. The first step in our protocol is to convert the fermionic Hamiltonian to a qubit Hamiltonian using the Bravyi-Kitaev transformation[73,74]. We show that using the Bravyi-Kitaev transformation instead of the Jordan-Wigner transformation is necessary for avoiding exponential control precision requirements in an experimental setting. Next, we show a new formulation of perturbative gadgets motivated by[77,80] that allows us to remove all terms involving YY couplings in a single gadget application (note that throughout this paper we use X, Y and Z to denote the Pauli matrices and these operators are defined to act as identity on unlabeled registers so that the dot product $Y_i Y_j$ is understood to represent the tensor product $Y_i \otimes Y_j$). Finally, we apply the gadgets described in[78] to produce a 2-local Hamiltonian with only ZZ, XX and ZX couplings.

The paper is organized as follows. In the first section we review the second quantized formulation of the electronic structure problem. Next we give the mapping of this problem to qubits. In the third section we introduce the gadgets that we will use for locality reduction. Finally, we apply our procedure to a simple example: molecular hydrogen in a minimal basis. We close the paper with some discussion and directions for future work.

Second Quantization

We begin by writing down the full configuration interaction (FCI) Hamiltonian in the occupation number basis. We define spin orbitals as the product of a spin function (representing either spin up or spin down) and a single-electron spatial function (usually molecular orbitals produced from a Hartree-Fock calculation). For example, in the case of molecular hydrogen there are two electrons and thus, two single-electron molecular orbitals, $|\psi_1\rangle$ and $|\psi_2\rangle$. Electrons have two possible spin states, $|\alpha\rangle$ (spin up) and $|\beta\rangle$ (spin down). The four spin orbitals for molecular hydrogen are therefore, $|\chi_0\rangle = |\psi_1\rangle |\alpha\rangle$, $|\chi_1\rangle = |\psi_1\rangle |\beta\rangle$, $|\chi_2\rangle = |\psi_2\rangle |\alpha\rangle$, and $|\chi_3\rangle = |\psi_2\rangle |\beta\rangle$.

The occupation number basis is formed from all possible configurations of n spin orbitals which are each either empty or occupied. We represent these vectors as a tensor product of individual spin orbitals written as $|f_{n-1} \ldots f_0\rangle$ where $f_j \in \mathbb{B}$ indicates the occupation of spin $|\chi_j\rangle$. Any interaction between electrons can be represented as some combination of creation and annihilation operators a_j^+ and a_j for $\{j \in \mathbb{Z} \mid 0 \leq j < n\}$. Because fermionic wave-functions must be anti-symmetric with respect to particle label exchange, these operators must obey the fermionic anti-commutation relations,

$$\left[a_j, a_k\right]_+ = \left[a_j^\dagger, a_k^\dagger\right]_+ = 0, \quad \left[a_j, a_k^\dagger\right]_+ = \delta_{jk}\mathbf{1}. \qquad (1)$$

$$(1)$$

With these definitions we write the second-quantized molecular electronic Hamiltonian,

$$H = \sum_{i,j} h_{ij} a_i^\dagger a_j + \frac{1}{2} \sum_{i,j,k,l} h_{ijkl} a_i^\dagger a_j^\dagger a_k a_l.$$

$$(2)$$

The coefficients h_{ij} and h_{ijkl} are single and double electron overlap integrals which are precomputed classically. The number of distinct integrals scale as O (n^4) in the number of molecular orbitals n.

QUBIT REPRESENTATION

The next step in our reduction will be to represent our fermionic wavefunction in terms of qubits. We use the direct mapping introduced in[1] that maps an occupancy state to a qubit basis state. Using Pauli operators we can represent qubit raising and lowering operators as,

$$Q_j^+ = |1\rangle\langle 0| = \frac{1}{2}\left(X_j - iY_j\right),$$

$$Q_j^- = |0\rangle\langle 1| = \frac{1}{2}\left(X_j + iY_j\right).$$

$$(3)$$

However, these operators do not obey the fermionic commutation relations given in Eq. 1. To write qubit operators that obey the commutation relations in Eq. 1, we could use the Jordan-Wigner transformation[1,75,76].

Unfortunately, the Jordan-Wigner transformation is not a scalable way to reduce electronic structure to an experimentally realizable Hamiltonian for AQC. This is because the Jordan-Wigner transformation introduces k-local interaction terms into the Hamiltonian and k grows linearly in the system size. Prima facie, this is not a major problem because there exist theoretical tools known as perturbative gadgets which allow for reductions in interaction order. However, in all known formulations of perturbative gadgets, control precision increases exponentially in k. Thus, the linear locality overhead introduced by the Jordan-Wigner transformation translates into an exponential control precision requirement in the reduction. An alternative mapping between the occupation number basis and qubit representation, known as the Bravyi-Kitaev transformation, introduces logarithmic locality overhead[73,74]. Two pieces of information are required in order to correctly construct creation and annihilation operators that act on qubits and obey the fermionic commutation relations. First, the occupancy of each orbital must be stored. Second, parity information must be stored so that for a pair of orbitals, it is possible to determine the parity of the occupancy of the orbitals that lie between them. This parity determines the phase which results from exchanging the occupancy of the two orbitals.

The occupation number basis stores the occupation directly in the qubit state (hence the name). This implies that occupancy is a fully local variable in this basis; one may determine the occupancy of an orbital by measuring a single qubit. However, this also implies that the parity information is completely non-local. It is this fact that determines the structure of the qubit creation and annihilation operators in the Jordan-Wigner transformation. Each such operator changes the state of a single qubit j (updating the occupancy information) but also acts on all qubits with indices less than j to determine the parity of their occupancy. This results in qubit operators, expressed as tensor products of Pauli matrices that contain strings of Z operators whose length grows with the number of qubits. One could consider storing the parity information locally, so that the qubit basis states store sums of orbital occupancies. Then determination of parity requires a single qubit operation. However, updating occupancy information requires updating the state of a number of qubits that again grows with the number of qubits. Hence this "parity basis" construction offers no advantage over the Jordan Wigner transformation[74].

The Bravyi-Kitaev transformation offers a middle ground in which both parity and occupancy information are stored non-locally, so neither can be determined by measurement of a single qubit[73,74].

Both parity and occupancy information can be accessed by acting on a number of qubits that scales as the logarithm of the number of qubits. This logarithmic scaling makes the proposed mapping of electronic structure to a 2-local qubit Hamiltonian efficient.

The consequences of this mapping, originally defined in[74], were computed for electronic structure in[73]. That work defines several subsets of qubits in which the parity and occupancy information is stored. The occupancy information is stored in the update set, whereas the parity information is stored in the parity set. These sets are distinct and their size is strictly bounded above by the logarithm base two of the number of qubits. The total number of qubits on which a qubit creation and annihilation operator may act can be a multiple of the logarithm base two of the number of qubits. However, this multiple is irrelevant from the point of view of the scalability of the construction. Using the Bravyi-Kitaev transformation, the spin Hamiltonian for molecular hydrogen in the minimal (STO-3G) basis, as reported in[73], is given by

$$H_{H_2} = f_0 \mathbf{1} + f_1 Z_0 + f_2 Z_1 + f_3 Z_2 + f_1 Z_0 Z_1$$
$$+ f_4 Z_0 Z_2 + f_5 Z_1 Z_3 + f_6 X_0 Z_1 X_2 + f_6 Y_0 Z_1 Y_2$$
$$+ f_7 Z_0 Z_1 Z_2 + f_4 Z_0 Z_2 Z_3 + f_3 Z_1 Z_2 Z_3$$
$$+ f_6 X_0 Z_1 X_2 Z_3 + f_6 Y_0 Z_1 Y_2 Z_3 + f_7 Z_0 Z_1 Z_2 Z_3 \tag{4}$$

where the integral values (in Hartree) are,

$f_0 = -0.81261$, $f_1 = 0.17120$,

$f_2 = 0.16862$, $f_3 = 0.22278$, $f_4 = 0.12055$,

$f_5 = 0.17435$, $f_6 = 0.04532$, $f_7 = 0.16587$: \hfill (5)

In general, the Bravyi-Kitaev transformation applied to electronic structure produces an n-qubit Hamiltonian which is (log n)-local, and has n4 real terms. This implies that each term has an even number of Y terms, or none.

Hamiltonian Gadgets

In order to embed electronic structure in an experimentally realizable Hamiltonian, we define a scalable methodology for transforming our (log n)-local qubit Hamiltonian into a 2-local Hamiltonian with only ZZ, XX and XZ interaction terms. In this section we will describe tools known as "gadgets" which allow us to simulate the target Hamiltonian with these interactions.

Hamiltonian gadgets provide a method for embedding the eigenspectra (and sometimes eigenvectors) of an n-qubit "target" Hamiltonian, denoted by H_{target}, in a restricted (typically lowenergy) subspace of a more constrained (N . n)-qubit "gadget" Hamiltonian, denoted by $\tilde{H}$. To illustrate the general idea of gadgets, we describe how a 2-local Hamiltonian can embed a k-local Hamiltonian. Suppose that we have a gadget Hamiltonian, $\tilde{H}$, which contains only 2-local terms which act on N 5 n 1 a qubits. Then,

$$\tilde{H} = \sum_{i=1} f_i O_i, \quad \tilde{H} \left| \tilde{\psi}_i \right\rangle = \tilde{\lambda}_i \left| \tilde{\psi}_i \right\rangle \tag{6}$$

where $\{f_i\}$ are scalar coefficients, $\tilde{\lambda}_j$ and $\left| \tilde{\psi}_i \right\rangle$ are the eigenvectors and eigenvalues of $\tilde{H}$, and $\{O_i\}$ are the 2-local interaction terms of the physical Hamiltonian. We choose our interaction terms to be Hilbert-Schmidt orthogonal so that $\mathrm{Tr}\,[O_i O_j] = 2^n \delta_{i,j}$. We now define an effective Hamiltonian which has support on the lowest 2^n states of the gadget, define an effective Hamiltonian which has support on the lowest 2^n states of the gadget,

$$H_{\mathrm{eff}} = \sum_{i=0}^{2^n-1} \tilde{\lambda}_i \left| \tilde{\psi}_i \right\rangle \left\langle \tilde{\psi}_i \right| = \sum_i f_i O_i \otimes \Pi.$$

(7)

Here Π is a projector onto a particular state (usually the lowest energy state) of the a ancilla qubits and the $\{O_i\}$ are a Hilbert-Schmidt orthogonal operator basis for operators on the space of the n logical qubits. In other words, the most general representation of H_{eff} is an expansion of all possible tensor products acting on the logical qubits. In general, there is no reason why $f_i = 0$ on all non-2-local terms. Therefore a 2-local gadget on $N = n + a$ qubits can embed a $(k > 2)$-local, n-qubit Hamiltonian using a ancilla bits.

The use of perturbation theory to derive Hamiltonian gadgets was introduced by Kempe et al. in their canonical proof showing that 2-Local Hamiltonian is QMA-Complete[77]. Their construction, which we refer to as the "bit-flip construction" for reasons that will become obvious later on, was analyzed by Jordan and Farhi using a formulation of perturbation theory due to Bloch[78]. Other perturbative gadget constructions were introduced by Oliveira and Terhal to prove the QMA-Completeness of Hamiltonian on a square lattice[79]. Following this work, Biamonte and Love used gadgets to show that XX and ZZ, or XZ couplings alone, suffice for the QMA-Completeness of 2-local Hamiltonian[92]. Several other papers improve these gadgets from an experimental perspective and introduce novel constructions which are compatible with the protocol developed here[80,91,93,94]. We note that different types of gadgets may have specific advantages when designing Hamiltonians for specific hardware. Results from[80] suggest that there is a rough tradeoff between the number of ancillae required and the amount of control precision required. For instance, Figure 2 indicates that bit-flip gadgets require less control precision than other gadget constructions (but generally more ancillae). In this paper we focus on the bit-flip family of gadgets.

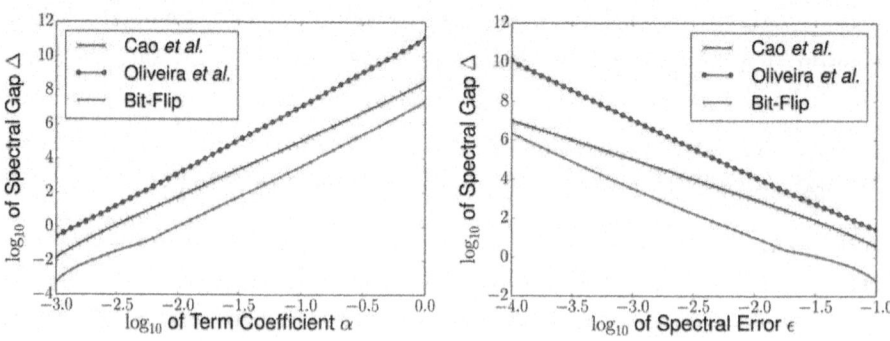

Figure 2: Numerics comparing the minimum spectral gaps required to reduce the term

$\alpha X_1 Y_2 Z_3$ to 2-local with an error in the eigenspectrum of at most. On the left, is fixed at 0.001 and gaps are plotted as a function of α. On the right, α is fixed at 0.1 and gaps are plotted as a function of. Here we compare the bit-flip construction[77,78], the Oliveira and Terhal construction[79] and an improved variant on Oliveira and Terhal by Cao et al.[80].

Although we employ the perturbation theory approach here, it does require a high degree of control precision and should be avoided when possible. We point out that when the Hamiltonian is entirely diagonal there are exact gadgets[94] which can embed the ground state with far less control precision and often far fewer ancillae but in a way that does not necessarily conserve the gap scaling. Moreover, "frustration-free" gadgets have been used extensively in proofs of the QMA-Completeness of various forms of quantum satisfiability, and in restricting the necessary Hamiltonian terms for universal adiabatic quantum computing[95,96,97,98].

While several types of perturbation theory have been used to derive these gadgets, we closely follow the approach and notation of Kempe et al.[77]. We wish to analyze the spectrum of the gadget Hamiltonian, $\tilde{H}=H+V$ for the case that the norm of the perturbation Hamiltonian, V, is small compared to the spectral gap between the ground state and first excited state of the unperturbed Hamiltonian, H. To accomplish this we use the Green's function of $\tilde{H}$,

$$\tilde{G}(z) \equiv (z\mathbf{1} - \tilde{H})^{-1} = \sum_j \frac{|\tilde{\psi}_j\rangle\langle\tilde{\psi}_j|}{z - \tilde{\lambda}_j}.$$

$$(8)$$

We also define $G(z)$ using the same expression except with H instead of $\tilde{H}$. Further, let $\mathcal{H} = \mathcal{L}_+ \oplus \mathcal{L}_-$ be the Hilbert space of $\tilde{H}$ where $\mathcal{L}_+$ is the "high-energy" subspace spanned by eigenvectors of $\tilde{H}$ with eigenvalues $\tilde{\lambda} \geq \lambda_*$ and $\mathcal{L}_-$ is the complementary "low-energy" subspace, spanned by eigenvectors of $\tilde{H}$ corresponding to eigenvalues of $\tilde{\lambda} \geq \lambda_*$. Let $\Pi_\pm$ correspond to projectors onto the support of $\mathcal{L}_+$. In a representation of $\mathcal{H} = \mathcal{L}_+ \oplus \mathcal{L}_-$, all the aforementioned operators V, H, $\tilde{H}$, $G(z)$, $\tilde{G}(z)$ are block-diagonal so we employ the notation that $A_{\pm\pm} = \Pi_\pm A \Pi_\pm$ and,

$$A = \begin{pmatrix} A_+ & A_{+-} \\ A_{-+} & A_- \end{pmatrix}.$$

$$(9)$$

Finally, we define the operator function known as the self-energy,

$$\Sigma_-(z) \equiv z\mathbf{1}_- - \tilde{G}_-^{-1}(z).$$

$$(10)$$

We use this notation to restate the "gadget theorem".

Theorem 1 Theorem 6.2 in[77]. Assume that H has a spectral gap Δ around the cutoff λ_*; i.e. all of its eigenvalues are in $(-\infty, \lambda_-] \cup [\lambda_+, +\infty)$ where $\lambda_+ = \lambda_* + \Delta/2$ and $\lambda_- = \lambda_* - \Delta/2$. Assume that $\|V\| \leq \Delta/2$. Let $\dot{o} > 0$ be arbitrary. Assume there exists an operator H_{eff} such that $\lambda\,(H_{eff}) \subset [c, d]$ for some $c \leq d < \lambda_* - \dot{o}$ and, moreover, the inequality $\left\| \Sigma_-(z) - H_{eff} \right\| \leq \dot{o}$ holds for all $z \in [c - \dot{o}, d + \dot{o}]$. Then each eigenvalue $\tilde{\lambda}_j$ of $\tilde{H}_-$ is $\dot{o} - close$ to the jth eigenvalue of H_{eff}.

Theorem 1 assures us that the eigenspectrum of the self-energy provides an arbitrarily good approximation to the eigenspectrum of the low-energy subspace of the gadget Hamiltonian. This is useful because the self-energy admits a series expansion,

$$\Sigma_-(z) = H_- + V_- + \sum_{k=2}^{\infty} V_{-+} G_+ (V_+ G_+)^{k-2} V_{+-}. \tag{11}$$

Using $G_+ = (z - \Delta)^{-1}\,\mathbf{1}_+$ and $H_- = 0$, we focus on the range $z = O(1) \ll \Delta$ and find that,

$$H_{eff} \approx V_- + \frac{1}{\Delta} \sum_{k=2}^{\infty} V_{-+} \left(\frac{V_+}{\Delta}\right)^{k-2} V_{+-}. \tag{12}$$

We use this effective Hamiltonian to approximate our k-local target Hamiltonian, which we now specify. The terms in our target Hamiltonian will have a locality that scales logarithmically with the number of orbitals. We may write such a term:

$$T = \bigotimes_{i=0}^{k-1} O_i : O_i \in \{X_i, Y_i, Z_i\} \quad \forall\, i. \tag{13}$$

One can always apply gadgets term by term to reduce locality; however, this may not be the optimal procedure. In addition, we are interested in replacing even tensor powers of the Y operator. For both these reasons we consider a slightly more general form of term as a target for gadgetization. We use the fact that it is only the commuting nature of the $\{O_i\}$ that is important for the gadget to function. We therefore write our target term as a product of k commuting operators, which includes the special case in which it is a product of k operators acting on distinct tensor factors,

$$T' = \prod_{i=0}^{k-1} O_i : [O_i, O_j] = 0 \quad \forall\, \{i, j\} \tag{14}$$

Hence, we can represent the target Hamiltonian as a sum of r terms which are the product of k commuting operators,

$$H_{\text{target}} = H_{\text{else}} + \sum_{s=1}^{r} \prod_{i=0}^{k-1} O_{s,i}$$

(15)

where all $\{O_{s,i}\}$ commute for a given s and H_{else} can be realized directly by the physical Hamiltonian. While previous formulations of bit-flip gadgets[77,78,80] have gadgetized operators acting on distinct tensor factors, it is only necessary that the operators commute. Their action on distinct tensor factors is sufficient but not necessary for the gadget construction. We take advantage of this property in order to realize YY terms without access to such couplings by making the substitution, $Y_i Y_j \rightarrow -X_i X_j Z_i Z_j$. Since $X_i X_j$ commutes with $Z_i Z_j$, we can create this effective interaction with a bit-flip gadget. For instance, suppose we have the term, $Z_0 Y_1 Y_2$. We gadgetize the term $A \cdot B \cdot C$ where $A = Z_0$, $B = -X_1 X_2$, and $C = Z_1 Z_2$ and all operators A, B, C commute. We note that another approach to removing YY terms is explained in[80].

We now introduce the form of the penalty Hamiltonian that acts only on the ancilla qubits. Bit-flip gadgets introduce an ancilla system which has two degenerate ground-states, usually taken to be $|111...\rangle_u$ and $|000...\rangle_u$ where u indicates that these kets refer to an ancilla space. For each of the r terms we use a separate ancilla system of the form,

$$H_s = \frac{\Delta}{2(k-1)} \sum_{0 \leq i < j \leq k-1} \left(1 - Z_{u_{s,i}} Z_{u_{s,j}}\right).$$

(16)

Again, we use u to indicate that operators act on an ancilla; e.g. the label $u_{3,2}$ indicates the ancilla corresponding to $O_{3,2}$ (the second operator in the third term). For each term we follow Farhi and Jordan in introducing an ancilla system connected by a complete graph with equal and negative edge weights. Thus, the ground state of the ancilla system is spanned by $|111...\rangle_u$ and $|000...\rangle_u$.

Next, we introduce the perturbation Hamiltonian,

$$V = H_{\text{else}} + \Lambda + \mu \sum_{s=1}^{r} \sum_{i=0}^{k-1} O_{s,i} X_{u_{s,i}},$$

(17)

where $\mu = \sqrt[k]{\frac{\Delta^{k-1}}{k!}}$ and Λ is a 2-local operator on logical bits which will be discussed later. The effect of this Hamiltonian on the low energy subspace is to introduce virtual excitations into the high energy space that modify the low energy effective Hamiltonian. Only terms which start and end in the ground state contribute to the perturbation series for the self-energy (see, for example, Figure 3). Thus, the gadget will produce the target term at order k in which a transition between the two degenerate ground states of the ancillae requires that each of the X_u terms in the perturbation act exactly once to flip all $r \cdot k$ bits from one ground state to the other. Crucially, the order in which the ancillae are flipped does not matter since the operators $O_{s,i}$ commute for a given s. The complete gadget is

$$\tilde{H} = \Lambda + H_{\text{else}} + \sum_{s=1}^{r} \left[\mu \sum_{i=0}^{k-1} O_{s,i} X_{u_{s,i}} + \frac{\Delta}{2(k-1)} \sum_{0 \le i < j \le k-1} (1 - Z_{u_{s,i}} Z_{u_{s,j}}) \right] \tag{18}$$

and is related to the target Hamiltonian and effective Hamiltonian by,

$$\tilde{H}_- = H_{\text{target}} \otimes \Pi_- = H_{\text{eff}} \tag{19}$$

for the appropriate choice of Λ and $\Delta \gg \|V\|$ where Π_- projects onto the ancillae ground space,

$$\Pi_- = |000\rangle\langle000|_u + |111\rangle\langle111|_u. \tag{20}$$

To illustrate the application of such a gadget and demonstrate how Λ is chosen, we scalably reduce the locality of molecular hydrogen and remove all Y terms in the next section.

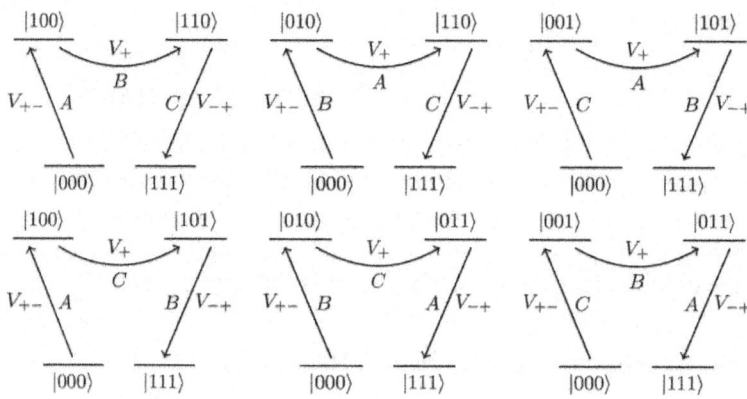

Figure 3: The six equivalent bit-flip processes at third order which produce the effective interaction $A \cdot B \cdot C$. Each of these diagrams also occurs backwards on the part of the ground state in $|111\rangle$.

For the example $H_{\text{target}} = A \cdot B \cdot C + H_{\text{else}}$, the perturbation is

$$V = \mu A X_a + \mu B X_b + \mu C X_c + H_{\text{else}} + \Lambda. \tag{21}$$

Its components in the low energy subspace, as in the block diagonal representation of Eq. 9 is:

$$V_- = (H_{\text{else}} + \Lambda) \otimes (|000\rangle\langle000|_u + |111\rangle\langle111|_u). \tag{22}$$

The projection into the high energy subspace is:

$$V_+ = (H_{else} + \Lambda) \otimes \left(\sum_{\{a,b,c\} \in \mathbb{B}^3} |a,b,c\rangle \langle a,b,c|_u \right) - V_-$$

$$+ \mu A \otimes \left(|0,1,0\rangle \langle 1,1,0|_u + |1,1,0\rangle \langle 0,1,0|_u \right.$$

$$+ |0,0,1\rangle \langle 1,0,1|_u + |1,0,1\rangle \langle 0,0,1|_u)$$

$$+ \mu B \otimes \left(|1,0,0\rangle \langle 1,1,0|_u + |1,1,0\rangle \langle 1,0,0|_u \right.$$

$$+ |0,0,1\rangle \langle 0,1,1|_u + |0,1,1\rangle \langle 0,0,1|_u)$$

$$+ \mu C \otimes \left(|1,0,0\rangle \langle 1,0,1|_u + |1,0,1\rangle \langle 1,0,0|_u \right.$$

$$|0,1,0\rangle \langle 0,1,1|_u + |0,1,1\rangle \langle 0,1,0|_u). \tag{23}$$

The projections coupling the low and high energy subspaces are:

$$V_{+-} = \mu A \otimes \left(|1,0,0\rangle \langle 0,0,0|_u + |0,1,1\rangle \langle 1,1,1|_u \right)$$

$$+ \mu B \otimes \left(|0,1,0\rangle \langle 0,0,0|_u + |1,0,1\rangle \langle 1,1,1|_u \right)$$

$$+ \mu C \otimes \left(|0,0,1\rangle \langle 0,0,0|_u + |1,1,0\rangle \langle 1,1,1|_u \right) \tag{24}$$

and $V_{-+} = (V_{+-})^\dagger$. Substituting these values into Eq. 12 we see that at order k = 3 a term appears with the following form,

$$\frac{1}{\Delta^2} V_{-+} V_+ V_{+-} = \frac{\mu^3}{\Delta^2} (ABC + ACB$$

$$+ BCA + CAB + BAC + CBA) \to ABC. \tag{25}$$

These terms arise because all ancilla qubits must be flipped and there are six ways of doing so, representing 3! (in general this will be k! for a gadget with kancillae) combinations of the operators. These six terms are represented diagrammatically in Figure 3. Note that it is the occurrence of all orderings of the operators A, B and C that imposes the requirement that these operators

commute. Hence, in order to realize our desired term we see that $\mu = \sqrt[k]{\frac{\Delta^{k-1}}{k!}}$. A few competing processes occur which contribute unwanted terms but these terms either vanish with increasing spectral gap Δ, or they can be removed exactly by introducing terms into the compensation term Λ. A simple way to compute Λ is to evaluate the perturbation series to order k and choose Λ so that problematic terms disappear.

At higher orders we encounter "cross-gadget contamination" which means that processes occur involving multiple ancilla systems, causing operators from different terms to interact. For a 3-operator gadget, such terms will always only contribute at order $O(\Delta^{-3})$. In reductions which require going to higher orders, these terms do not necessarily depend on Δ, and so may introduce unwanted terms into the effective Hamiltonian. For instance, Figure 4 shows an example of the four processes which occur at fourth order for a multiple term, 4-operator reduction. The diagrams involving multiple ancilla registers are examples of cross-gadget contamination.

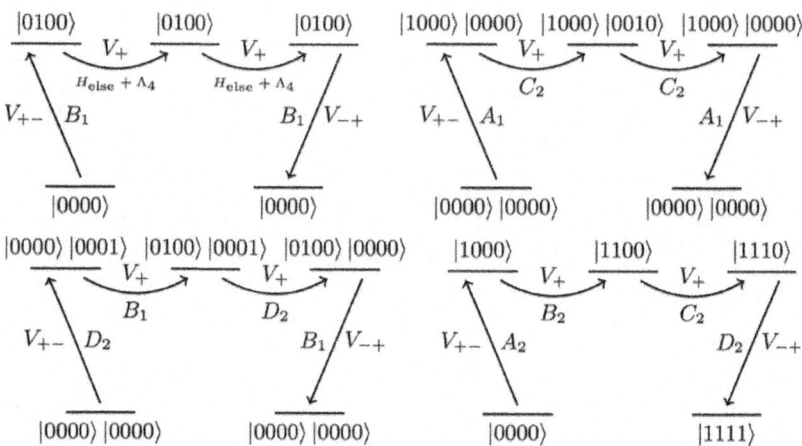

Figure 4: Diagrams showing an example of each of the four processes at fourth order. In the upper left is the process $B_1 (H_{else} + \Lambda)^2 B_1$. In the upper right is the process $A_1 C_2^2 A_1$. In the lower left is the process $D_2 B_1 D_2 B_1$. In the lower right is the process $A_2 B_2 C_2 D_2$.

However, if terms are factored into tensor products of operators that square to the identity (as is the case for products of Pauli operators, which is always possible), cross-gadget contamination can only contribute a constant shift to the energy which can be compensated for in Λ. This is because any process contributing to the perturbation series which does not transition between the two different ground states must contain an even multiple of each operator and if we choose to act on the non-ancilla qubits with operators that square to identity we obtain only a constant shift. Consider the two cross-gadget terms represented in these diagrams: $A_1 C_2^2 A_1 = A_1 1 A_1 = 1$ and $D_2 B_1 D_2 B_1 = (D_2 B_1)^2 = 1$. At even higher orders, individual cross-gadget terms might not equal a constant shift (i.e. the sixth order term $A_1 A_2 A_3 A_2 A_1 A_3$) but the occurrence of all combinations of operators and the fact that all Pauli terms either commute or anti-commute will guarantee that such terms disappear. In the sixth order example, if $[A_1, A_2] = 0$ then $A_1 A_2 A_3 A_2 A_1 A_3 = A_1 A_2 A_3 A_1 A_2 A_3 = (A_1 A_2 A_3)^2 = 1$, otherwise $[A_1, A_2]_+ = 0$ which implies that $A_1 A_2 A_3 A_2 A_1 A_3 + A_1 A_2 A_3 A_1 A_2 A_3 = 0$.

Example Problem: Molecular Hydrogen

We begin by factoring and rewriting the k-local molecular hydrogen Hamiltonian from Eq. 4 into a 4-local part and a 2-local part so that $H_{H_2} = H_{4L} + H_{2L}$ where,

$$H_{4L} = (f_4 Z_0 + f_3 Z_1) Z_2 Z_3 + (Z_1 + Z_1 Z_3)(f_6 X_0 X_2 + f_6 Y_0 Y_2 + f_7 Z_0 Z_0) \quad (26)$$

$$H_{2L} = f_0 1 + f_2 Z_1 + f_3 Z_2 + f_4 Z_0 Z_2 + f_5 Z_1 Z_3 + f_1 Z_0 (1 + Z_1) \quad (27)$$

In order to reduce H_{H_2} to a 2-local ZZ/XX/XZ-Hamiltonian we further factor H_{4L} to remove YY terms,

$$H_{4L} = \underbrace{(f_4 Z_0 + f_3 Z_1)}_{A_1} \underbrace{Z_2}_{B_1} \underbrace{Z_3}_{C_1} + \underbrace{f_7 Z_0}_{A_2} \underbrace{Z_2}_{B_2} \underbrace{(Z_1 + Z_1 Z_3)}_{C_2}$$

$$+ \underbrace{f_6 X_0 X_2}_{A_3} \underbrace{(1 - Z_0 Z_2)}_{B_3} \underbrace{(Z_1 + Z_1 Z_3)}_{C_3}$$

$$= A_1 B_1 C_1 + A_2 B_2 C_2 + A_3 B_3 C_3. \quad (28)$$

Within each term, the operators all commute so that $[A_i, B_i] = [A_i, C_i] = [B_i, C_i] = 0$. We emphasize that factoring terms into commuting operators is always possible and necessary in order for bit-flip gadgets to work correctly.

Each of the operators defined in Eq. 28 will have a corresponding ancilla qubit labelled to indicate the operator with which it is associated, e.g. the ancilla for operator B_2 has label b_2. Our unperturbed Hamiltonian is a sum of fully connected ancilla systems in which each ancilla system corresponds to a term,

$$H_1 = \frac{9\Delta_1}{4} 1 - \frac{\Delta_1}{4} (Z_{a_1} Z_{b_1} + Z_{a_1} Z_{c_1} + Z_{b_1} Z_{c_1}$$

$$+ Z_{a_2} Z_{b_2} + Z_{a_2} Z_{c_2} + Z_{b_2} Z_{c_2} + Z_{a_3} Z_{b_3} + Z_{a_3} Z_{c_3} + Z_{b_3} Z_{c_3}). \quad (29)$$

The spectral gap and Hamiltonian have the subscript "1" to associate them with the first of two applications of perturbation theory. We perturb the ancilla system with the Hamiltonian,

$$V_1 = \mu_1 (A_1 X_{a_1} + B_1 X_{b_1} + C_1 X_{c_1} + A_2 X_{a_2} + B_2 X_{b_2}$$

$$+ C_2 X_{c_2} + A_3 X_{a_3} + B_3 X_{b_3} + C_3 X_{c_3}) + H_{2L} + \Lambda_1 \quad (30)$$

where $\mu_1 = \sqrt{\dfrac{\Delta_1^2}{6}}$ and Λ_1 is a 2-local compensation Hamiltonian acting on the logical qubits only. Later on, Λ_1 will be chosen to cancel extraneous terms from the perturbative expansion. The interaction terms involving A, B, and C will arise at third order $(V_{-+} V_+ V_{+-})$ from processes which involve a transition between the two degenerate ground states of the ancilla systems. This occurs at third order because to make the transition $|000\rangle \rightleftharpoons |111\rangle$, we must flip all three ancilla bits in each term by applying the operators X_a, X_b, and X_c. Since these operators are coupled to A, B, and C, sequential action of bit flip operators yields our desired term. Because the operators commute, the order of the bit flipping does not matter. We now calculate the effective Hamiltonian using the perturbative expansion of the self-energy from Eq. 12.

Second Order

The only processes which start in the ground state and return to the ground state at second order are those which flip a single bit and then flip the same bit back. Thus, effective interactions are created between each operator and itself,

$$-\frac{1}{\Delta_1}V_{-+}V_{+-} = -\frac{\mu_1^2}{\Delta_1}\left(A_1^2 + B_1^2 + C_1^2 + A_2^2 + B_2^2 + C_2^2 + A_3^2 + B_3^2 + C_3^2\right)$$

$$= -\sqrt[3]{\frac{\Delta_1}{36}}[(9+f_3^2+f_4^2+f_6^2+f_7^2)\mathbf{1}$$

$$+2f_3f_4Z_0Z_1 - 2Z_0Z_2 + 4Z_3]. \tag{31}$$

These processes are shown in Figure 5.

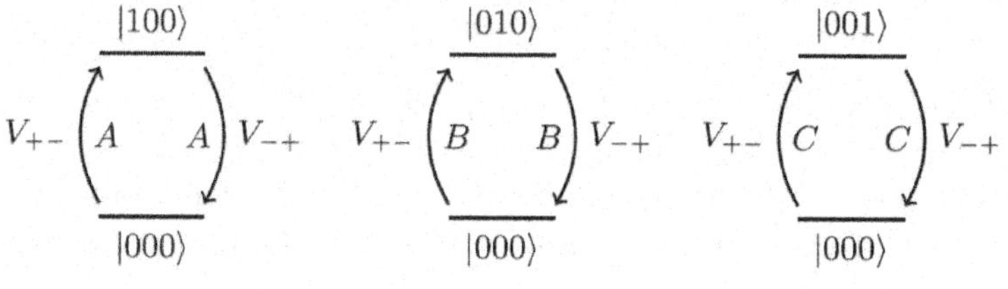

Figure 5: The three bit-flip processes at second order. These occur for each term. Note that each of these diagrams occurs in reverse for the part of the ground state in $|111\rangle$.

The second order effective Hamiltonian at large Δ_1 is,

$$H_{\text{eff}}^{(2)} = H_{2L} + \Lambda_1 - \sqrt[3]{\frac{\Delta_1}{36}}[(9+f_3^2+f_4^2+f_6^2+f_7^2)\mathbf{1}$$

$$+2f_3f_4Z_0Z_1 - 2Z_0Z_2 + 4Z_3] + O(\Delta_1^{-2}). \tag{32}$$

Third Order

The target Hamiltonian terms appears at third order from processes that transition between degenerate ground states. However, there is also an additional, unwanted process which occurs at this order. This competing process involves one interaction with H_{2L} and Λ_1 in the high-energy subspace,

$$\frac{1}{\Delta_1^2} V_{-+} V_+ V_{+-}^{(1)} = \frac{\mu_1^2}{\Delta_1^2} [A_1 (H_{2L} + \Lambda_1) A_1$$

$$+ B_1(H_{2L} + \Lambda_1)B_1 + C_1(H_{2L} + \Lambda_1)C_1 + A_2(H_{2L} + \Lambda_1)A_2$$

$$B_2(H_{2L} + \Lambda_1)B_2 + C_2(H_{2L} + \Lambda_1)C_2 + A_3(H_{2L} + \Lambda_1)A_3$$

$$+ B_3(H_{2L} + \Lambda_1)B_3 + C_3(H_{2L} + \Lambda_1)C_3]. \tag{33}$$

These processes are illustrated diagrammatically in Figure 6.

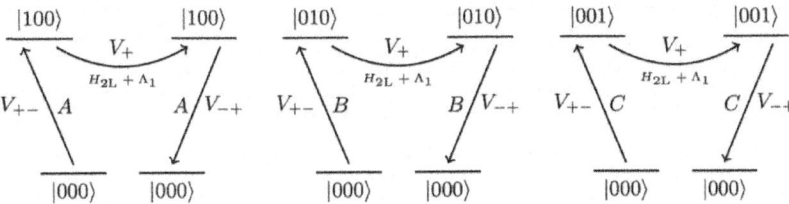

Figure 6: Diagrams for the competing process encountered at third order. Note that each of these diagrams can also occur backwards if the system starts in $|111\rangle$.

The process we want occurs with the ancilla transition $|000\rangle \rightleftharpoons |111\rangle$ which flips all three bits (for each term separately since they have different ancillae). There are $3! = 6$ possible ways to flip the bits for each term, (these processes are illustrated in Figure 3),

$$\frac{1}{\Delta_1^2} V_{-+} V_+ V_{+-}^{(2)} = 6 \frac{\mu_1^3}{\Delta_1^2} (A_1 B_1 C_1 + A_2 B_2 C_2 + A_3 B_3 C_3)$$

$$= A_1 B_1 C_1 + A_2 B_2 C_2 + A_3 B_3 C_3. \tag{34}$$

Because H_{2L} has no Δ_1 dependence and μ_1 is order $O(\Delta_1^{2/3})$, terms such as $(\mu_1^2 / \Delta_1^2) A_1 H_{2L} A_1$ will vanish in the limit of large Δ_1; therefore, the third order effective Hamiltonian is,

$$H_{eff}^{(3)} = H_{2L} + \Lambda_1 - - = -\sqrt[3]{\frac{\Delta_1}{36}} [(9 + f_3^2 + f_4^2 + f_6^2 + f_7^2)\mathbf{1}$$

$$+ 2f_3 f_4 Z_0 Z_1 - 2Z_0 Z_2 + 4Z_3] + \frac{\mu_1^2}{\Delta_1^2} (A_1 \Lambda_1 A_1 + B_1 \Lambda_1 B_1$$

$$+ C_1 \Lambda_1 C_1 + A_2 \Lambda_1 A_2 + B_2 \Lambda_1 B_2 + C_2 \Lambda_1 C_2 + A_3 \Lambda_1 A_3$$

$$+ B_3 \Lambda_1 B_3 + C_3 \Lambda_1 C_3) + A_1 B_1 C_1 + A_2 B_2 C_2 + A_3 B_3 C_3 \tag{35}$$

with error $O(\Delta_1^{-3})$. We see that if $\Lambda_1 = \frac{1}{\Lambda}V_{-+}V_{+-}$ then the unwanted contribution at third order will go to zero in the limit of large Δ_1 and the second order term will cancel exactly with Λ_1. Thus,

$$H_{\text{eff}}^{(3)} \approx H_{2L} + A_1 B_1 C_1 + A_2 B_2 C_2 + A_3 B_3 C_3 \tag{36}$$

$$H_{H_2} \to H_1 + V_1 \tag{37}$$

where "$\to$" denotes an embedding. There are still 3-local terms remaining in V_1,

$$
\begin{aligned}
V_1 = {} & \mu_1 (f_4 Z_0 + f_3 Z_1) X_{a_1} + \mu_1 X_2 (X_{b_1} + X_{b_2}) \\
& + \mu_1 Z_3 X_{c_1} + \mu_1 f_7 Z_0 X_{a_2} + \mu_1 Z_1 (Z_{c_2} + X_{c_3}) + \mu_1 X_{b_3} \\
& + \underbrace{\mu_1 Z_1}_{A_4} \underbrace{Z_3}_{B_4} \underbrace{(X_{c_2} + X_{c_3})}_{C_4} + \underbrace{\mu_1 f_6 X_0}_{A_5} \underbrace{X_2}_{B_5} \underbrace{X_{a_3}}_{C_5} \\
& + \underbrace{(-\mu_1) Z_0}_{A_6} \underbrace{Z_2}_{B_6} \underbrace{X_{b_3}}_{C_6} + H_{2L} + \Lambda_1 .
\end{aligned}
\tag{38}
$$

With this notation we reorganize our Hamiltonian a final time, so that $H_{H_2} \to H_{2L} \to H_{3L}$,

$$H_{3L} = A_4 B_4 C_4 + A_5 B_5 C_5 + A_6 B_6 C_6 \tag{39}$$

$$
\begin{aligned}
H_{2L} = {} & \left(f_0 + \frac{9\Delta_1}{4} \right) 1 + f_2 Z_1 + f_3 Z_2 + f_4 Z_0 Z_2 \\
& + f_5 Z_1 Z_3 + f_1 Z_0 (1 + Z_1) - \frac{\Delta_1}{4} (Z_{a_1} Z_{b_1} + Z_{a_1} Z_{c_1} \\
& + Z_{b_1} Z_{c_1} + Z_{a_2} Z_{b_2} + Z_{a_2} Z_{c_2} + Z_{b_2} Z_{c_2} + Z_{a_3} Z_{b_3} \\
& + Z_{a_3} Z_{c_3} + Z_{b_3} Z_{c_3}) + \sqrt[3]{\frac{\Delta_1^2}{6}} [(f_4 Z_0 + f_3 Z_1) X_{a_1} \\
& + Z_3 X_{c_1} + f_7 Z_0 X_{a_2} + X_2 (X_{b_1} + X_{b_2}) + X_{b_3} \\
& + Z_1 (X_{c_2} + X_{c_3})] + - \sqrt[3]{\frac{\Delta_1}{36}} [(9 + f_3^2 + f_4^2 + f_6^2 + f_7^2) 1 \\
& + 2 f_3 f_4 Z_0 Z_1 - 2 Z_0 Z_2 + 4 Z_3].
\end{aligned}
\tag{40}
$$

The third order gadget we need to reduce H_{3L} takes exactly the same form as before except with the term labels 1, 2, 3 exchanged for the term labels 4, 5, 6. The components of the final gadget are

$$
\begin{aligned}
H_2 = {} & \frac{9\Delta_2}{4} 1 - \frac{\Delta_2}{4} (Z_{a_4} Z_{b_4} + Z_{a_4} Z_{c_4} + Z_{b_4} Z_{c_4} \\
& + Z_{a_5} Z_{b_5} + Z_{a_5} Z_{c_5} + Z_{b_5} Z_{c_5} + Z_{a_6} Z_{b_6} + Z_{a_6} Z_{c_6} + Z_{b_6} Z_{c_6})
\end{aligned}
\tag{41}
$$

and

$$
\begin{aligned}
V_2 = {} & \mu_2 (A_4 X_{a_4} + B_4 X_{b_4} + C_4 X_{c_4} + A_5 X_{a_5} \\
& + B_5 X_{b_5} + C_5 X_{c_5} + A_6 X_{a_6} + B_6 X_{b_6} + C_6 X_{c_6}) + H_{2L} + \Lambda_2
\end{aligned}
\tag{42}
$$

where $\mu_2 = \sqrt[3]{\dfrac{\Delta_2^2}{6}}$ and

$$\Lambda_2 = \frac{\mu_2^2}{\Delta_2}\left(A_4^2 + B_4^2 + C_4^2 + A_5^2 + B_5^2 + C_5^2 + A_6^2 + B_6^2 + C_6^2\right)$$

$$= \sqrt[3]{\frac{\Delta_2}{6}}\left[\frac{7}{\sqrt[3]{6}} + \Delta_1^{4/3}\left(\frac{1}{3} + \frac{f_6^2}{6}\right)\right]\mathbf{1} + \sqrt[3]{\frac{2\Delta_2}{9}}X_{c_2}X_{c_3}. \qquad (43)$$

This time the spectral gap and Hamiltonian have the subscript "2" to associate them with our second application of perturbation theory. We have thus shown the embedding $H_{H_2} = H_2 + V_2$. We present an interaction graph for the embedded Hamiltonian in Figure 7.

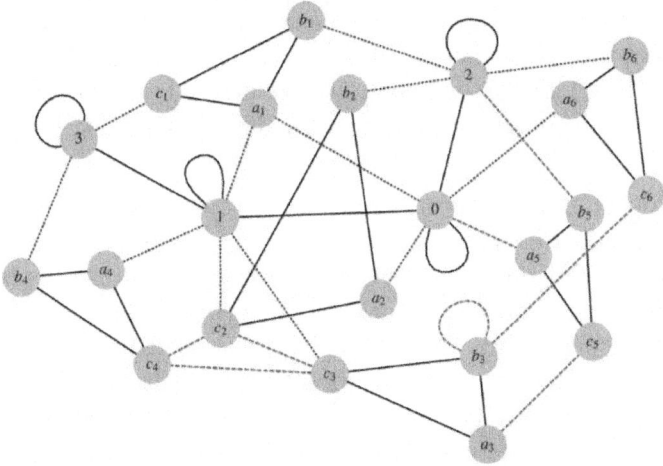

Figure 7: Interaction graph for embedded molecular hydrogen Hamiltonian. Each node represents a qubit. The solid, black edges represent ZZ terms and the black loops represent local Z terms. The dashed, red edges represent XX terms and the red loops represent local X terms. The dotted, blue edges represent XZ terms. It is easy to see the unperturbed Hamiltonians corresponding to the six 3-operator terms (the black triangles).

CONCLUSION

We have presented a fully general method for mapping any molecular electronic structure instance to a 2-local Hamiltonian containing only ZZ, XX and XZ terms. Our method is scalable in the sense that all experimental resources (qubits, control precision, graph degree) scale polynomially in the number of orbitals. We used perturbative gadgets which embed the entire target Hamiltonian (as opposed to just the ground state), thus guaranteeing that the eigenvalue gap is conserved under our reduction. Furthermore, we showed that bit-flip gadgets can be applied to remove experimentally challenging YY terms. The resulting Hamiltonian is suitable for implementation in superconducting systems, quantum dots and other systems of artificial spins with the correct engineered interactions.

Further reduction of the types of interactions present is possible, to either ZZ and XX terms or ZZ and XZ terms, using the techniques of[92]. This makes the required interactions for simulating electronic structure Hamiltonians equivalent to the requirements of universal adiabatic quantum computation[92]. However, repeated reduction of the Hamiltonian results in more stringent precision requirements. The chosen target set of interactions strikes a balance between control precision and a reasonable set of distinct types of controllable interaction. The techniques developed here could also be applied to interacting fermion problems on the lattice. However, in that case it is possible to improve beyond the Bravyi-Kitaev mapping and exploit the locality of the interactions to directly obtain Hamiltonians whose locality is independent of the number of orbitals[99].

We propose to read out energy eigenvalues using the tunneling spectroscopy of a probe qubit. This technique has already been demonstrated experimentally with rf SQUID flux qubits in32. In this scheme, a probe qubit is coupled to a single qubit of the simulation. Tunneling transitions allow the probe qubit to flip when the energy bias of the probe is close to an eigenvalue of the original system. Hence detection of these transitions reveals the eigenspectrum of the original system. In this way, we would be able to directly measure the eigenspectra of the molecular systems embedded into the spin Hamiltonian using the techniques developed in the present paper. Alternatively, one could evaluate the energy by determining the expectation value of each term in the Hamiltonian via projective measurements.

There has been rapid recent progress in new classical algorithms, such as DMRG (density matrix renormalization group) and related tensor network methods, and proving complexity and approximability results pertaining to minimal resource model Hamiltonians. By using and understanding the techniques we have introduced in this paper, problems in chemistry can be reduced to such models and these discoveries can be leveraged to make advances in electronic structure theory. However, we note that the spin Hamiltonians that result from the mapping developed here will be non-stoquastic, and classical simulation techniques will therefore suffer from the fermionic sign problem100. This further motivates the construction of quantum hardware to address the electronic structure problem by quantum simulation of these spin Hamiltonians.

REFERENCE

1.   Aspuru-Guzik, A., Dutoi, A. D., Love, P. J. & Head-Gordon, M. Simulated Quantum Computation of Molecular Energies. Science. 309, 20; 10.1126/science.1113479 (2006).

2.   Wecker, D., Bauer, B., Clark, B. K., Hastings, M. B. & Troyer, M. Gate-count estimates for performing quantum chemistry on small quantum computers. Phys. Rev. A 90, 022305; 10.1103/PhysRevA.90.022305 (2014).

3.   Whitfield, J. D., Biamonte, J. & Aspuru-Guzik, A. Simulation of Electronic Structure Hamiltonians Using Quantum Computers. Mol. Phys. 2, 106–111;10.1080/00268976.2011.552441 (2010).

4.   Poulin, D. et al. The Trotter Step Size Required for Accurate QuantumSimulation of Quantum Chemistry. e-print arXiv: 1406.4920; (2014). URL http://arxiv.org/abs/1406.4920.

5.   Hastings, M. B., Wecker, D., Bauer, B. & Troyer, M. Improving Quantum Algorithms for Quantum Chemistry. e-print arXiv: 1403.1539; (2014).

6.   McClean, J. R., Babbush, R., Love, P. J. & Aspuru-Guzik, A. Exploiting locality in quantum computation for quantum chemistry. e-print arXiv: 1407.7863; (2014). URL http://arxiv.org/abs/1407.7863.

7.   Farhi, E., Goldstone, J., Gutmann, S. & Sipser, M. Quantum Computation by Adiabatic Evolution. e-print arXiv: 0001106; (2000). URL http://arxiv.org/abs/quant-ph/0001106.

8.   Farhi, E. et al. A Quantum Adiabatic Evolution Algorithm Applied to Random Instances of an NP-Complete Problem. Science 292, 472–475; 10.1126/science.1057726 (2001).

9.   Born, M. & Fock, V. Beweis des Adiabatensatzes. Zeitschrift fu"r Phys. A 51, 165–180 (1928).

10.   Boixo, S. & Somma, R. D. Necessary Condition for the Quantum Adiabatic Approximation. Phys. Rev. A 81, 5; 10.1103/PhysRevA.81.032308 (2009).

11.   Hogg, T. Adiabatic quantum computing for random satisfiability problems. Phys. Rev. A 67, 22314; 10.1103/Phys-RevA.67.022314 (2003).

12.   Choi, V. Adiabatic Quantum Algorithms for the NP-Complete Maximum-Weight Independent Set, Exact Cover and 3SAT Problems. e-print arXiv: 1004.2226; (2010). URL http://arxiv.org/abs/1004.2226.

13.   Neuhaus, T., Peschina, M., Michielsen, K. & De Raedt, H. Classical and quantum annealing in the median of three-satisfiability. Phys. Rev. A 83, 12309; 10.1103/PhysRevA.83.012309 (2011).

14. Farhi, E. et al. Performance of the quantum adiabatic algorithm on random instances of two optimization problems on regular hypergraphs. Phys. Rev. A 86; 10.1103/PhysRevA.86.052334 (2012).

15. Boixo, S. et al. Quantum annealing with more than one hundred qubits. e-print arxiv: 1304.4595; (2013).

16. Perdomo-Ortiz, A., Dickson, N., Drew-Brook, M., Rose, G. & Aspuru-Guzik, A. Finding low-energy conformations of lattice protein models by quantum annealing. Sci. Rep. 2; 10.1038/srep00571 (2012).

17. Babbush, R. et al. Construction of Energy Functions for Lattice Heteropolymer Models: Efficient Encodings for Constraint Satisfaction Programming and Quantum Annealing. Adv. Chem. Phys. 155, 201–243; 10.1002/9781118755815.ch05 (2014).

18. Babbush, R., Denchev, V., Ding, N., Isakov, S. & Neven, H. Construction of nonconvex polynomial loss functions for training a binary classifier with quantum annealing. e-print arXiv: 1406.4203; (2014). URL http://arxiv.org/abs/1406.4203.

19. Denchev, V. S., Ding, N., Vishwanathan, S. V. N. & Neven, H. Robust Classification with Adiabatic Quantum Optimization. e-print arXiv: 1205.1148; (2012). URL http://arxiv.org/abs/1205.1148.

20. Roland, J. & Cerf, N. J. Quantum search by local adiabatic evolution. Phys. Rev. A 65, 42308; 10.1103/Phys-RevA.65.042308 (2002).

21. Roland, J. & Cerf, N. J. Adiabatic quantum search algorithm for structured problems. Phys. Rev. A 68, 62312; 10.1103/PhysRevA.68.062312 (2003).

22. Garnerone, S., Zanardi, P. & Lidar, D. A. Adiabatic Quantum Algorithm for Search Engine Ranking. Phys. Rev. Lett. 108, 230506; 10.1103/PhysRevLett.108.230506 (2012).

23. Smelyanskiy, V. N. et al. A Near-Term Quantum Computing Approach for Hard Computational Problems in Space Exploration. Electr. Eng. 68; (2012).

24. Barahona, F. On the computational complexity of Ising spin glass models. J. Phys. A. Math. Gen. 15, 3241; 10.1088/0305-4470/15/10/028 (1982).

25. Bernstein, E. & Vazirani, U. Quantum complexity theory. SIAM J. Comput. 26, 1411–1473 (1997).

26. Harris, R. et al. Sign-and magnitude-tunable coupler for superconducting flux qubits. Phys. Rev. Lett. 98, 177001; 10.1103/PhysRevLett.98.177001 (2007).

27. Harris, R. et al. Probing Noise in Flux Qubits via Macroscopic Resonant Tunneling. Phys. Rev. Lett. 101, 117003; 10.1103/

PhysRevLett.101.117003 (2008).

28. Harris, R. et al. Synchronization of multiple coupled rf-SQUID flux qubits. New J. Phys. 11, 123022; 10.1088/1367-2630/11/12/123022 (2009).

29. Lanting, T. et al. Geometrical dependence of the low-frequency noise in superconducting flux qubits. Phys. Rev. B 79, 60509; 10.1103/PhysRevB.79.060509 (2009).

30. Johansson, J. et al. Landau-Zener transitions in a superconducting flux qubit. Phys. Rev. B 80, 12507; 10.1103/Phys-RevB.80.012507 (2009).

31. Berkley, A. J. et al. A scalable readout system for a superconducting adiabatic quantum optimization system. Supercond. Sci. Technol. 23, 105014; 10.1088/0953-2048/23/10/105014 (2010).

32. Berkley, A. J. et al. Tunneling spectroscopy using a probe qubit. Phys. Rev. B 87, 020502; 10.1103/PhysRevB.87.020502 (2013).

33. Johnson, M.W. et al. Quantum annealing with manufactured spins. Nature 473, 194–198; 10.1038/nature10012 (2011).

34. Dickson, N. G. et al. Thermally assisted quantum annealing of a 16-qubit problem. Nat. Commun. 4, 1903; 10.1038/ncomms2920 (2013).

35. Pudenz, K. L., Albash, T. & Lidar, D. A. Error-corrected quantum annealing with hundreds of qubits. Nat. Commun. 5, 3243; 10.1038/ncomms4243 (2014).

36. Bian, Z., Chudak, F., Macready, W. G., Clark, L. & Gaitan, F. Experimental Determination of Ramsey Numbers. Phys. Rev. Lett. 111, 130505; 10.1103/PhysRevLett.111.130505 (2013).

37. Wang, L. et al. Comment on: "Classical signature of quantum annealing" e-print arXiv: 1305.5837; (2013). URL http://arxiv.org/abs/1305.5837.

38. Smolin, J. A.& Smith, G. Classical signatures of quantum annealing. e-print arXiv: 1305.4904; (2013). URL http://arxiv.org/abs/1305.4904.

39. Shor, P. W. Polynomial-time algorithms for prime factorization and discrete logarithms on a quantum computer. SIAM J. Comput. 26, 1484–1509; 10.1137/S0097539795293172 (1997).

40. Childs, A. M. et al. Exponential algorithmic speedup by a quantum walk. Proc. thirty-fifth Annu. ACM Symp. Theory Comput. 35, 59–68; 10.1145/780542.780552 (2003).

41. Grover, L. K. A fast quantum mechanical algorithm for database search. In Proc. twenty-eighth Annu. ACM Symp. Theory Comput., STOC 996, 212–219; 10.1145/237814.237866 (1996).

42. Feynman, R. P. Simulating physics with computers. Int. J. Theor. Phys.

21, 467–488; 10.1007/BF02650179 (1982).

43. Meyer, D. A. From quantum cellular automata to quantum lattice gases. J. Stat. Phys. 85, 551–574; 10.1007/BF02199356 (1996).

44. Wiesner, S. Simulations ofmany-body quantum systems by a quantum computer. e-print arXiv: 9603028; (1996). URL http://arxiv.org/abs/quant-ph/9603028.

45. Abrams, Daniel S. & Lloyd, Seth. Quantum Algorithm Providing Exponential Speed Increase for Finding Eigenvalues and Eigenvectors. Phys. Rev. Lett. 83, 5162–5165; 10.1103/PhysRevLett.83.5162 (1999).

46. Lidar, D. A. & Biham, O. Simulating Ising spin glasses on a quantum computer. Phys. Rev. E (Statistical Phys. 56, 3661–3681; 10.1103/PhysRevE.56.3661 (1997).

47. Boghosian, B. M. & Taylor, W. Simulating quantum mechanics on a quantum computer. Phys. D-Nonlinear Phenom. 120, 30–42; 10.1016/S0167-2789(98)00042-6 (1998).

48. Zalka, C. Efficient Simulation of Quantum Systems by Quantum Computers. Fortschritte der Phys. 46, 877–879; 10.1002/(SICI)1521-3978(199811)46:6/8877::AID-PROP877.3.0.CO;2-A (1998).

49. Abrams, D. S. & Lloyd, S. Quantum Algorithm Providing Exponential Speed Increase for Finding Eigenvalues and Eigenvectors. Phys. Rev. Lett. 83, 5162–5165; 10.1103/PhysRevLett.83.5162 (1999).

50. Berry, D. W.,Ahokas, G., Cleve, R.&Sanders, B. C. EfficientQuantumAlgorithms for Simulating Sparse Hamiltonians. Commun. Math. Phys. 270, 359–371; 10.1007/s00220-006-0150-x (2007).

51. Kassal, I., Jordan, S. P., Love, P. J., Mohseni, M. & Aspuru-Guzik, A. Polynomialtime quantum algorithm for the simulation of chemical dynamics. Proc. Natl. Acad. Sci. 105, 18681–18686; 10.1073/pnas.0808245105 (2008).

52. Wiebe, N., Berry, D. W.,Hoyer, P.& Sanders, B. C. Higher Order Decompositions of Ordered Operator Exponentials. J. Phys. A Math. Theor. 43, 1–21; 10.1088/1751-8113/43/6/065203 (2010).

53. Ward, N. J., Kassal, I. & Aspuru-Guzik, A. Preparation of many-body states for quantum simulation. J. Chem. Phys. 130, 194105–194114; 10.1063/1.3115177 (2008).

54. Raeisi, S., Wiebe, N. & Sanders, B. C. Quantum-circuit design for efficient simulations of many-body quantum dynamics. New J. Phys. 14, 3017; 10.1088/1367-2630/14/10/103017 (2012).

55. Sanders, B. C. EfficientAlgorithms for Universal Quantum Simulation. Lect. Notes Comput. Sci. 7948, 1–10; 10.1007/978-3-642-38986-31 (2013).

56. Weimer, H., Müller, M., Lesanovsky, I., Zoller, P. & Büchler, H. P. A Rydberg quantum simulator. Nat. Phys. 6, 382–388; 10.1038/nphys1614 (2010).

57. Ma, X.-S., Dakić, B., Naylor, W., Zeilinger, A. &Walther, P. Quantum simulation of the wavefunction to probe frustrated Heisenberg spin systems. Nat. Phys. 7, 399–405; 10.1038/nphys1919 (2011).

58. Hague, J. P., Downes, S., MacCormick, C. & Kornilovitch, P. E. Cold Rydberg atoms for quantum simulation of exotic condensed matter interactions. J. Supercond. Nov. Magn.; 10.1007/s10948-013-2414-y (2013).

59. Cohen, I. & Retzker, A. Proposal for Verification of the Haldane Phase Using Trapped Ions. Phys. Rev. Lett. 112, 040503; 10.1103/ PhysRevLett.112.040503 (2014).

60. Hauke, P., Marcos, D., Dalmonte, M. & Zoller, P. Quantum simulation of a lattice Schwinger model in a chain of trapped ions. Phys. Rev. X 3, 18; 10.1103/PhysRevX.3.041018 (2013).

61. Simon, J. et al. Quantum simulation of antiferromagnetic spin chains in an optical lattice. Nature 472, 307–312; 10.1038/nature09994 (2011).

62. Gillen, J. I. et al. Two-dimensional quantum gas in a hybrid surface trap. Phys. Rev. A 80, 21602; 10.1103/Phys-RevA.80.021602 (2009).

63. Leibfried, D. et al. Trapped-Ion Quantum Simulator: Experimental Application to Nonlinear Interferometers. Phys. Rev. Lett. 89, 247901; 10.1103/PhysRevLett.89.247901 (2002).

64. Friedenauer, A., Schmitz,H., Glueckert, J. T., Porras, D.&Schaetz, T. Simulating a quantum magnet with trapped ions. Nat. Phys. 4, 757–761; 10.1038/nphys1032(2008).

65. Johanning, M., Varón, A. F. & Wunderlich, C. Quantum simulations with cold trapped ions. J. Phys. B At. 42, 4009; 10.1088/0953-4075/42/15/154009 (2009).

66. Richerme, P. et al. Experimental performance of a quantum simulator: Optimizing adiabatic evolution and identifying many-body ground states. Phys. Rev. A 88, 12334; 10.1103/PhysRevA.88.012334 (2013).

67. Kassal, I., Whitfield, J. D., Perdomo-Ortiz, A., Yung, M.-H. & Aspuru-Guzik, A. Simulating chemistry using quantum computers. Annu. Rev. Phys. Chem. 62, 185–207; 10.1146/annurev-physchem-032210-103512

(2010).

68. Welch, J., Greenbaum, D., Mostame, S. & Aspuru-Guzik, A. Efficient quantum circuits for diagonal unitaries without ancillas. New J. Phys. 16, 033040; 10.1088/1367-2630/16/3/033040 (2014).

69. Whitfield, J. D. Spin-free quantum computational simulations and symmetry adapted states. J. Chem. Phys. 139; 10.1063/1.4812566 (2013).

70. Lu, D. et al. Simulation of chemical isomerization reaction dynamics on a NMR quantum simulator. Phys. Rev. Lett. 107, 020501; 10.1103/PhysRevLett.107.020501 (2011).

71. Peruzzo, A. et al. A variational eigenvalue solver on a photonic quantum processor. Nat. Commun. 5; 10.1038/ncomms5213 (2014).

72. Yung, M. H. et al. From transistor to trapped-ion computers for quantum chemistry. e-print arXiv: 1311.3297; (2013). URL http://arxiv.org/abs/1307.4326.

73. Seeley, J. T., Richard, M. J. & Love, P. J. The Bravyi-Kitaev transformation for quantum computation of electronic structure. J. Chem. Phys. 137; 10.1063/1.4768229 (2012).

74. Bravyi, S. & Kitaev, A. Fermionic quantum computation. Ann. Phys. (N. Y). 298, 18; 10.1006/aphy.2002.6254 (2000).

75. Jordan, P. & Wigner, E. u¨ber das paulische a¨quivalenzverbot. Zeitschrift fu¨r Phys. 47, 631–651 (1928).

76. Somma, R., Ortiz, G., Gubernatis, J., Knill, E. & Laflamme, R. Simulating physical phenomena by quantum networks. Phys. Rev. A 65, 17; 10.1103/PhysRevA.65.042323 (2002).

77. Kempe, J., Kitaev, A. & Regev, O. The Complexity of the Local Hamiltonian Problem. SIAM J. Comput. 35, 30; 10.1137/S0097539704445226 (2004).

78. Jordan, S. P. & Farhi, E. Perturbative Gadgets at Arbitrary Orders. Phys. Rev. A 77, 1–8 (2008).

79. Oliveira, R. & Terhal, B. M. The complexity of quantum spin systems on a twodimensional square lattice. Quant Inf Comp 8, 19 (2005).

80. Cao, Y., Babbush, R., Biamonte, J. & Kais, S. Towards Experimentally Realizable Hamiltonian Gadgets. e-print arXiv: 1311.3297; (2013). URL http://arxiv.org/abs/1311.2555.

81. Biamonte, J., Bergholm, V., Fitzsimons, J. & Aspuru-Guzik, A. Adiabatic quantum simulators. AIP Adv. 1; 10.1063/1.3598408 (2011).

82. Wang, H., Kais, S., Aspuru-Guzik, A. & Hoffmann, M. R. Quantum algorithm for obtaining the energy spectrum of molecular systems. Phys. Chem. Chem. Phys. 10, 5388–5393; 10.1039/B804804E (2008).

83. Veis, L. & Pittner, J. Quantum computing applied to calculations of molecular energies: CH2 benchmark. J. Chem. Phys. 133, 194106; 10.1063/1.3503767 (2010).

84. Toloui, B. & Love, P. J. Quantum Algorithms for Quantum Chemistry based on the sparsity of the CI-matrix. e-print arXiv: 1311.3297; (2013). URL http://arxiv.org/abs/1312.2579.

85. Lanyon, B. P. et al. Towards Quantum Chemistry on a Quantum Computer. Nat. Chem. 2, 20; 10.1038/nchem.483 (2009).

86. Li, Z. et al. Solving Quantum Ground-State Problems with Nuclear Magnetic Resonance. Sci. Rep. 1; doi:10.1038/srep00088 (2011).

87. Du, J. et al. NMR implementation of a molecular hydrogen quantum simulation with adiabatic state preparation. Phys. Rev. Lett. 104, 030502; 10.1103/PhysRevLett.104.030502 (2010).

88. Veis, L.&Pittner, J. Adiabatic state preparation study of methylene. J. Chem. Phys. 140, 214111; 10.1063/1.4880755 (2014).

89. Schuch, N. & Verstraete, F. Computational complexity of interacting electrons and fundamental limitations of density functional theory. Nat. Phys. 5, 732–735;10.1038/nphys1370 (2009).

90. Whitfield, J. D., Love, P. J. & Aspuru-Guzik, A. Computational complexity in electronic structure. Phys. Chem. Chem. Phys. 15, 397–411; 10.1039/c2cp42695a (2013).

91. Bravyi, S., DiVincenzo, D. P., Loss, D. & Terhal, B. M. Quantum simulation of many-body Hamiltonians using perturbation theory with bounded-strength interactions. Phys. Rev. Lett. 101, 070503; 10.1103/PhysRevLett.101.070503(2008).

92. Biamonte, J. D. & Love, P. J. Realizable Hamiltonians for Universal Adiabatic Quantum Computers. Phys. Rev. A 78, 1–7; 10.1103/PhysRevA.78.012352 (2007).

93. Duan, Q.-H. & Chen, P.-X. Realization of Universal Adiabatic Quantum Computation with Fewer Physical Resources. Phys. Rev. A 84, 4; 10.1103/PhysRevA.84.042332 (2011).

94. Babbush, R., O'Gorman, B. & Aspuru-Guzik, A. A. Resource Efficient Gadgets for Compiling Adiabatic QuantumOptimization Problems. Ann. Phys. 525, 877–888;10.1002/andp.201300120 (2013).

95. Nagaj, D. & Mozes, S. New construction for a QMA complete three-local Hamiltonian. J. Math. Phys. 48, 2104; 10.1063/1.2748377 (2007).

96. Nagaj, D. Fast universal quantum computation with railroad-switch local Hamiltonians. J. Math. Phys. 51, 2201; 10.1063/1.3384661 (2010).

97. Gosset, D. & Nagaj, D. Quantum 3-SAT is QMA1-complete. e-print arXiv: 1302.0290; (2013). URL http://arxiv.org/abs/1302.0290.

98. Childs, A. M.,Gosset, D.&Webb, Z. The Bose-Hubbard model is QMA-complete. e-print arXiv: 1311.3297; (2013).

99. Verstraete, F. & Cirac, J. I. Mapping local Hamiltonians of fermions to local Hamiltonians of spins. J. Stat. Mech. Theory Exp. P09012; 10.1088/1742-5468/2005/09/P09012 (2005).

100. Bravyi, S., DiVincenzo, D. P., Oliveira, R. I. & Terhal, B. M. The Complexity of Stoquastic Local Hamiltonian Problems. Quantum Inf. Comput. 8, 361–385(2008).

Chapter 9

FROM TRANSISTOR TO TRAPPED-ION COMPUTERS FOR QUANTUM CHEMISTRY

M.-H. Yung [1,2], J. Casanova [3], A. Mezzacapo [3], J. McClean [2], L. Lamata [3], A. Aspuru-Guzik [2] & E. olano [3,4]

[1] Center for Quantum Information, Institute for Interdisciplinary Information Sciences, Tsinghua University, Beijing, 100084, P. R. China

[2] Department of Chemistry and Chemical Biology, Harvard University, Cambridge MA, 02138, USA

[3] Department of Physical Chemistry, University of the Basque Country UPV/EHU, Apartado 644, 48080 Bilbao, Spain

[4] IKERBASQUE, Basque Foundation for Science, Alameda Urquijo 36, 48011 Bilbao, Spain

ABSTRACT

Over the last few decades, quantum chemistry has progressed through the development of computational methods based on modern digital computers. However, these methods can hardly fulfill the exponentially-growing resource requirements when applied to large quantum systems. As pointed out by Feynman, this restriction is intrinsic to all computational models based on classical physics. Recently, the rapid advancement of trapped-ion technologies has opened new possibilities for quantum control and quantum simulations. Here, we present an efficient toolkit that exploits both the internal and motional degrees of freedom of trapped ions for solving problems in quantum chemistry, including molecular electronic structure, molecular dynamics, and vibronic coupling. We focus on applications that go beyond the capacity of classical computers, but may be realizable on state-of-the-art trapped-ion systems. These results allow us to envision a new paradigm of quantum chemistry that shifts from the current transistor to a near-future trapped-ion-based technology.

INTRODUCTION

Quantum chemistry represents one of the most successful applications of quantum mechanics. It provides an excellent platform for understanding matter from atomic to molecular scales, and involves heavy interplay of experimental and theoretical methods. In 1929, shortly after the completion of the basic structure of the quantum theory, Dirac speculated[1] that the fundamental laws for chemistry were completely known, but the application of the fundamental laws led to equations that were too complex to be solved. About ninety years later, with the help of transistor-based digital computers, the development of quantum chemistry continues to flourish, and many powerful methods, such as Hartree-Fock, configuration interaction, density functional theory, coupled-cluster, and quantum Monte Carlo, have been developed to tackle the complex equations of quantum chemistry (see e.g. for a historical review[2]). However, as the system size scales up, all of the methods known so far suffer from limitations that make them fail to maintain accuracy with a finite amount of resources[3]. In other words, quantum chemistry remains a hard problem to be solved by the current computer technology.

As envisioned by Feynman[4], one should be able to efficiently solve problems of quantum systems with a quantum computer. Instead of solving the complex equations, this approach, known as quantum simulation (see the recent reviews in Refs. [5,6,7]), aims to solve the problems by simulating target systems with another controllable quantum system, or qubits. Indeed, simulating many-body systems beyond classical resources will be a cornerstone of quantum computers. Quantum simulation is a very active field of study and various methods have been developed. Quantum simulation methods have been proposed for preparing specific states such as ground[8,9,10,11,12,13] and thermal states[14,15,16,17,18,19,20], simulating time evolution[21,22,23,24,25,26,27], and the measurement of physical observables[28,29,30,31].

Trapped-ion systems (see Fig. 1) are currently one of the most sophisticated technologies developed for quantum information processing[32]. These systems offer an unprecedented level of quantum control, which opens new possibilities for obtaining physico-chemical information about quantum chemical problems. The power of trapped ions for quantum simulation is manifested by the high-precision control over both the internal degrees of freedom of the individual ions and the phonon degrees of freedom of the collective motions of the trapped ions, and the high-fidelity initialization and measurement[32,33]. Up to 100 quantum logic gates have been realized for six qubits with trapped ions[22], and quantum simulators involving 300 ions have been demonstrated[34].

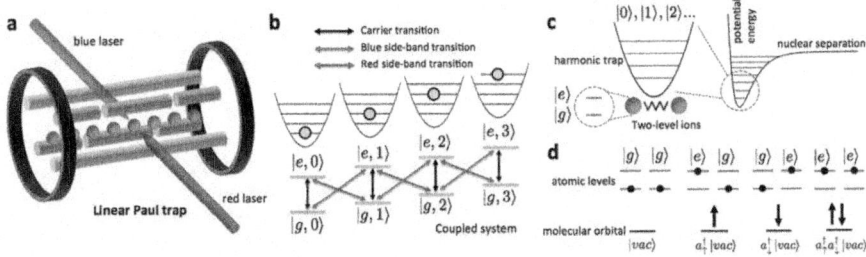

Figure 1: Simulating quantum chemistry with trapped ions. (a) Scheme of a trapped-ion setup for quantum simulation, which contains a linear chain of trapped ions confined by a harmonic potential, and external lasers that couple the motional and internal degrees of freedom. (b) Transitions between internal and motional degrees of freedom of the ions in the trap. (c) The normal modes of the trapped ions can simulate the vibrational degrees of freedom of molecules. (d) The internal states of two ions can simulate all four possible configurations of a molecular orbital.

In this work, we present an efficient toolkit for solving quantum chemistry problems based on the state-of-the-art in trapped-ion technologies. The toolkit comprises two components i) First, we present a hybrid quantum-classical variational optimization method, called quantum-assisted optimization, for approximating both ground-state energies and the ground-state eigenvectors for electronic problems. The optimized eigenvector can then be taken as an input for the phase estimation algorithm to project out the exact eigenstates and hence the potential-energy surfaces (see Fig. 2). Furthermore, we extend the application of the unitary coupled-cluster method[35]. This allows for the application of a method developed for classical numerical computations in the quantum domain. ii) The second main component of our toolkit is the optimized use of trapped-ion phonon degrees of freedom not only for quantum-gate construction, but also for simulating molecular vibrations, representing a mixed digital-analog quantum simulation. The phonon degrees of freedom in trapped-ion systems provide a natural platform for addressing spin-boson or fermion-boson-type problems through quantum simulation[23,36,37,38,39,40]. It is noteworthy to mention that, contrary to the continuous of modes required for full-fledged quantum field theories, quantum simulations of quantum chemistry problems could reach realistic conditions for finite bosonic and fermionic mode numbers. Consequently, trapped ions can be exploited to solve dynamical problems involving linearly or non-linearly coupled oscillators, e.g., spin-boson models[41,42], that are difficult to solve either analytically or numerically with a classical computer. Furthermore, we have also developed a novel protocol to measure correlation functions of observables in trapped ions that will be crucial for the quantum simulation of quantum chemistry.

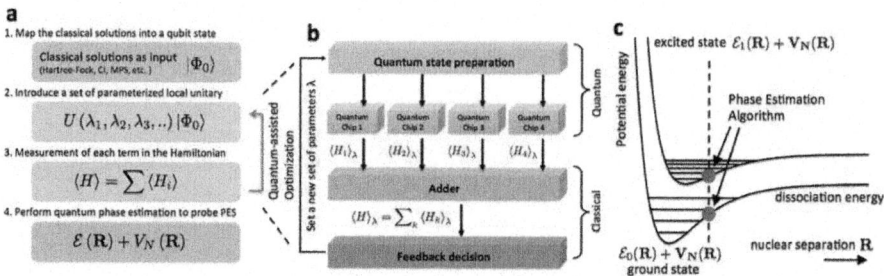

Figure 2: Outline of the quantum-assisted optimization method. (a) The key steps for quantum assisted optimization, which starts from classical solutions. For each new set of parameters λs, determined by a classical optimization algorithm, the expectation value $\langle H \rangle$ is calculated. The potential energy surface is then obtained by quantum phase estimation. (b) Quantum measurements are performed for the individual terms in H, and the sum is obtained classically. (c) The same procedure is applied for each nuclear configuration R to probe the energy surface.

RESULTS AND DISCUSSION

Trapped ions for quantum chemistry

Quantum chemistry deals with the many-body problem involving electrons and nuclei. Thus, it is very well suited for being simulated with trapped-ion systems, as we will show below. The full quantum chemistry Hamiltonian, H $= T_e + V_e + T_N + V_N + V_{eN}$, is a sum of the kinetic energies of the electrons

$$T_e \equiv -\frac{\hbar^2}{2m}\Sigma_i \nabla_{e,i}^2 \text{ and nuclei } T_N \equiv -\Sigma_i \frac{\hbar^2}{2M_i} \nabla_{N,i}^2, \text{ and the electron-electron}$$
$V_e \equiv \Sigma_{j>i} e^2 \mid r_i - r_j \mid$, nuclei-nuclei $V_N \equiv \Sigma_{j>i} Z_i Z_j e^2 \mid R_i - R_j \mid$, and electron-nuclei $V_{eN} \equiv -\Sigma_{j,i} Z_j e^2 \mid r_i - R_j \mid$ potential energies, where $\mathbf{r}$ and $\mathbf{R}$ respectively refer to the electronic and nuclear coordinates.

In many cases, it is more convenient to work on the second-quantization representation for quantum chemistry. The advantage is that one can choose a good fermionic basis set of molecular orbitals, $\mid p \rangle = c_p^\dagger \mid vac \rangle$, which can compactly capture the low-energy sector of the chemical system. This kind of second quantized fermionic Hamiltonians are efficiently simulatable in trapped ions[23]. To be more specific, we will choose first M > N orbitals for an N-electron system. Denote $\varphi_p(\mathbf{r}) \equiv \langle r \mid p \rangle$ as the single-particle wavefunction corresponding to mode p. The electronic part, $H_e(\mathbf{R}) \equiv T_e + V_{eN}(\mathbf{R}) + V_e$, of the Hamiltonian H can be expressed as follows:

$$H_e(\mathbf{R}) = \sum_{pq} h_{pq} c_p^\dagger c_q + \frac{1}{2} \sum_{pqrs} h_{pqrs} c_p^\dagger c_q^\dagger c_r c_s,$$

(1)

where h_{pq} is obtained from the single-electron integral $h_{pq} \equiv -\int dr \phi_p^*(r)(T_e + V_{eN}) \phi_q(r)$, and h_{pqrs} comes from the electron-electron Coulomb interaction, $h_{pqrs} \equiv \int dr_1 dr_2 \phi_p^*(r_1) \phi_q^*(r_2) V_e(|r_1 - r_2|) \phi_r(r_2) \phi_s(r_1)$. We note that the total number of terms in H_e is $O(M^4)$; typically M is of the same order asN. Therefore, the number of terms in H_e scales polynomially in N, and the integrals $\{h_{pq}, h_{pqrs}\}$ can be numerically calculated by a classical computer with polynomial resources[9].

To implement the dynamics associated with the electronic Hamiltonian in Eq. (1) with a trapped-ion quantum simulator, one should take into account the fermionic nature of the operators c_p and $c_q^\dagger$. We invoke the Jordan-Wigner transformation (JWT), which is a method for mapping the occupation representation to the spin (or qubit) representation[43]. Specifically, for each fermionic mode p, an unoccupied state $|0\rangle_p$ is represented by the spin-down state $|\downarrow\rangle_p$, and an occupied state $|1\rangle_p$ is represented by the spin-up state $|\uparrow\rangle_p$. The exchange symmetry is enforced by the Jordan-Wigner transformation:

$c_q^\dagger = \left(\Pi_{m<p} \sigma_m^z\right) \sigma_p^+$ and $c_p = \left(\Pi_{m<p} \sigma_m^z\right) \sigma_p^-$, where $\sigma^\pm \equiv (\sigma^x \pm i\sigma^y)/2$. Consequently, the electronic Hamiltonian in Eq. (1) becomes highly nonlocal in terms of the Pauli operators $\{\sigma^x, \sigma^y, \sigma^z\}$, i.e.,

$$H_e \xrightarrow[\mathrm{JWT}]{} \sum_{i,j,k... \in \{x,y,z\}} g_{ijk...} \left(\sigma_1^i \otimes \sigma_2^j \otimes \sigma_3^k ...\right).$$

(2)

Nevertheless, the simulation can still be made efficient with trapped ions, as we shall discuss below.

In trapped-ion physics two metastable internal levels of an ion are typically employed as a qubit. Ions can be confined either in Penning traps or radio frequency Paul traps[33], and cooled down to form crystals. Through sideband cooling the ions motional degrees of freedom can reach the ground state of the quantum Harmonic oscillator that can be used as a quantum bus to perform gates among the different ions. Using resonance fluorescence with a cycling transition quantum non demolition measurements of the qubit can be performed. The fidelities of state preparation, single- and two-qubit gates, and detection, are all above 99%[32].

The basic interaction of a two-level trapped ion with a single-mode laser is given by[32], $H = \hbar \Omega \sigma_+ e^{-i(\Delta t - \phi)} \exp\left(in\left[ae^{-i\omega_t t} + a^\dagger e^{i\omega_t t}\right]\right) + H.c$, where $\sigma_\pm$ are the atomic raising and lowering operators, a ($a^\dagger$) is the annihilation (creation) operator of the considered motional mode, and Ω is the Rabi frequency associated to the laser strength. $\eta = kz_0$ is the Lamb-Dicke parameter, with k the wave vector of the laser and $z_0 = \sqrt{\hbar/(2m\omega_t)}$ the ground state width of the motional mode. φ is a controllable laser phase and Δ the laser-atom detuning.

In the Lamb-Dicke regime where $\eta\sqrt{\langle(a+a^\dagger)^2\rangle} \ll 1$, the basic interaction of a two-level trapped ion with a laser can be rewritten as

$$H = \hbar\Omega\left[\sigma_+ e^{-i(\Delta t - \phi)} + i\eta\sigma_+ e^{-i(\Delta t - \phi)}\left(ae^{-i\omega_t t} + a^\dagger e^{i\omega_t t}\right) + \text{H.c.}\right]$$

By adjusting the laser detuning Δ, one can generate the three basic ion-phonon interactions, namely: the carrier interaction ($\Delta = 0$) $H_c = \hbar\Omega(\sigma_+ e^{i\phi} + \sigma_- e^{-i\phi})$, the red sideband interaction ($\Delta = -\omega_t$) $H_r = i\hbar\eta\Omega(\sigma_+ ae^{i\phi} - \sigma_- a^\dagger e^{-i\phi})$, and the blue sideband interaction ($\Delta = \omega_t$) $H_b = i\hbar\eta\Omega(\sigma_+ a^\dagger e^{i\phi} - \sigma_- ae^{-i\phi})$. By combining detuned red and blue sideband interactions, one obtains the Mølmer-Sørensen gate[44], which is the basic building block for our methods. With combinations of this kind of gates, one can obtain dynamics as the associated one to H_e in Eq. (2), that will allow one to simulate arbitrary quantum chemistry systems.

Quantum-assisted optimization

Quantum-assisted optimization[45] (see also Fig. 2) for obtaining ground-state energies aims to optimize the use of quantum coherence by breaking down the quantum simulation through the use of both quantum and classical processors; the quantum processor is strategically employed for expensive tasks only.

To be more specific, the first step of quantum-assisted optimization is to prepare a set of quantum states $\{|\psi_\lambda\rangle\}$ that are characterized by a set of parameters $\{\lambda\}$. After the state is prepared, the expectation value $E_\lambda \equiv \langle\psi_\lambda| H |\psi_\lambda\rangle$ of the Hamiltonian H will be measured directly, without any quantum evolution in between. Practically, the quantum resources for the measurements can be significantly reduced when we divide the measurement of the Hamiltonian $\sum_i H_i$ into a polynomial number of small pieces $\langle H_i\rangle$ (cf Eq. (2)). These measurements can be performed in a parallel fashion, and no quantum coherence is needed to maintain between the measurements (see Fig. 2a and 2b). Then, once a data point of E_λ is obtained, the whole procedure is repeated for a new state $\{|\psi'_\lambda\rangle\}$ with another set of parameters $\{\lambda'\}$. The choice of the new parameters is determined by a classical optimization algorithm that aims to minimize E_λ (see Methods). The optimization procedure is terminated after the value of E_λ converges to some fixed value.

Finally, for electronic Hamiltonians $H_e(\mathbf{R})$, the optimized state can then be sent to a quantum circuit of phase estimation algorithm to produce a set of data point for some $\mathbf{R}$ on the potential energy surfaces (Fig. 2c shows the 1D case). After locating the local minima of the ground and excited states, vibronic coupling for the electronic structure can be further studied (see Supplementary Material).

The performance of quantum-assisted optimization depends crucially on (a) the choice of the variational states, and (b) efficient measurement methods. We found that the unitary coupled-cluster (UCC) states35 are particularly suitable for being the input state for quantum-assisted optimization, where each quantum state $|\psi_\lambda\rangle$ can be prepared efficiently with standard techniques in trapped ions. Furthermore, efficient measurement methods for H_e are also available for trapped ion systems. We shall discuss these results in detail in the following sections.

Unitary coupled-cluster (UCC) ansatz

The unitary coupled-cluster (UCC) ansatz[35] assumes electronic states $|\psi\rangle$ have the following form, $|\psi\rangle = e^{T-T^\dagger}|\Phi\rangle$, where $|\Phi\rangle$ is a reference state, which can be, e.g., a Slater determinant constructed from Hartree-Fock molecular orbitals. The particle-hole excitation operator, or cluster operator T, creates a linear combination of excited Slater determinants from $|\Phi\rangle$. Usually, T is divided into subgroups based on the particle-hole rank. More precisely, $T = T_1 + T_2 + T_3 + \ldots + T_N$ for an N-electron system, where

$$T_1 = \sum_{i,a} t_i^a c_a^\dagger c_i, \quad T_2 = \sum_{i,j,a,b} t_{ij}^{ab} c_a^\dagger c_b^\dagger c_j c_i,$$ and so on.

Here $c_a^\dagger$ creates an electron in the orbital a. The indices a, b label unoccupied orbitals in the reference state $|\Phi\rangle$, and i, j label occupied orbitals. The energy obtained from UCC, namely $E = \langle \Phi | e^{T^\dagger - T} H e^{T - T^\dagger} | \Phi \rangle$ is a variational upper bound of the exact ground-state energy.

The key challenge for implementing UCC on a classical computer is that the computational resource grows exponentially. It is because, in principle, one has to expand the expression $\tilde{H} \equiv e^{T^\dagger - T} H e^{T - T^\dagger}$ into an infinity series, using the Baker-Campbell-Hausdorff expansion. Naturally, one has to rely on approximate methods[35,46] to truncate the series and keep track of finite numbers of terms. Therefore, in order to make good approximations by perturbative methods, i.e., assuming T is small, one implicitly assumes that the reference state $|\Phi\rangle$ is a good solution to the problem. However, in many cases, such an assumption is not valid and the use of approximate UCC breaks down. We explain below how implementing UCC on a trapped-ion quantum computer can overcome this problem.

Implementation of UCC through time evolution

We can generate the UCC state by simulating a pseudo time evolution through Suzuki-Trotter expansion on the evolution operator $e^{T-T^\dagger}$[21]. To proceed, we consider an N-electron system with M, where M > N, molecular orbitals (including spins). We need totally M qubits; the reference state is the Hartree-Fock state where N orbitals are filled, and M − N orbitals are empty, i.e, $|\Phi\rangle$ = $|000...0111...1\rangle$. We also define an effective Hamiltonian $K \equiv i(T - T^\dagger)$, which means that we should prepare the state $e^{-iK}|\Phi\rangle$.

We decompose K into subgroups $K = K_1 + K_2 + K_3 + ... + K_P$, where $P \leq$ N, and $K_i \equiv i\left(T_i - T_i^\dagger\right)$. We now write $e^{-iK} = (e^{-iK\delta})^{1/\delta}$ for some dimensionless constantδ. For small δ, we have $e^{-iK\delta} \approx e^{-iK_P\delta} \ldots e^{-iK_2\delta} e^{-iK_1\delta}$. Since each K_j contains$N^j(M-N)^j$ terms of the creation $c^\dagger$ and annihilation c operators, we will need to individually simulate each term separately, e.g., $e^{-i\left(tc_a^\dagger c_i - t^* c_i^\dagger c_a\right)}$ and $e^{-i\left(tc_a^\dagger c_b^\dagger c_j c_i - t^* c_i^\dagger c_j^\dagger c_b c_a\right)}$, which can be implemented by transforming into spin operators through Jordan-Wigner transformation. The time evolution for each term can be simulated with a quantum circuit involving many nonlocal controlled gates, which can be efficiently implemented with trapped ions as we shall see below.

Implementation of UCC and simulation of time evolution with trapped-ions

Our protocol for implementing the UCC ansatz requires the simulation of the small-time t/n evolution of non-local product of Pauli matrices of the form: $e^{-iH_l t/n}$, where $H_l = g_l\sigma_1^i\sigma_2^j\sigma_3^k$ ... for i, j, k ∈ {x, y, z}. Note that for any N-spin interaction, the $e^{-iH_l t/n}$ terms are equivalent to $e^{i\phi\sigma_1^z\sigma_2^x\sigma_3^x\cdots\sigma_N^x}$ through local spin rotations, which are simple to implement on trapped ions. Such a non-local operator can be implemented using the multi-particle Mølmer-Sørensen gate[23,39]: $U_{MS}(\theta, \varphi) \equiv \exp[-i\theta(\cos\varphi S_x + \sin\varphi S_y)^2/4]$, where $S_{x,y} \equiv \sum_i \sigma_i^{x,y}$ is a collective spin operator. Explicitly,

$$e^{i\phi\sigma_1^z \sigma_2^x \sigma_3^x \cdots \sigma_N^x} = U_{MS}\left(\frac{-\pi}{2},0\right) R_N(\phi) U_{MS}\left(\frac{\pi}{2},0\right) \tag{3}$$

Here $R_N(\phi)$ is defined as follows: for any $m \in \mathbb{N}$, $R_N(\phi) = e^{\pm i\phi\sigma_1^z}$ for $N = 4m \pm 1$, and (ii) $R_N(\phi) = e^{i\phi\sigma_1^y}$ for $N = 4m$, and (iii) $R_N(\phi) = e^{i\phi\sigma_1^y}$ for $N = 4m - 2$.

It is remarkable that the standard quantum-circuit treatment (e.g. see Ref.[47]) for implementing each $e^{-iH_l t/n}$ involves as many as 2N two-qubit gates for simulating N fermionic modes; in our protocol one needs only two Mølmer-Sørensen gates, which are straightforwardly implementable with current trapped-ion technology. Furthermore, the local rotation $R_N(\phi)$ can also include motional degrees of freedom of the ions for simulating arbitrary fermionic Hamiltonians coupled linearly to bosonic operators a_k and $a_k^\dagger$.

Measurement of arbitrarily-nonlocal spin operators

For any given state $|\psi\rangle$, we show how to encode expectation value of products of Pauli matrices $\left\langle \sigma_1^i \otimes \sigma_2^j \otimes \sigma_3^k \otimes \cdots \right\rangle \equiv \langle\psi| \sigma_1^i \otimes \sigma_2^j \otimes \sigma_3^k \otimes \cdots |\psi\rangle$, where i, j, k $\in \{x, y, z\}$, onto an expectation value of a single qubit. The idea is to first apply the unitary evolution of the form: $e^{-i\theta\left(\sigma_1^i \otimes \sigma_2^j \otimes \cdots\right)}$, which as we have seen (cf Eq. 3) can be generated by trapped ions efficiently, to the state $|\psi\rangle$ before the measurement. For example, defining $|\psi_\theta\rangle \equiv e^{-i\theta\left(\sigma_1^x \otimes \sigma_2^x \otimes \cdots\right)}|\psi\rangle$, we have the relation

$$\langle\psi_\theta|\sigma_1^z|\psi_\theta\rangle = \cos(2\theta)\langle\sigma_1^z\rangle + \sin(2\theta)\langle\sigma_1^y \otimes \sigma_2^x \otimes \cdots\rangle \tag{4}$$

which equals $\langle\psi| \left(\sigma_1^y \otimes \sigma_2^x \otimes \cdots\right) |\psi\rangle$ for $\theta = \pi/4$. Note that the application of this method requires the measurement of one qubit only, making this technique especially suited for trapped ion systems where the fidelity of the measurement of one qubit is 99.99%[48].

This method can be further extended to include bosonic operators in the resulting expectation values. For example, re-define $|\psi_\theta\rangle \equiv e^{-i\theta\left(\sigma_1^i \otimes \sigma_2^j \otimes \cdots\right) \otimes \left(a + a^\dagger\right)}|\psi\rangle$ and consider $\theta \rightarrow \theta(a + a^\dagger)$ in Eq. (4). We can obtain the desired correlation through the derivative of the single-qubit measurement: $\partial_\theta\langle\psi_\theta|\sigma_1^z|\psi_\theta\rangle|_{\theta=0} = -2\left\langle\left(\sigma_1^y \otimes \sigma_2^x \otimes \cdots\right)\left(a + a^\dagger\right)\right\rangle$. Note that the evolution operator of the form $e^{-i\theta\left(\sigma_1^i \otimes \sigma_2^j \otimes \cdots\right) \otimes \left(a+a^\dagger\right)}$ can be generated by replacing the local

operation $R_N(\varphi)$ in Eq. 3 with $e^{\pm i\phi\sigma_1^i\left(a+a^\dagger\right)}$. This technique allows us to obtain a diverse range of correlations between bosonic and internal degrees of freedom.

Probing potential energy surfaces

In the Born-Oppenheimer (BO) picture, the potential energy surface $\mathcal{E}_k(R) + V_N(R)$ associated with each electronic eigenstate $|\varphi_k\rangle$ is obtained by scanning the eigenvalues $\mathcal{E}_k(R)$ for each configurations of the nuclear coordinates $\{R\}$. Of course, we can apply the standard quantum phase estimation algorithm[49] that allows us to extract the eigenvalues. However, this can require many ancilla qubits. In fact, locating these eigenvalues can be achieved by the phase estimation method utilizing one extra ancilla qubit[12]corresponding, in our case, to one additional ion.

This method works as follows: suppose we are given a certain quantum state $|\psi\rangle$ (which may be obtained from classical solutions with quantum-assisted optimization) and an electronic Hamiltonian $H_e(R)$ (cf. Eq. (1)). Expanding the input state, $|\psi\rangle = \sum_k \alpha_k |\phi_k\rangle$, by the eigenstate vectors $|\varphi_k\rangle$ of $H_e(R)$, where $H_e(R)|\phi_k\rangle = \mathcal{E}_k(R)|\phi_k\rangle$, then for the input state $|0\rangle |\psi\rangle$, the quantum circuit of the quantum phase estimation produces the following output state, $\left(1/\sqrt{2}\right)\sum_k \alpha_k \left(|0\rangle + e^{-i\omega_k t}|1\rangle\right)|\phi_k\rangle$, where $\omega_k = \mathcal{E}_k/\hbar$. The corresponding reduced density matrix,

$$\frac{1}{2}\begin{pmatrix} 1 & \sum_k |\alpha_k|^2 e^{i\omega_k t} \\ \sum_k |\alpha_k|^2 e^{-i\omega_k t} & 1 \end{pmatrix} \tag{5}$$

of the ancilla qubit contains the information about the weight (amplitude-square) $|\alpha_k|^2$ of the eigenvectors $|\varphi_k\rangle$ in $|\psi\rangle$ and the associated eigenvalues ω_k in the off-diagonal matrix elements. All $|\alpha_k|^2$'s and ω_k's can be extracted by repeating the quantum circuit for a range of values of t and performing a (classical) Fourier transform to the measurement results. The potential energy surface is obtained by repeating the procedure for different values of the nuclear coordinates $\{R\}$.

Numerical investigation

In order to show the feasibility of our protocol, we can estimate the trapped-ion resources needed to simulate, e.g., the prototypical electronic Hamiltonian

$H_e = \sum h_{pq} a_p^\dagger a_q + (1/2) \sum h_{pqrs} a_p^\dagger a_q^\dagger a_r a_s$ as described in Eq. (1), for the specific case of the H_2 molecule in a minimal STO-3G basis. This is a two-electron system represented in a basis of four spin-orbitals. The hydrogen atoms were separated by 0.75 Å, near the equilibrium bond distance of the molecule. The Hamiltonian is made up of 12 terms, that include 4 local ion operations and 8 non-local interactions. Each of the non-local terms can be done as a combination of two Mølmer-Sørensen (MS) gates and local rotations, as described in Table 1. Therefore, to implement the dynamics, one needs 16 MS gates per Trotter step and a certain number of local rotations upon the ions. Since $\pi/2$ MS gates can be done in ~ 50 μs, and local rotations can be performed in negligible times (~ 1 μs)[22,32], the total simulation time can be assumed of about 800 μs for the n = 1 protocol, 1.6 ms and 2.4 ms for the n = 2 and n = 3 protocols. Thus total simulation times are within the decoherence times for trapped-ion setups, of about 30 ms[32]. In a digital protocol performed on real quantum systems, each gate is affected by an error. Thus, increasing the number of Trotter steps leads to an accumulation of the single gate error. To implement an effective quantum simulation, on one hand one has to increase the number of steps to reduce the error due to the digital approximation, on the other hand one is limited by the accumulation of the single gate error. We plot inFig. 3a, 3b, 3c, the fidelity loss $1 - |\langle \Psi_S | \Psi_E \rangle|^2$ of the simulated state $|\Psi_S\rangle$ versus the exact one $|\Psi_E\rangle$, for the hydrogen Hamiltonian, starting from the initial state with two electrons in the first two orbitals. We plot, along with the digital error, three horizontal lines representing the accumulated gate error, for n= 1, 2, 3 in each plot, considering a protocol with an error per Trotter step of $\epsilon = 10^{-3}$ (a), $\epsilon = 10^{-4}$ (b) and $\epsilon = 10^{-5}$ (c). To achieve a reasonable fidelity, one has to find a number of steps that fits the simulation at a specific time. The vertical lines and arrows in the figure mark the time regions in which the error starts to be dominated by the digital error. Trapped-ion two-qubit gates are predicted to achieve in the near future infidelities of 10^{-4}, thus making the use of these protocols feasible[50]. In Fig. 3d we plot the behavior of the energy of the system for the initial state $|\uparrow\uparrow\downarrow\downarrow\rangle$ for the exact dynamics, versus the digitized one. Again, one can observe how the energy can be retrieved with a small error within a reduced number of digital steps.

Table 1: Using trapped ions to simulate quantum chemistry

	Simulating Quantum Chemistry	Implementation with Trapped Ions
Hamiltonian transformation:	The fermionic (electronic) Hamiltonian H_e is transformed into a spin Hamiltonian through the Jordan-Wigner transformation. $$H_e \rightarrow \sum_{i,j,k\in\{x,y,z\}} g_{i,j\ldots}\left(\sigma_1^i \otimes \sigma_2^j \otimes \sigma_3^k \cdots\right) \equiv \sum_{l=1}^m H_l$$	The spin degrees of freedom in H_e are represented by the internal degrees of freedom of the trapped ions.
Simulation of time evolution:	The time evolution operator $e^{-iH_e t}$ is split into n small-time $[t/n]$ pieces $e^{-iH_e t/n}$ through the Suzuki-Trotter expansion. $$e^{-i\sum_{l=1}^m H_l t} \approx \left(e^{-iH_1 t/n} e^{-iH_2 t/n} \cdots e^{-iH_m t/n}\right)^n$$	Each individual term $e^{-iH_l t/n}$ can be simulated with trapped ions through the use of Mølmer-Sørensen gates U_{MS}. Explicitly, $$e^{-i\theta\sigma_z^i/2} = U_{MS}\left(\frac{-\pi}{2}, 0\right) U_{\sigma_z}(\phi) U_{MS}\left(\frac{\pi}{2}, 0\right).$$
Obtaining average energy:	The average energy $\langle H_e\rangle$ of the Hamiltonian can be obtained through the sum of the individual terms $\langle H_l\rangle$, which reduces to the measurement of products of Pauli matrices.	For any prepared state $|\psi\rangle$, average values of the products of Pauli matrices $J_{jk} \equiv \sigma_1^i \otimes \sigma_2^j \otimes \sigma_3^k \cdots$ can be measured by first applying the pseudo time evolution operator $e^{-i\pi J/4}|\psi_a\rangle$ to $|\psi\rangle$ and then measuring $\langle \sigma_z^i\rangle$.
Measuring eigenvalues:	The eigenvalues of the Hamiltonian can be obtained through the phase estimation algorithm. Good trial states can be obtained through classical computing, or the unitary coupled-cluster method.	The phase estimation algorithm can be implemented through the simulation of controlled time evolutions.
Molecular vibrations:	The inclusion of vibrational degrees of freedom is necessary for corrections on the Born-Oppenheimer picture in the electronic structure of molecules.	The vibrational degrees of freedom are represented by the quantized vibrational motion of the trapped ions.

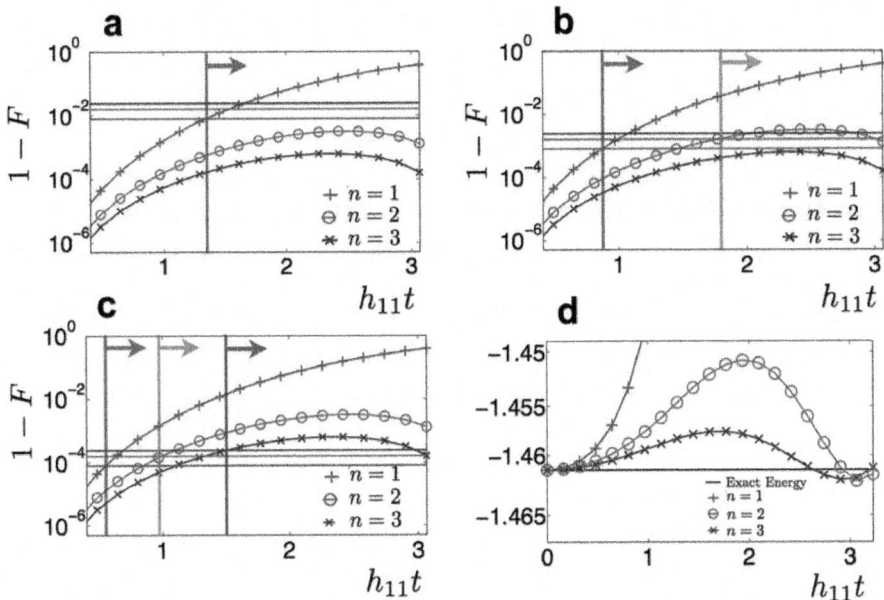

Figure 3: Digital error $1 - F$ (curves) along with the accumulated gate error (horizontal lines) versus time in h_{11} energy units, for n = 1, 2, 3 Trotter steps in each plot, considering a protocol with an error per Trotter step of $\dot{o} = 10^{-3}$ (a), $\dot{o} = 10^{-4}$ (b) and $\dot{o} = 10^{-5}$ (c). The initial state considered is $|\uparrow\uparrow\downarrow\downarrow\rangle$, in the qubit representation of the Hartree-Fock state in a molecular orbital basis with one electron on the first and second orbital. Vertical lines and arrows define the time domain in which the dominant part of the error is due to the digital approximation. d) Energy of the system, in h_{11} units, for the initial state $|\uparrow\uparrow\downarrow\downarrow\rangle$ for the exact dynamics, versus the digitized one. For a protocol with three Trotter steps the energy is recovered up to a negligible error

CONCLUSIONS

Summarizing, we have proposed a quantum simulation toolkit for quantum chemistry with trapped ions. This paradigm in quantum simulations has several advantages: an efficient electronic simulation, the possibility of interacting electronic and vibrational degrees of freedom, and the increasing scalability provided by trapped-ion systems. This approach for solving quantum chemistry problems aims to combine the best of classical and quantum computation.

Methods

To implement the optimization with the UCC wavefunction ansatz on a trapped-ion quantum simulator, our proposal is to first employ classical algorithms to obtain approximate solutions[35,46]. Then, we can further improve the quality of the solution by searching for the true minima with an ion trap. The idea is as follows: first we create a UCC ansatz by the Suzuki-Trotter method described in the previous section. Denote this choice of the cluster operator as $T^{(0)}$, and other choices as $T^{(k)}$ with k = 1, 2, 3, ... The corresponding energy $E_0 = \langle \Phi | e^{T^{(0)\dagger} - T^{(0)}} H e^{T^{(0)} - T^{(0)\dagger}} | \Phi \rangle$ of the initial state is obtained by a classical computer.

Next, we choose another set of cluster operator $T^{(1)}$ which is a perturbation around $T^{(0)}$. Define the new probe state $| \phi_k \rangle \equiv e^{T^{(k)} - T^{(k)\dagger}} | \Phi \rangle$. Then, the expectation value of the energy $E_1 = \langle \Phi | e^{T^{(1)\dagger} - T^{(1)}} H e^{T^{(1)} - T^{(1)\dagger}} | \Phi \rangle$ can be obtained by measuring components of the second quantized Hamiltonian, $\langle \phi_1 | H | \phi_1 \rangle = \sum_{pqrs} \tilde{h}_{pqrs} \langle \phi_1 | c_p^\dagger c_q^\dagger c_r c_s | \phi_1 \rangle$. Recall that the coefficients $\tilde{h}_{pqrs}$ are all precomputed and known.

In order to obtain measurement results for the operators $\langle \phi_1 | c_p^\dagger c_q^\dagger c_r c_s | \phi_1 \rangle$, we will first convert the fermion operators into spin operators via Jordan-Wigner transformation; the same procedure is applied for creating the state $| \varphi_1 \rangle$. The quantum measurement for the resulting products of Pauli matrices can be achieved efficiently with trapped ions, using the method we described.

The following steps are determined through a classical optimization algorithm. There can be many choices for such an algorithm, for example gradient descent method, Nelder-Mead method, or quasi-Newton methods. For completeness, we summarize below the application of gradient descent method to our optimization problem.

First we define the vector $T^{(k)} = \left(t_i^{a(k)}, t_{ij}^{ab(k)}, \ldots \right)^T$ to contain all coefficients in the cluster operator $T^{(k)}$ at the k-th step. We can also write the expectation value $E(\mathbf{T}^{(k)}) \equiv \langle \varphi_k | H | \varphi_k \rangle$ for each step as a function of $\mathbf{T}^{(k)}$. The main idea of the gradient descent method is that E $(\mathbf{T}^{(k)})$ decreases fastest along the direction of the negative gradient of E $(\mathbf{T}^{(k)})$, $-\nabla$E $(\mathbf{T}^{(k)})$. Therefore, the (k + 1)-th step is determined by the following relation:

$$T^{(k+1)} = T^{(k)} - a_k \nabla E \left(\mathbf{T}^{(k)} \right)$$

(6)

where a_k is an adjustable parameter; it can be different for each step. To obtain values of the gradient ∇E $(\mathbf{T}^{(k)})$, one may use the finite-difference method to approximate the gradient. However, numerical gradient techniques are often susceptible to numerical instability. Alternatively, we can invoke the Hellman-Feynman theorem and get, e.g., $\left(\partial / \partial t_i^a \right) E \left(\mathbf{T}^{(k)} \right) = \langle \phi_k | [H, c_a^\dagger c_i] | \phi_k \rangle$

, which can be obtained with a method similar to that for obtaining $E(\mathbf{T}^{(k)})$. Finally, as a valid assumption for general cases, we assume our parametrization of UCC gives a smooth function for E $(\mathbf{T}^{(k)})$. Thus, it follows

that $E\left(T^{(0)}\right) \geq E\left(T(1)\right) \geq E\left(T^{(2)}\right) \geq \cdots$, and eventually E $(\mathbf{T}^{(k)})$ converges to a minimum value for large k. Finally, we can also obtain the optimized UCC quantum state.

REFERENCE

1. Dirac, P. A. M. Quantum Mechanics of Many-Electron Systems. Proc. R. Soc. A 123, 714–733 (1929).

2. Love, P. J. Back to the Future: A roadmap for quantum simulation from vintage quantum chemistry. eprint arXiv:1208.5524 (2012). (To Appear in Advances in Chemical Physics).

3. Head-Gordon,M. & Artacho, E. Chemistry on the computer. Phys. Today, 61, 58 (2008).

4. Feynman, R. P. Simulating physics with computers. Int. J. Theor. Phys. 21,467–488 (1982).

5. Kassal, I., Whitfield, J. D., Perdomo-Ortiz, A., Yung, M.-H. & Aspuru-Guzik, A. Simulating chemistry using quantum computers. Annu. Rev. Phys. Chem. 62, 185207 (2011).

6. Yung, M.-H., Whitfield, J. D., Boixo, S., Tempel, D. G. & Aspuru-Guzik, A. Introduction to Quantum Algorithms for Physics and Chemistry. EprintarXiv,1203.1331 (2012).

7. Aspuru-Guzik, A. & Walther, P. Photonic quantum simulators. Nat. Phys. 8, 285291 (2012).

8. Abrams, D. & Lloyd, S. Quantum Algorithm Providing Exponential Speed Increase for Finding Eigenvalues and Eigenvectors. Phys. Rev. Lett. 83, 5162–5165 (1999).

9. Aspuru-Guzik, A., Dutoi, A. D., Love, P. J. & Head-Gordon, M. Simulated quantum computation of molecular energies. Science 309, 1704–7 (2005).

10. Lanyon, B. P. et al. Towards quantum chemistry on a quantum computer. Nat. Chem. 2, 106–11 (2010).

11. Poulin, D. & Wocjan, P. Preparing Ground States of Quantum Many-Body Systems on a Quantum Computer. Phys. Rev. Lett. 102, 130503 (2009).

12. Li, Z. et al. Solving quantum ground-state problems with nuclear

magnetic resonance. Sci. Rep. 1, 88 (2011).

13. Xu, J.-S. et al. Demon-like Algorithmic Quantum Cooling and its Realization with Quantum Optics.eprint arXiv,1208.2256 (2012).

14. Lidar, D.& Biham,O. Simulating Ising spin glasses on a quantum computer. Phys. Rev. E 56, 3661–3681 (1997).

15. Poulin, D. & Wocjan, P. Sampling from the Thermal Quantum Gibbs State and Evaluating Partition Functions with a Quantum Computer. Phys. Rev. Lett. 103, 220502 (2009).

16. Yung, M.-H., Nagaj, D., Whitfield, J. & Aspuru-Guzik, A. Simulation of classical thermal states on a quantum computer: A transfer-matrix approach. Phys. Rev. A 82, 060302 (2010).

17. Bilgin, E. & Boixo, S. Preparing Thermal States of Quantum Systems by Dimension Reduction. Phys. Rev. Lett. 105, 170405 (2010).

18. Temme, K., Osborne, T. J., Vollbrecht, K. G., Poulin, D. & Verstraete, F. Quantum Metropolis sampling. Nature 471, 87–90 (2011).

19. Yung, M.-H. & Aspuru-Guzik, A. A quantum-quantum Metropolis algorithm. PNAS 109, 754–9 (2012).

20. Zhang, J., Yung, M.-H., Laflamme, R., Aspuru-Guzik, A. & Baugh, J. Digital quantum simulation of the statistical mechanics of a frustrated magnet. Nat. Comm. 3, 880 (2012).

21. Lloyd, S. Universal Quantum Simulators. Science 273, 1073–1078 (1996).

22. Lanyon, B. P. et al. Universal digital quantum simulation with trapped ions. Science 334, 57–61 (2011).

23. Casanova, J., Mezzacapo, A., Lamata, L. & Solano, E. Quantum Simulation of Interacting Fermion Lattice Models in Trapped Ions. Phys. Rev. Lett. 108, 190502 (2012).

24. Zalka, C. Simulating quantum systems on a quantum computer. Proc R. Soc. A 454, 313–322 (1998).

25. Wu, L.-A., Byrd, M. & Lidar, D. Polynomial-Time Simulation of Pairing Models on a Quantum Computer. Phys. Rev. Lett. 89, 057904 (2002).

26. Kassal, I., Jordan, S. P., Love, P. J., Mohseni, M. & Aspuru-Guzik, A. Polynomialtime quantum algorithm for the simulation of chemical dynamics. PNAS 105, 18681–6 (2008).

27. Childs, A. M. & Kothari, R. Simulating sparse Hamiltonians with star decompositions. Theory of Quantum Computation Communication and Cryptography TQC 2010 6519:94–103, 2011.

28. Lidar, D. & Wang, H. Calculating the thermal rate constant with exponential speedup on a quantum computer. Phys. Rev. E 59, 2429–2438 (1999).

29. Master, C., Yamaguchi, F. & Yamamoto, Y. Efficiency of free-energy calculations of spin lattices by spectral quantum algorithms. Phys. Rev. A 67, 032311 (2003).

30. Kassal, I. & Aspuru-Guzik, A. Quantum algorithm for molecular properties and geometry optimization. J. Chem. Phys. 131, 224102 (2009).

31. Wocjan, P., Chiang, C.-F., Nagaj, D. & Abeyesinghe, A. Quantum algorithm for approximating partition functions. Phys. Rev. A 80, 022340 (2009).

32. Ha¨ffner, H., Roos, C. F. & Blatt, R. Quantum computing with trapped ions. Phys. Rep. 469, 155–203 (2008).

33. Leibfried, D., Blatt, R., Monroe, C. & Wineland, D. Quantum dynamics of single trapped ions. Rev. Mod. Phys. 75, 281–324 (2003).

34. Britton, J. W. et al. Engineered two-dimensional Ising interactions in a trappedion quantum simulator with hundreds of spins. Nature 484, 489–492 (2012).

35. Taube, A. G. & Bartlett, R. J. New perspectives on unitary coupled-cluster theory. Int. J. Quant. Chem. 106, 3393–3401 (2006).

36. Lamata, L., Leo´n, J., Scha¨tz, T. & Solano, E. Dirac Equation and Quantum Relativistic Effects in a Single Trapped Ion. Phys. Rev. Lett. 98, 253005 (2007).

37. Gerritsma, R. et al.Quantum simulation of the Dirac equation. Nature 463, 68–71 (2010).

38. Casanova, J. et al. Quantum Simulation of Quantum Field Theories in Trapped Ions. Phys. Rev. Lett. 107, 260501 (2011).

39. Mu¨ller, M., Hammerer, K., Zhou, Y. L., Roos, C. F. & Zoller, P. Simulating open quantum systems: from many-body interactions to stabilizer pumping. New J. Phys. 13, 085007 (2011).

40. Mezzacapo, A., Casanova, J., Lamata, L. & Solano, E. Digital Quantum Simulation of the Holstein Model in Trapped Ions. Phys. Rev. Lett. 109, 200501 (2012).

41. Leggett, A. et al. Dynamics of the dissipative two-state system. Rev. Mod. Phys. 59, 185 (1987).

42. Mostame, S. et al. Quantum simulator of an open quantum system using superconducting qubits: exciton transport in photosynthetic complexes.

New J. Phys. 14, 105013 (2012).

43. Ortiz, G., Gubernatis, J., Knill, E. & Laflamme, R. Quantum algorithms for fermionic simulations. Phys. Rev. A 64, 022319 (2001).

44. Mølmer, K. & Sørensen, A. Multiparticle Entanglement of Hot Trapped Ions. Phys. Rev. Lett. 82, 1835 (1999).

45. Peruzzo, A. et al. A variational eigenvalue solver on a quantum processor. EprintarXiv,1304.3061 (2013).

46. Kutzelnigg, W. Error analysis and improvements of coupled-cluster theory. Theor. Chim. Acta 80, 349–386 (1991).

47. Whitfield, J. D., Biamonte, J. & Aspuru-Guzik, A. Simulation of electronic structure Hamiltonians using quantum computers. Mol. Phys. 109, 735750 (2011).

48. Myerson, A. H. et al. High-Fidelity Readout of Trapped-Ion Qubits. Phys. Rev. Lett. 100, 200502 (2008).

49. Kaye, P., Laflamme, R. & Mosca, M. An introduction to quantum computing. (Oxford University Press, USA, 2007).

50. Kirchmair, G. et al. Deterministic entanglement of ions in thermal states of motion. New J. Phys. 11, 023002 (2009).

Chapter 10

FROM MOLECULAR PHYLOGENETICS TO QUANTUM CHEMISTRY: DISCOVERING ENZYME DESIGN PRINCIPLES THROUGH COMPUTATION

Troy Wymore[a], Charles L. Brooks III[b]

[a]Pittsburgh Supercomputing Center, 300 South Craig Street, Pittsburgh, PA 15213 USA

[b]University of Michigan, Department of Chemistry and Biophysics, 930 North University Avenue, Ann Arbor, MI 48109 USA

INTRODUCTION

Enzymes have obtained their amazing transformational capabilities through a colossal experiment in the optimization and diversification of protein structure-function relationships carried out over enormous stretches of time [1]. As such, it is critical to understand both the physicochemical properties of a biomolecular system and its history. Thus, the subject of this mini-review is on how both sequence-based bioinformatics and molecular dynamics (MD) simulations, particularly those using hybrid Quantum Chemical/Molecular Mechanical (QC/MM, the term QM/MM is also often used) potential energy functions, can be employed to discover the most critical amino acids relating to the enzyme's specific function as well as the atomic details of an enzymatic mechanism. We will highlight some results obtained from our own lab studying sesquiterpene synthases [2] and class D b-lactamases[3], two enzyme families known for their evolvability.

Comparative analysis of sequence and structural features of extant, mechanistically diverse, enzyme family members can result in the construction of powerful structure-function relationships and continues to dominate enzymatic studies today[4]. Yet, there remains an immense lack of knowledge at the atomic level on how enzymes evolve novel functions especially when residues outside the active site play a central role in this evolution. Some of the now classical studies of molecular adaptation, on topics such as insecticide resistance[5], color vision [6], antibiotic resistance [7-9], cofactor selectivity [10,11], hormone receptor selectivity [12-14] as well as those from directed evolution [15], have uncovered a complex network of interactions often involving residues outside the active site. These studies have elucidated structure-function relationships that otherwise would have remained hidden with conventional analyses [1]. In addition, they have revealed stability-function trade-offs [16-18], promiscuity of ancestral proteins [19], and the profound concept of functional epistasis [20]. Functional epistasis is defined here as the phenotypic consequences of a mutation depending on the genetic sequence in which it occurs [21]. Epistasis restricts the evolutionary pathway to novel functions so that adaptive walks through sequence space must be acquired in a particular order and through a rugged functional landscape to avoid non-functional intermediates; an idea pioneered by Linus Pauling and Emile Zuckerkandl in a summary of their research [22] and by the evolutionary biologist John Maynard Smith in response to an attack on the concept of natural selection[23]. These studies obligate us to seek answers to broader, more fundamental questions [24] such as, "Does functional evolution proceed by a few mutations of large effect or by many mutations each of small effect?"; "Could alternative solutions to the same problem have evolved and if so how might they differ in sequence, structure and mechanism?"; "What role does epistasis have in structuring evolutionary trajectories?" and "Can we predict the course of genetic evolution?"[25]. The answers to these questions have deep implications for unlocking nature's fundamental design principles, understanding chemical allostery [26] and for developing totally new strategies for the rational design of molecules to control biochemical processes [27]. Achieving these remarkable insights into protein structure and function evolution will require a multi-disciplinary approach involving cutting-edge molecular biology, structural biology, bioinformatics and molecular modeling methods.

Enzymatic catalytic cycles often involve several chemical steps requiring the stabilization of multiple intermediate and transition states during catalysis. In principle, their active sites could be pre-organized to catalyze all of the chemical reactions involved (a principle championed by Arieh Warshel [28]) minimizing any reorganizational motions required to meet the demands of subsequent steps. This principle was demonstrated in a study of serine esterases [29]. However, if the active site reorganization needed to achieve catalysis of a subsequent step involves, for example, a conformational change, then evolution could have acted to 1) lower the energetic cost of reorganization and/or 2) evolve a novel function. In the latter case, protein fluctuations [30] play an important role in the mechanism and in understanding its evolution of novel function. In addition, the study of enzymes that show changes in catalytic activity upon mutation of residues distant from the active site presents a clear opportunity to better understand 1) allosteric principles [31,32] to control, not just the binding of a substrate to the enzyme but also its reactivity with implications for the design of new natural products and the chemical rescue of disease proteins[33,34], and 2) the vast neutral sequence space of proteins that endows them with two seemingly conflicting properties; robustness and evolvability [35].

MOLECULAR PHYLOGENETICS

Molecular phylogenetics is now ubiquitous in most branches of biology [36] but can also be leveraged to make decisions on the type and locations of specific residues (those that are highly conserved or conserved only within an orthologous group) that would yield the most insight into structure-function relationships as well as aid in the design of species-specific inhibitors. This section describes the individual steps in carrying out a phylogenetic analysis and analyses based on a multiple sequence alignment (MSA). The initial step in a molecular phylogenetic analysis of an enzyme family is to gather enough protein sequences of sufficient variety in order to generate robust hypotheses from the subsequent analyses. For example, if a researcher would like to determine what residues are critical for specificity, then the data set must obviously contain several known paralagous members or sub-groups. Often, a BLAST [37] or PSI-BLAST [38] search is performed on the non-redundant protein database from NCBI using an appropriate query sequence to find orthologous and paralogous sequences. Alternatively, one could start from the iProClass database[39] that contains collections of sequences grouped according to protein function. Almost without exception, a researcher will then have to prune this dataset manually (to remove identical or nearly identical sequences and fragments) or with the aid of programs like CD-HIT [40]. This latter program is an excellent tool for clustering sequences and the subsequent selection of cluster representatives, often resulting in a diverse and more manageable set of protein sequences. The next step is to develop an accurate MSA for which several programs and online accessible interfaces are available. The most popular programs are T-Coffee [41], MUSCLE [42], ProbCons [43], and Clustal[44] and have been the subject of multiple reviews [45]. Yet, only in the relatively simple cases, do MSA programs get the entire alignment "correct" (as judged by comparison to a 3D structural alignment). Careful inspection and adjustment of the MSA should then be performed in some editor, like Genedoc [46] or Jalview [47] since even a highly conserved residue can be misaligned depending upon the degree of conservation in adjacent residues. Still, editing a large MSA can be a daunting task even with the help of an editor. Therefore, our group uses the Meme program [48] with the zoops model (zero or one motif per sequences) to search for the most conserved patterns or motifs ranging in length between six and 50 residues over the entire list of sequences. We have found that these parameters to MEME generally return motifs that are immensely useful for efficiently refining MSAs with the Genedoc editor (patterns/motifs are assigned a color which is highlighted over the MSA wherever the motif occurs, sb.nrbsc.org) as well as assisting in assigning sequences to groups (described in more detail below). The information content of a motif is determined by the conservation of residues along the motif, the length of the motif and the distribution of the motif residues in the submitted dataset [48]. In addition, a structural alignment of several 3-dimensional structures from the enzyme family can be performed with programs like STAMP within the MultiSeq module [49] contained within the VMD program [50] and used to assist in the MSA refinement. There are several very good multiple structure alignment programs [51] that all have similar performance when trying to align two proteins of similar length that share the same fold. The final MSA can then be designated "high-resolution" to distinguish it from the raw MSA program output.

The quality of a phylogenetic tree is highly dependent on the quality of the MSA and the regions included for tree construction [52,53] . Therefore, the final MSA must be trimmed by deleting sections where the alignment is equivocal; primarily this trimming occurs at the N- and C-termini. Several automated programs, such as trimAl [54]and GBlocks [55], can be used to perform alignment trimming, though our manual trimming exercises have always resulted in trees with higher bootstrap values. The trimmed MSA file can then be used to create a distance-based phylogenetic tree or one based on maximum parsimony, Bayesian or maximum likelihood methods [36]. Once complete, several viewers are available for phylogenetic tree visualization [46]. Finally, organizing all of this sequence information and communicating the results to colleagues can still be problematic and tedious. Therefore, our group has developed a suite of utilities called HarvestSeq (sb.nrbsc.org) that will retrieve the functional characteristics for all sequences, order the MSA file based on the phylogenetic tree and perform other information gathering tasks and analyses.

The resulting phylogenetic tree can be used as a guide along with the gathered metadata on the sequences to partition the sequences into separate groups, typically those that cluster together with high bootstrap support. When the set of sequences is of a protein superfamily, the sequences cluster according to shared biochemical function. Sequences that have been incorrectly annotated are readily identified and with maximum likelihood or

Bayesian methods, ancestral sequences can be inferred [56]. The construction of ancestral sequences has provided a wealth of information on how enzymes evolve new functions [56]. Other programs based on principle component analysis[57] and n-gram analysis [58] offer different but complimentary ways of grouping sequences. The MSA and a partitioning of sequences into defined groups can then serve as input to the GEnt program [3] that identifies amino acid residues characteristic of sub-groups within a set of orthologous proteins or characteristic of individual protein families within a collection of paralogous proteins. These characteristic residues are identified as having 1) low overall family relative entropy defined as:

$$\Sigma p_i \log_2(p_i/q_i)$$

where for each of the 20 amino acids p_i is the fraction of residue type i at that alignment position, and q_i is the fraction of residue type i expected in a random sequence. q_i is usually taken from an appropriate non-redundant database and 2) high group cross-entropy computed as:

$$\Sigma(p_i - q_i) \log_2(p_i/q_i)$$

where p_i is the fraction of residue type i at a particular position in the alignment for sequences in the predefined group while q_i is the fraction of residue type i at that position for sequences not in the predefined group. Often these group-specific residues are most responsible for changes in biochemical properties like substrate selectivity. We have observed in several analyses of protein families that group-specific residues often cluster around highly conserved active site residues, but that some can be quite distant which strongly suggests some stability-function tradeoff relationship.

By probing the order that these residues may have appeared in their respective lineage yields insights into the how enzymes may evolve novel functions. Other programs that identify specificity determining positions though the algorithmic details are different include SDPFox [59], SPEER [60], and multi-Harmony [61]. The programs also differ in the way they treat columns in the MSA that contain gaps. Another program often highlighted in these discussions is Evolutionary Trace [62] though this program does not appear to distinguish between identifying strictly conserved and specificity-determining residues. It should be emphasized that these programs should be viewed as hypothesis generation devices since what properties constitute a group can change depending on the question you ask (see section on Sesquiterpene Syntheses).

HYBRID QUANTUM CHEMICAL/MOLECULAR ME-CHANICAL METHODS

Hybrid QC/MM potentials were first described by Warshel and Levitt [63] in a simulation of the lysozyme reaction. They were developed and continue to be extensively used to investigate the atomic and sub-atomic details of enzymatic reactions due to the high computational cost of QC methods, which are necessary for accurately modeling the electronic reorganization that occurs upon chemical bonds being broken and created, but with a goal of accurately treating the heterogeneous environment of the active site. The total potential energy in these simulations is the sum of three terms; one for the atoms in the QC region, one for those in the surrounding MM region and a term that describes the interactions between the two (see Figure 1). The methods have been the subject of numerous excellent reviews [64-67]. Despite their more widespread use, numerous challenges remain, including finding an appropriate model chemistry that can accurately represent the large (by QC standards) enzyme "active site" and a definition of the reaction coordinate. Furthermore, if free energy profiles/surfaces obtained by umbrella-sampled molecular dynamics simulations along a reaction coordinate(s) are desired, then some compromises in the QC potential must be made. Usually this requires the employment of semi empirical molecular orbital (SMO) methods [68] that are computationally efficient due to approximations of many two-electron integrals and the representation of valence electrons only. Other approximate methods based on Density Functional Theory (DFT) are also often employed[69]. All of these methods can suffer from the lack of quantitative accuracy that may be needed in order to distinguish one mechanism from another [70]. A well-established method for correcting the free energy profiles/surfaces derived from hybrid SMO/MM simulations that employs higher level QC results as a reference and spline functions that interpolate between the low and high level methods [71,72] and which has been demonstrated to improve the free energy profiles/surfaces and result in calculated rate constants in good agreement with experiment.

Two of the major challenges in determining an enzymatic mechanism through QC/MM simulations, besides those already mentioned, is 1) representing the correct protonation states of active site residues in the Michaelis-Menten complex and 2) simulating low-barrier proton transfer reactions that may have a nuclear dynamical or tunneling component [73] and that may occur either step-wise or concerted with nucleophilic attacks. To overcome this challenge, neutron diffraction experiments of protein crystals can aid in the assignment of protonation states and the dissection of enzyme mechanistic details surrounding the location of hydrogen atoms [74]. Implicit solvent models such as MM-Poisson-Boltzmann, MM-Generalized Born [75] as well as knowledge based potentials [76] can be used for assigning protonation states. If the goal in enzyme engineering is to enhance the capabilities of natural enzymes, then dissecting the atomic details of enzymatic mechanisms will be critical to the effort.

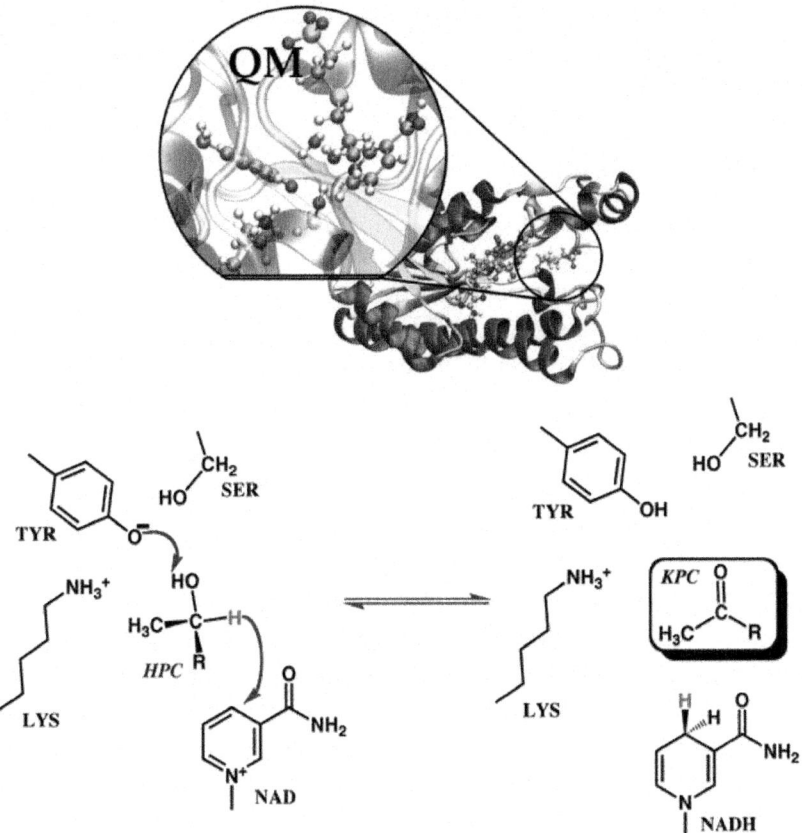

Figure 1: (left) Depiction of the R-hydroxypropylthioethanesulfonate dehydrogenase (R-HPCDH) structure with those atoms most likely to undergo significant electronic reorganization or significantly contribute to this reorganization represented by a QM method (Ser142, Tyr155, Lys140, the R-HPC substrate and the nicotinamide moiety of nicotinamide adenine dinucleotide (NAD)) while the surroundings are represented by a MM force field. Solvent molecules not shown for clarity. (right) Chemical mechanism for the oxidation of R-HPC by R-HPCDH.

For QC/MM simulations, our lab (TW) uses pDynamo [77] (www. pdynamo.org). This program and its Fortran predecessor was developed by Martin J. Field (Institut de Biologie Structurale) and has been utilized by our group for over 10 years to study detailed enzyme reactions. The program can directly read molecular systems constructed within CHARMM [78], AMBER [79] and GROMACS [80]. The program contains all of the standard and new semiempirical molecular orbital (SMO) methods. In addition, it contains an intuitive interface to the QC program ORCA [81] enabling higher level QC methods to be utilized for high resolution refinement or corrections to surfaces calculated with less accurate methods. Nowadays, most modern QC and MM software packages contain the capabilities to perform QC/MM simulations in some form.

Sesquiterpene Synthases

Plant sesquiterpene synthases, a subset of the terpene synthase superfamily, are a mechanistically diverse family of enzymes capable of synthesizing hundreds of complex compounds with high regio- and stereospecificity and are of biological importance due to their role in plant defense mechanisms. Several excellent reviews on the larger terpene synthase family are available covering the essential enzymatic transformations, structural biology and phylogenetics[82-85]. Sesquiterpene synthases bind farnesyl diphosphate and three divalent $Mg2+$ ions. Sesquiterpene biosynthesis is initiated by ionization of the C1-OPP bond generating a reactive carbocation (see Figure 2 for a depiction of the unfolded form) and the first major important branching of a mechanistic network. At this point, the diphosphate moiety can then bind to C3, isomerize about the C2-C3 bond, and then ionize the C3-OPP bond to form the reactive nerolidyl carbocation. Carbocations can be quenched by proton transfer from the intermediate to the enzyme or by addition of water molecules at any point along the mechanistic decision network. Otherwise, the synthesis can proceed through an intermolecular electrophilic attack on one of the two double bonds of the substrate to form a cyclic species. Possible subsequent reactions include hydride shifts, proton transfers, methyl or methylene shifts, and further intermolecular electrophilic additions. Sesquiterpene synthases span a large range of specificity. Increasingly, similar catalytic versatility is being discovered in several protein families and may be an inherent property of enzymes [86,87]. Such secondary catalytic activities can become the primary activity through gene duplication and subsequent divergence as well as the starting points for directed evolution in protein engineering applications [15]. Understanding these secondary activities may help lower attrition rates in drug discovery programs and identify drug interaction surfaces less susceptible to escape mutations [88].

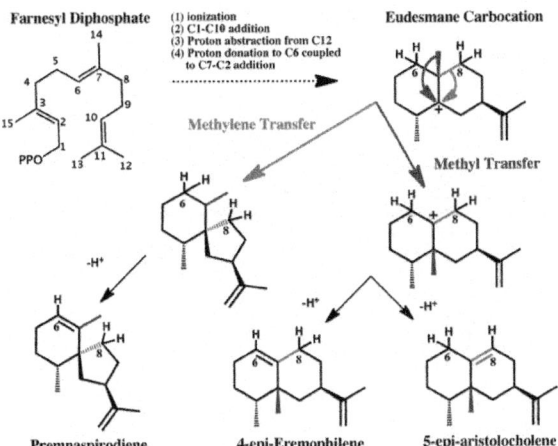

Figure 2: Chemical reactions progressing from the shared eudesmane carbocation intermediate to premnaspirodiene (methylene transfer followed by proton abstraction from C6), 4-epi-eremophilene and 5-epi-aristolocholene (methyl transfer followed by proton abstraction from C6 or C8 respectively).

In a recent study of sesquiterpene synthases (that encompasses all the hallmarks of the classical molecular evolution studies previously noted) in which a 5-epi-aristolocholene synthase (5EAS) was transformed to a premnaspirodiene synthase (PSDS) through mutational swaps of nine residues (none of which make specific contacts with the substrate) as well as experimental classification of all 512 proteins with different combinations of these nine residues, a functional landscape underlying the evolution of sesquiterpene chemical diversity was revealed (see Figures 2 and 3) [89]. The catalytic cycle of both synthases passes through several common intermediate states and only diverges in the last few chemical steps with some mutants along this putative evolutionary swath producing 4-epi-eremophilene, a product with hybrid activity; possibly ancestral to both enzymes. Also of significance, the mutants exhibited functional epistasis by the fact that 1) no single amino acid correlated with the product distribution and 2) the effect of a mutation depended on the state of the other eight residues. While these studies on sesquiterpene synthases are highly innovative and relevant to the task of seeking fundamental enzyme design principles, a decisive physicochemical explanation for the evolution of novel sesquiterpene synthase function is lacking. Therefore, our lab began investigating these details by first performing an extensive molecular phylogenetic analysis and then leveraging this information to support mechanistic conclusions obtained from atomic MD simulations [2]. Uncovering the biophysical principles governing a functional landscape can assist in narrowing the many possible evolutionary pathways to novel functions [9]; information that deepens our knowledge of structure-function relationships.

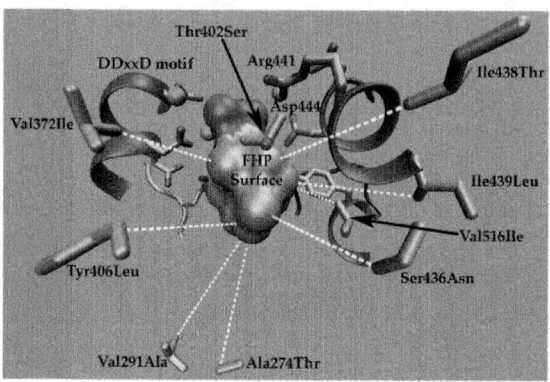

Figure 3: The nine residues and their substitutions located outside the active site designated here by the farnesylhydroxyphosphonate (FHP) surface and other highly conserved residues that functionally convert a 5-epi-aristolocholene synthase to a premnaspirodiene synthase.

Our recently published molecular phylogenetic analysis of the plant sesquiterpene synthase family has been, and can in the future, be utilized to support conclusions from the simulations and to leverage experimental results on sesquiterpene synthases from other branches of the phylogenetic tree [2]. The number and complexity of sesquiterpene products had in the past inhibited the use of phylogenetic analysis for constructing sequence-function relationships because of uncertainties in the statistical relevance of the resulting inferences. However, through a carefully-crafted multiple sequence alignment of ~200 plant sesquiterpene synthases using the procedures described in the Molecular Phylogenetics section, we observed that all sequences that cluster together on the phylogenetic tree into well-defined groups share at least the first reaction in the catalytic mechanism subsequent to the initial ionization step and many share steps beyond this, down to proton transfers between the enzyme and substrate. The multiple sequence alignment showed 15 highly conserved residues (95% and above) in the C-terminal catalytic domain, five of which are outside the active site. Most significant was the previously unreported high conservation of a Tyr520-Asp444-Asp525 triad (numbering according to the Nicotiana tobacum sequence for which there is a structure, see Figure 4). Given the high conservation of the Asp444-Tyr-520-Asp-525 triad, its position relative to a folded substrate analogue [90], the demonstration of its importance in generating (+)-germacrene A either as an intermediate or product [91], and our own atomistic MD simulations of the eudesmane carbocation in 5EAS, and finally the absence of likely proton donors/acceptors in other parts of many plant sTS active sites, we proposed that this triad is an important functional element responsible for many proton transfers to and from the substrate and intermediates along the plant sesquiterpene synthase catalytic cycle. Though this triad is obviously not the key to understanding all of plant sTS enzymatic chemistry, we nevertheless proposed that the triad can be tuned in a variety of ways to generate a diversity of products. These include 1) substituting residues on the opposite side of the active site forcing the farnesyl diphosphate to fold in different ways so that through both mechanisms the triad will donate/abstract protons to and from different carbons and 2) substituting residues surrounding the triad to shift their position relative to the substrate and/or intermediates. Finally, these results highlight what insights can be gained and a wealth of hypotheses that can arise from a phylogenetic analysis coupled to molecular modeling of the substrate/intermediate-enzyme complex. Developing these hypotheses was critical to our efforts to dissect the 5EAS mechanism further by hybrid QC/MM methods.

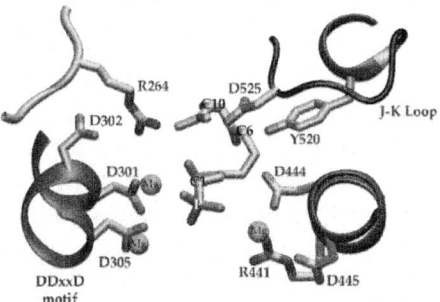

Figure 4: Highly conserved (<95%) residues highlighted on the 5-epi-aristolocholene synthase active site structure (PDB entry: 5eat). Also shown are the co-crystallized Mg^{2+} ions (green) and farnesylhydroxyphosphanate substrate mimic (middle). Helices are in purple.

The first structural models of a plant sesquiterpene synthase [90] based on x-ray crystallography presented enough information to propose a full mechanism for the many chemical steps in transforming farnesyl pyrophosphate to its product, 5-epi-aristolocholene (5EA). Part of this mechanism has gained experimental support [91], though the latter steps that form the basis of functional divergence remain to be fully explained. Furthermore, much is still unknown about the atomic details of many sesquiterpene synthase mechanisms, even though the intrinsic reactivity of the substrate has in several cases been elucidated through quantum chemical calculations [92,93]. A QC/MM simulation study on a fungal aristolocholene synthase discovered an energetically feasible intramolecular proton transfer reaction leading to formation of the eudesmane carbocation as well as determining the functional roles of active site residues [94]. Another QC/MM study on a closely related bornyl diphosphate synthase revealed how the electrostatic environment of the enzyme's active site steered the substrate to its' final product [95]. Due to the transient nature of so many intermediate states along terpene synthase catalytic cycles, it is extremely difficult to experimentally determine reaction rates or thermodynamic data for the individual reactions. In addition, simulation studies on terpene synthases are quite challenging due to the fact that the enzymes' contribution to catalysis may be relatively subtle with residues appearing to act in concert and ones outside the active site affecting product distributions [89]. Past QC calculations on the methyl and methylene transfer reactions emanating from the eudesmane carbocation in 5EAS [96] have primarily served to amplify the mystery behind the physicochemical properties that control sesquiterpene synthase product distributions.

These reactions could be under thermodynamic or kinetic control. The enzyme may preferentially stabilize the non-classical carbocation (carbonium ions) structures that occur along both of these reaction paths. Yet, our latest simulation results (manuscript in preparation) employing QC/MM methods and representation of the fully solvated 5EAS-eudesmane carbocation intermediate reveals the basis for 5EAS specificity is the preferential stabilization of the 4-epi-eremophilenyl intermediate (see Figure 2) over the premnaspirodienyl intermediate. The calculated free energy barrier for both reactions is very low, around 5-6 kcal/mol. The free energy differences, on the other hand, are exergonic for the methyl transfer reaction and endergonic for the methylene transfer. The methyl transfer reaction from the eudesmane carbocation is favored in 5EAS in part because the active site forms a "cage" around the substrate's isoprenyl group which is not "disturbed" by the methyl transfer reaction. In contrast, the methylene transfer reaction results in a more substantial change in the shape of the intermediate and several steric clashes occur between the isoprenyl group of the substrate and surrounding residues. Furthermore, the catalytic triad of Asp444-Tyr520-Asp525 is favorably positioned to abstract the proton from C8 of the 4-epi-eremophilenyl carbocation resulting in 5EA product formation. Thus, there are multiple structural and functional requirements for specificity to arise in 5EAS. Further examination of this functional landscape [89] through hybrid QC/MM simulations could provide direction on how to engineer these enzymes to be less promiscuous and biosynthesize completely new natural products.

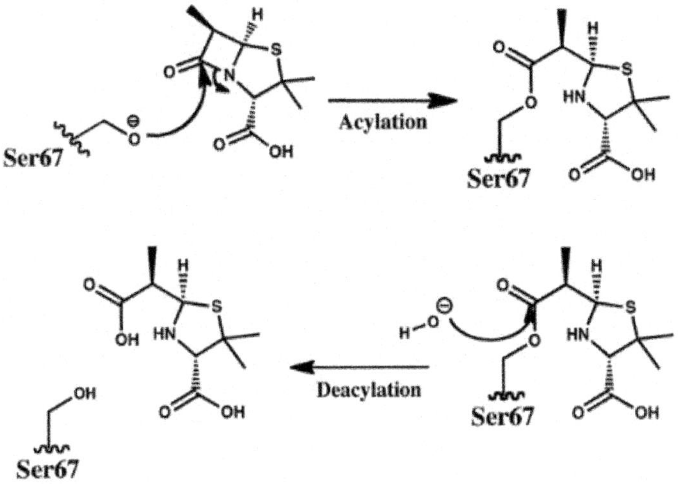

Figure 5: Two stage process for the hydrolysis of antibiotic molecule by β-lactamases.

Class D b-Lactamases

Because of their safety and efficacy, β-lactams still constitute the most widely used antibiotics that inhibit bacterial cell-wall transpeptidases (PBPs – Penicilin Binding Proteins). Bacteria's most important resistance mechanism relies on the scavenging potential of β-lactamases to break the amide bond before antibiotics reach their cellular target. The reaction is a two-step process: 1) acylation of the enzyme Ser residue followed by 2) deacylation and release of the cleaved antibiotic (see Figure 5). Phylogenetic analysis divides the β-lactamases into groups A, C, D (serine hydrolases) and B (metallo- enzymes). The remarkable variety of β-lactamases, the rapid rate of their evolution and acquisition of resistance towards newly developed drugs make it of prime importance for us to understand fully their sequence–structure–function relationships and evolutionary mechanisms towards antibiotic resistance in order to break the current cycle of drug development followed by resistance. Class D, often referred to as OXA, after several initially classified members demonstrated unusually high hydrolysis efficacy against oxacillin, is the most diverse group of β-lactamases, most recently identified, and also the least studied [97,98]. They possess a remarkably diverse range of hydrolytic profiles encompassing penicillins, cephems (including third generation drugs: cefotaxime and ceftazidime) as well as carbapenems [98]. Interestingly, however, to date no class D enzyme has shown ability to hydrolyze both extended spectrum cephalosporins and carbapenems [98,99].

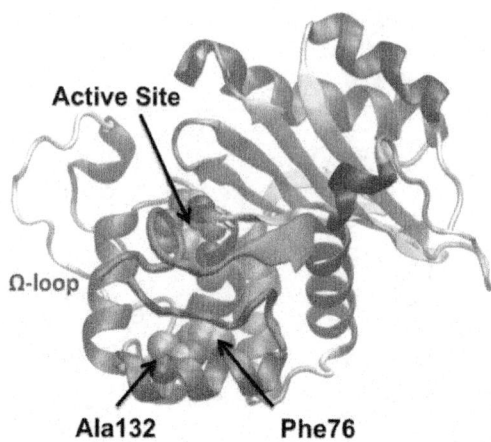

Figure 6: "OXA-2" group-specific residues (Phe76 and Ala132) within the OXA-2 structure (PDB entry 1K38)

Class D is one of two known enzymes that employ a post-translational carboxylation of a lysine residue as part of their catalytic machinery. This carboxylated lysine (Lys70) is crucial for activating a water molecule for the deacylation reaction and likely acts as a general base to abstract a proton from the absolutely conserved Ser67 that initiates nucleophilic attack on the β-lactam ring. Our recently published phylogenetic analysis of class D enabled the division/classifying of the sequences into distinct groups [3]. Pairwise sequence identities within these groups were very small (as low as 22% in the largest group), yet there were distinct signatures or group-specific residues that allowed us to uniquely demarcate the groups. These group-specific residues are mainly located adjacent to the active site with their van der Waals surfaces in contact with the binding pocket. Thus they are likely to modulate the properties of the conserved active site catalytic residues [21,100] and may correlate with the enzyme's spectrum of activity (see Figure 6). In addition, their location outside the active site may have only a minimal effect on protein stability [16]. The largest sub-group of class D sequences (OXA-1 subgroup) made up of α-, β-, and γ-proteobacteria also contains a group-specific Cys pair. In the OXA-1 crystal structure (PDB entry 3ISG [101]) this Cys pair is positioned to form a disulfide bond and our analysis of the deposited electron density map reveals that both the reduced and oxidized forms are present in apo OXA-1 (PDB entry 1M6K [102]). The reducing cytoplasmic environment of the *E. coli* expression system is responsible for partial or complete reduction of the disulfide bond and its apparent absence in the OXA-1 3ISG crystal structure. Patterns of evolutionary constraint seen in our analyses support the prediction that formation of this disulfide bond in an oxidizing environment may be important for survival by maintaining the hydrophobic core of the enzyme. Additional support to this hypothesis is provided by the fact that *P. aeruginosa* group (OXA- 10 group) also contains a group-specific disulfide-bonded Cys pair – located on the opposite end of the same segment of β-sheet (PDB entry 1K57 [103]). Alternatively, the increased stabilization of this region may have an impact on the dynamic behavior of the active site. Long timescale (several microseconds) MD simulations using classical MM force fields showed that when the Cys pair is in a reduced state, the active site becomes disorganized which if accurate would also lead to decarboxylation of the Lys sidechain and destruction of the enzyme's function (manuscript in preparation) while in the oxidized state the active site remained stable.

SUMMARY AND OUTLOOK

The mechanistic details of enzymatic reactions obtained from hybrid QC/MM simulations combined with identification of amino acid residues critical for the maintenance and diversity of function within a enzyme (super)family is a powerful approach for elucidating enzyme design principles. This information can then be leveraged to design enzymes for environmental remediation, energy production, the prediction of antibiotic resistance and the production of therapeutic compounds. The use of QC/MM methods to investigate enzymatic reactions still requires some expertise. The resulting free energy profiles or surfaces can be very sensitive to the initial coordinates, protonation states of surrounding residues, the QC method, and the length of umbrella sampling simulations [104]. Nevertheless, there have been significant improvements in software employing QC/MM methods over the last ten years since the authors began using them. We can expect to see further improvement in semiempirical molecular orbital parameterizations and methods which will enable faster exploration of possible reactive configurations as well as the generation of initial reaction paths that can be refined with higher level QC methods. But possibly more importantly is that in the future we will increasingly see the atomic and subatomic level details of enzyme reaction mechanisms interpreted within the context of its evolutionary history using molecular phylogenetics. One of the main limitations in this endeavor today is the fact that so many sequences are uncharacterized or even mischaracterized [105]. Assignment of function by homology may only be successful at relatively high sequence identities. Beyond this, the enzyme may carry out essentially the same catalytic chemistry but on very different substrates. Until this problem is remedied, the value of the sequence data will not reach its full potential benefit for use in enzyme design. For example, if a much larger percentage of plant sesquiterpene synthase sequences had experimentally determined functions, then analyses could be performed on the MSA to determine what residues are most responsible for traversing the nerolidyl pathway versus the trans pathway. Nevertheless, these are exciting times for computational biochemists who have an amazing opportunity to utilize both atomic-scale MD simulations, using both MM and QC/MM potential energy functions, and molecular phylogenetics in their research that will provide results and subsequent stories (publications), one could argue, that can stand aside some of the past classical studies of adaptation and begin to provide robust answers to the challenging questions posed in this review.

ACKNOWLEDGEMENTS

We would like to acknowledge our fellow colleagues, Martin J. Field (Institut de Biologie Structurale), Ricardo R. Gonzalez Mendez (University of Puerto Rico School of Medicine), Hugh B. Nicholas Jr. (PSC), Joseph Noel (Salk Institute), Luis Vazquez Quinones (Metropolitano University), Alexander J. Ropelewski (PSC) and Agnieszka Szarecka (Grand Valley State University) who have played a part in our thinking about enzyme design principles and the methods we have employed to discover them. This work was supported by the National Center for Research Resources at the National Institutes of Health (grant number RR06009).

REFERENCES

1.   Harms MJ, Thornton JW (2010) Analyzing protein structure and function using ancestral gene reconstruction. Curr Opin Struct Biol. http://dx.doi.org/10.1016/j.sbi.2010.03.005 PMid:20413295 PMCid:291695

2.   Wymore T, Chen BY, Nicholas HB, Ropelewski AJ, Brooks CL (2011) A Mechanism for Evolving Novel Plant Sesquiterpene Synthase Function. Molecular Informatics 30: 896-906. http://dx.doi.org/10.1002/minf.201100087

3.   Szarecka A, Lesnock KR, Ramirez-Mondragon CA, Nicholas HB, Wymore T (2011) The Class D β-lactamase family: residues governing the maintenance and diversity of function. Protein Engineering Design and Selection 24: 801-809. http://dx.doi.org/10.1093/protein/gzr041 PMid:21859796 PMCid:3170078

4.   Glasner ME, Gerlt JA, Babbitt PC (2006) Evolution of enzyme superfamilies. Current opinion in chemical biology 10: 492-497. http://dx.doi.org/10.1016/j.cbpa.2006.08.012 PMid:16935022

5.   Hartley CJ, Newcomb RD, Russell RJ, Yong CG, Stevens JR, et al.

(2006) Amplification of DNA from preserved specimens shows blowflies were preadapted for the rapid evolution of insecticide resistance. Proc Natl Acad Sci USA 103: 8757-8762. http://dx.doi.org/10.1073/pnas.0509590103 PMid:16723400 PMCid:1482651

6. Shi Y, Yokoyama S (2003) Molecular analysis of the evolutionary significance of ultraviolet vision in vertebrates. Proc Natl Acad Sci USA 100: 8308-8313. http://dx.doi.org/10.1073/pnas.1532535100 PMid:12824471 PMCid:166225

7. Bershtein S, Goldin K, Tawfik DS (2008) Intense neutral drifts yield robust and evolvable consensus proteins. J Mol Biol 379: 1029-1044. http://dx.doi.org/10.1016/j.jmb.2008.04.024 PMid:18495157

8. Wang X, Minasov G, Shoichet BK (2002) Evolution of an antibiotic resistance enzyme constrained by stability and activity trade-offs. J Mol Biol 320: 85-95. http://dx.doi.org/10.1016/S0022-2836(02)00400-X

9. Weinreich DM, Delaney NF, Depristo MA, Hartl DL (2006) Darwinian evolution can follow only very few mutational paths to fitter proteins. Science 312: 111-114. http://dx.doi.org/10.1126/science.1123539 PMid:16601193

10. Lunzer M, Miller SP, Felsheim R, Dean AM (2005) The biochemical architecture of an ancient adaptive landscape. Science 310: 499-501. http://dx.doi.org/10.1126/science.1115649 PMid:16239478

11. Miller SP, Lunzer M, Dean AM (2006) Direct demonstration of an adaptive constraint. Science 314: 458-461. http://dx.doi.org/10.1126/science.1133479 PMid:17053145

12. Bridgham JT, Carroll SM, Thornton JW (2006) Evolution of hormone-receptor complexity by molecular exploitation. Science 312: 97-101. http://dx.doi.org/10.1126/science.1123348 PMid:16601189

13. Ortlund EA, Bridgham JT, Redinbo MR, Thornton JW (2007) Crystal structure of an ancient protein: evolution by conformational epistasis. Science 317: 1544-1548. http://dx.doi.org/10.1126/science.1142819 PMid:17702911 PMCid:2519897

14. Bridgham JT, Ortlund EA, Thornton JW (2009) An epistatic ratchet constrains the direction of glucocorticoid receptor evolution. Nature 461: 515-519. http://dx.doi.org/10.1038/nature08249 PMid:19779450

15. Romero PA, Arnold FH (2009) Exploring protein fitness landscapes by directed evolution. Nat Rev Mol Cell Biol 10: 866-876. http://dx.doi.org/10.1038/nrm2805 PMid:19935669 PMCid:2997618

16. Depristo MA, Weinreich DM, Hartl DL (2005) Missense meanderings in

sequence space: a biophysical view of protein evolution. Nat Rev Genet 6: 678-687. http://dx.doi.org/10.1038/nrg1672 PMid:16074985

17. Tokuriki N, Tawfik DS (2009) Stability effects of mutations and protein evolvability. Current Opinion in Structural Biology 19: 596-604. http://dx.doi.org/10.1016/j.sbi.2009.08.003 PMid:19765975

18. Tokuriki N, Stricher F, Serrano L, Tawfik DS (2008) How protein stability and new functions trade off. PLoS Computational Biology 4: e1000002. http://dx.doi.org/10.1371/journal.pcbi.1000002   PMid:18463696 PMCid:2265470

19. Conant GC, Wolfe KH (2008) Turning a hobby into a job: how duplicated genes find new functions. Nat Rev Genet 9: 938-950. http://dx.doi.org/10.1038/nrg2482 PMid:19015656

20. Poelwijk FJ, Kiviet DJ, Weinreich DM, Tans SJ (2007) Empirical fitness landscapes reveal accessible evolutionary paths. Nature 445: 383-386. http://dx.doi.org/10.1038/nature05451 PMid:17251971

21. Phillips PC (2008) Epistasis--the essential role of gene interactions in the structure and evolution of genetic systems. Nat Rev Genet 9:   855-867.   http://dx.doi.org/10.1038/nrg2452   PMid:18852697 PMCid:2689140

22. ZUCKERKANDL E (1976) Evolutionary processes and evolutionary noise at the molecular level. I. Functional density in proteins. J Mol Evol 7: 167-183. PMid:933174

23. Smith JM (1970) Natural selection and the concept of a protein space. Nature 225: 563-564. http://dx.doi.org/10.1038/225563a0 PMid:5411867

24. Dean AM, Thornton JW (2007) Mechanistic approaches to the study of evolution: the functional synthesis. Nat Rev Genet 8: 675-688. http://dx.doi.org/10.1038/nrg2160 PMid:17703238 PMCid:2488205

25. Stern DL, Orgogozo V (2009) Is genetic evolution predictable? Science 323: 746-751. http://dx.doi.org/10.1126/science.1158997 PMid:19197055 PMCid:3184636

26. Laskowski R, Gerick F, Thornton J (2009) The structural basis of allosteric regulation in proteins. FEBS Letters. http://dx.doi.org/10.1016/j.febslet.2009.03.019 PMid:19303011

27. Yoshikuni Y, Dietrich JA, Nowroozi FF, Babbitt PC, Keasling JD (2008) Redesigning enzymes based on adaptive evolution for optimal function in synthetic metabolic pathways. Chemistry & Biology 15: 607-618. http://dx.doi.org/10.1016/j.chembiol.2008.05.006 PMid:18559271

28. Warshel A (2003) Computer simulations of enzyme catalysis: methods, progress, and insights. Annu Rev Biophys Biomol Struct 32: 425-443. http://dx.doi.org/10.1146/annurev.biophys.32.110601.141807 PMid:12574064

29. Smith AJT, Müller R, Toscano MD, Kast P, Hellinga HW, et al. (2008) Structural reorganization and preorganization in enzyme active sites: comparisons of experimental and theoretically ideal active site geometries in the multistep serine esterase reaction cycle. Journal of the American Chemical Society 130: 15361-15373. http://dx.doi.org/10.1021/ja803213p PMid:18939839 PMCid:2728765

30. Boehr DD, Nussinov R, Wright PE (2009) The role of dynamic conformational ensembles in biomolecular recognition. Nat Chem Biol 5: 789-796. http://dx.doi.org/10.1038/nchembio.232 PMid:19841628 PMCid:2916928

31. Goodey NM, Benkovic SJ (2008) Allosteric regulation and catalysis emerge via a common route. Nat Chem Biol 4: 474-482. http://dx.doi.org/10.1038/nchembio.98 PMid:18641628

32. Cui Q, Karplus M (2008) Allostery and cooperativity revisited. Protein Sci 17: 1295-1307. http://dx.doi.org/10.1110/ps.03259908 PMid:18560010 PMCid:2492820

33. Hassan A, Koh J (2008) Selective Chemical Rescue of a Thyroid-Hormone-Receptor Mutant, TRbeta(H435Y), Identified in Pituitary Carcinoma and Resistance to Thyroid Hormone. Angew Chem Int Ed Engl 47: 7280-7283. http://dx.doi.org/10.1002/anie.200801742 PMid:18683837

34. Chen C-H, Budas GR, Churchill EN, Disatnik M-H, Hurley TD, et al. (2008) Activation of aldehyde dehydrogenase-2 reduces ischemic damage to the heart. Science 321: 1493-1495. http://dx.doi.org/10.1126/science.1158554 PMid:18787169 PMCid:2741612

35. Wagner A (2008) Robustness and evolvability: a paradox resolved. Proc Biol Sci 275: 91-100. http://dx.doi.org/10.1098/rspb.2007.1137 PMid:17971325 PMCid:2562401

36. Whelan S, Liò P, Goldman N (2001) Molecular phylogenetics: state-of-the-art methods for looking into the past. Trends Genet 17: 262-272. http://dx.doi.org/10.1016/S0168-9525(01)02272-7

37. Altschul SF, Gish W, Miller W, Myers EW, Lipman DJ (1990) Basic local alignment search tool. Journal of Molecular Biology 215: 403-410. PMid:2231712

38. Altschul SF, Madden TL, Schaffer AA, Zhang J, Zhang Z, et al. (1997)

Gapped BLAST and PSI-BLAST: a new generation of protein database search programs. Nucleic acids research 25: 3389-3402. http://dx.doi.org/10.1093/nar/25.17.3389 PMid:9254694 PMCid:146917

39. Wu CH, Huang H, Nikolskaya A, Hu Z, Barker WC (2004) The iProClass integrated database for protein functional analysis. Comput Biol Chem 28:   87-96.   http://dx.doi.org/10.1016/j.compbiolchem.2003.10.003 PMid:15022647

40. Huang Y, Niu B, Gao Y, Fu L, Li W (2010) CD-HIT Suite: a web server for clustering and comparing biological sequences. Bioinformatics 26: 680-682. http://dx.doi.org/10.1093/bioinformatics/btq003 PMid:20053844 PMCid:2828112

41. Notredame C, Higgins DG, Heringa J (2000) T-Coffee: A novel method for fast and accurate multiple sequence alignment. Journal of Molecular Biology   302:   205-217.   http://dx.doi.org/10.1006/jmbi.2000.4042 PMid:10964570

42. Edgar RC (2004) MUSCLE: multiple sequence alignment with high accuracy and high throughput. Nucleic Acids Res 32: 1792-1797. http://dx.doi.org/10.1093/nar/gkh340 PMid:15034147 PMCid:390337

43. Do CB, Mahabhashyam MS, Brudno M, Batzoglou S (2005) ProbCons: Probabilistic   consistency-based   multiple   sequence   alignment. Genome Research 15: 330-340. http://dx.doi.org/10.1101/gr.2821705 PMid:15687296 PMCid:546535

44. Larkin MA, Blackshields G, Brown NP, Chenna R, McGettigan PA, et al. (2007) Clustal W and Clustal X version 2.0. Bioinformatics 23: 2947-2948. http://dx.doi.org/10.1093/bioinformatics/btm404 PMid:17846036

45. Edgar RC, Batzoglou S (2006) Multiple sequence alignment. Curr Opin Struct Biol 16: 368-373. http://dx.doi.org/10.1016/j.sbi.2006.04.004 PMid:16679011

46. Procter JB, Thompson J, Letunic I, Creevey C, Jossinet F, et al. (2010) Visualization of multiple alignments, phylogenies and gene family evolution. Nature methods 7: S16-25. http://dx.doi.org/10.1038/nmeth.1434 PMid:20195253

47. Waterhouse AM, Procter JB, Martin DMA, Clamp M, Barton GJ (2009) Jalview Version 2--a multiple sequence alignment editor and analysis workbench. Bioinformatics 25: 1189-1191. http://dx.doi.org/10.1093/bioinformatics/btp033 PMid:19151095 PMCid:2672624

48. Bailey TL (1994) Fitting a mixture model by expectation maximization to discover motifs in biopolymers. Proceedings of 2nd International Conference on ISMB: 1-7.

49. Roberts E, Eargle J, Wright D, Luthey-Schulten Z (2006) MultiSeq: unifying sequence and structure data for evolutionary analysis. BMC Bioinformatics 7: 382. http://dx.doi.org/10.1186/1471-2105-7-382 PMid:16914055 PMCid:1586216

50. Humphrey W, Dalke A, Schulten K (1996) VMD: visual molecular dynamics. Journal of molecular graphics 14: 33-38, 27-38.

51. Hasegawa H, Holm L (2009) Advances and pitfalls of protein structural alignment. Current Opinion in Structural Biology 19: 341-348. http:// dx.doi.org/10.1016/j.sbi.2009.04.003 PMid:19481444

52. Sjölander K (2004) Phylogenomic inference of protein molecular function: advances and challenges. Bioinformatics 20: 170-179. http:// dx.doi.org/10.1093/bioinformatics/bth021 PMid:14734307

53. Sjölander K (2010) Getting started in structural phylogenomics. PLoS Computational Biology 6: e1000621. http://dx.doi.org/10.1371/journal. pcbi.1000621 PMid:20126522 PMCid:2813252

54. Capella-Gutierrez S, Silla-Martinez JM, Gabaldon T (2009) trimAl: a tool for automated alignment trimming in large-scale phylogenetic analyses. Bioinformatics 25: 1972-1973.

55. http://dx.doi.org/10.1093/bioinformatics/btp348

56. PMid:19505945 PMCid:2712344

57. Talavera G, Castresana J (2007) Improvement of phylogenies after removing divergent and ambiguously aligned blocks from protein sequence alignments. Systematic biology 56: 564-577. http://dx.doi. org/10.1080/10635150701472164 PMid:17654362

58. Thornton JW (2004) Resurrecting ancient genes: experimental analysis of extinct molecules. Nat Rev Genet 5: 366-375. http://dx.doi.org/10.1038/ nrg1324 PMid:15143319

59. Casari G, Sander C, Valencia A (1995) A method to predict functional residues in proteins. Nature structural biology 2: 171-178. http://dx.doi. org/10.1038/nsb0295-171 PMid:7749921

60. Maetschke SR, Kassahn KS, Dunn JA, Han S-P, Curley EZ, et al. (2010) A visual framework for sequence analysis using n-grams and spectral rearrangement. Bioinformatics 26: 737-744. http://dx.doi.org/10.1093/ bioinformatics/btq042 PMid:20130028

61. Mazin PV, Gelfand MS, Mironov AA, Rakhmaninova AB, Rubinov AR, et al. (2010) An automated stochastic approach to the identification of the protein specificity determinants and functional subfamilies. Algorithms for molecular biology : AMB 5: 29. http://dx.doi.org/10.1186/1748-

7188-5-29 PMid:20633297 PMCid:2914642

62. Chakraborty A, Mandloi S, Lanczycki CJ, Panchenko AR, Chakrabarti S (2012) SPEER-SERVER: a web server for prediction of protein specificity determining sites. Nucleic acids research 40: W242-248. http://dx.doi.org/10.1093/nar/gks559 PMid:22689646 PMCid:3394334

63. Brandt BW, Feenstra KA, Heringa J (2010) Multi-Harmony: detecting functional specificity from sequence alignment. Nucleic acids research 38: W35-40. http://dx.doi.org/10.1093/nar/gkq415 PMid:20525785 PMCid:2896201

64. Lichtarge O, Bourne HR, Cohen FE (1996) An evolutionary trace method defines binding surfaces common to protein families. Journal of Molecular Biology 257: 342-358. http://dx.doi.org/10.1006/jmbi.1996.0167 PMid:8609628

65. Warshel A, Levitt M (1976) Theoretical studies of enzymic reactions: dielectric, electrostatic and steric stabilization of the carbonium ion in the reaction of lysozyme. Journal of Molecular Biology 103: 227-249. http://dx.doi.org/10.1016/0022-2836(76)90311-9

66. Gao J, Ma S, Major DT, Nam K, Pu J, et al. (2006) Mechanisms and free energies of enzymatic reactions. Chem Rev 106: 3188-3209. http://dx.doi.org/10.1021/cr050293k PMid:16895324

67. Hu H, Yang W (2008) Free energies of chemical reactions in solution and in enzymes with ab initio quantum mechanics/molecular mechanics methods. Annu Rev Phys Chem 59: 573-601. http://dx.doi.org/10.1146/annurev.physchem.59.032607.093618 PMid:18393679

68. Mulholland AJ (2005) Modelling enzyme reaction mechanisms, specificity and catalysis. Drug Discov Today 10: 1393-1402. http://dx.doi.org/10.1016/S1359-6446(05)03611-1

69. Senn HM, Thiel W (2009) QM/MM methods for biomolecular systems. Angew Chem Int Ed Engl 48: 1198-1229. http://dx.doi.org/10.1002/anie.200802019 PMid:19173328

70. Stewart JJP (2007) Optimization of parameters for semiempirical methods V: modification of NDDO approximations and application to 70 elements. Journal of molecular modeling 13: 1173-1213. http://dx.doi.org/10.1007/s00894-007-0233-4 PMid:17828561 PMCid:2039871

71. Elstner M (2006) The SCC-DFTB method and its application to biological systems. Theor Chem Acc 116: 316-325. http://dx.doi.org/10.1007/s00214-005-0066-0

72. Sattelmeyer KW, Tirado-Rives J, Jorgensen WL (2006) Comparison of

SCC-DFTB and NDDO-based semiempirical molecular orbital methods for organic molecules. The journal of physical chemistry A, Molecules, spectroscopy, kinetics, environment & general theory 110: 13551-13559.

73. Chuang Y, Corchado J, Truhlar D (1999) Mapped interpolation scheme for single-point energy corrections in reaction rate calculations and a critical evaluation of dual-level reaction path dynamics methods. J Phys Chem A 103: 1140-1149. http://dx.doi.org/10.1021/jp9842493

74. Ruiz-Pernia J, Silla E, Tunon I, Marti S, Moliner V (2004) Hybrid QM/MM potentials of mean force with interpolated corrections. J Phys Chem B 108: 8427-8433. http://dx.doi.org/10.1021/jp049633g

75. Truhlar DG, Gao J, Alhambra C, Garcia-Viloca M, Corchado J, et al. (2002) The incorporation of quantum effects in enzyme kinetics modeling. Acc Chem Res 35: 341-349. http://dx.doi.org/10.1021/ar0100226 PMid:12069618

76. Glusker JP, Carrell HL, Kovalevsky AY, Hanson L, Fisher SZ, et al. (2010) Using neutron protein crystallography to understand enzyme mechanisms. Acta Crystallographica Section D-Biological Crystallography 66: 1257-1261. http://dx.doi.org/10.1107/S0907444910027915 PMid:21041947 PMCid:2967424

77. Chen J, Brooks CL, 3rd, Khandogin J (2008) Recent advances in implicit solvent-based methods for biomolecular simulations. Current Opinion in Structural Biology 18: 140-148. http://dx.doi.org/10.1016/j.sbi.2008.01.003 PMid:18304802 PMCid:2386893

78. Olsson MHM, Sondergaard CR, Rostkowski M, Jensen JH (2011) PROPKA3: Consistent Treatment of Internal and Surface Residues in Empirical pK(a) Predictions. Journal of Chemical Theory and Computation 7: 525-537. http://dx.doi.org/10.1021/ct100578z

79. Field M (2008) The pDynamo Program for Molecular Simulations using Hybrid Quantum Chemical and Molecular Mechanical Potentials. Journal of Chemical Theory and Computation 4: 1151-1161. http://dx.doi.org/10.1021/ct800092p

80. Brooks BR, Brooks CL, Mackerell AD, Nilsson L, Petrella RJ, et al. (2009) CHARMM: the biomolecular simulation program. Journal of computational chemistry 30: 1545-1614. http://dx.doi.org/10.1002/jcc.21287 PMid:19444816 PMCid:2810661

81. Case DA, Cheatham TE, Darden T, Gohlke H, Luo R, et al. (2005) The Amber biomolecular simulation programs. Journal of computational chemistry 26: 1668-1688. http://dx.doi.org/10.1002/jcc.20290 PMid:16200636 PMCid:1989667

82. Hess B, Kutzner C, van der Spoel D, Lindahl E (2008) GROMACS 4: Algorithms for highly efficient, load-balanced, and scalable molecular simulation. Journal of Chemical Theory and Computation 4: 435-447. http://dx.doi.org/10.1021/ct700301q

83. Neese F (2012) The ORCA program system. Wiley Interdisciplinary Reviews-Computational Molecular Science 2: 73-78. http://dx.doi.org/10.1002/wcms.81

84. Christianson DW (2008) Unearthing the roots of the terpenome. Current Opinion in Chemical Biology 12: 141-150. http://dx.doi.org/10.1016/j.cbpa.2007.12.008 PMid:18249199 PMCid:2430190

85. Christianson DW (2006) Structural biology and chemistry of the terpenoid cyclases. Chem Rev 106: 3412-3442. http://dx.doi.org/10.1021/cr050286w PMid:16895335

86. CANE D (1990) ENZYMATIC FORMATION OF SESQUITERPENES. Chem Rev 90: 1089-1103. http://dx.doi.org/10.1021/cr00105a002

87. Degenhardt J, Köllner TG, Gershenzon J (2009) Monoterpene and sesquiterpene synthases and the origin of terpene skeletal diversity in plants. Phytochemistry 70: 1621-1637. http://dx.doi.org/10.1016/j.phytochem.2009.07.030 PMid:19793600

88. Babtie A, Tokuriki N, Hollfelder F (2010) What makes an enzyme promiscuous? Current opinion in chemical biology 14: 200-207. http://dx.doi.org/10.1016/j.cbpa.2009.11.028 PMid:20080434

89. Khersonsky O, Tawfik DS (2010) Enzyme promiscuity: a mechanistic and evolutionary perspective. Annu Rev Biochem 79: 471-505. http://dx.doi.org/10.1146/annurev-biochem-030409-143718 PMid:20235827

90. Nobeli I, Favia AD, Thornton JM (2009) Protein promiscuity and its implications for biotechnology. Nat Biotechnol 27: 157-167. http://dx.doi.org/10.1038/nbt1519 PMid:19204698

91. O›maille PE, Malone A, Dellas N, Andes Hess B, Smentek L, et al. (2008) Quantitative exploration of the catalytic landscape separating divergent plant sesquiterpene synthases. Nat Chem Biol 4: 617-623. http://dx.doi.org/10.1038/nchembio.113 PMid:18776889 PMCid:2664519

92. Starks CM, Back K, Chappell J, Noel JP (1997) Structural basis for cyclic terpene biosynthesis by tobacco 5-epi-aristolochene synthase. Science 277: 1815-1820. http://dx.doi.org/10.1126/science.277.5333.1815 PMid:9295271

93. Rising K, Starks C, Noel J, Chappell J (2000) Demonstration of germacrene A as an intermediate in 5-epi-aristolochene synthase catalysis.

Journal of the American Chemical Society 122: 1861-1866. http://dx.doi. org/10.1021/ja993584h

94. Tantillo DJ (2010) The carbocation continuum in terpene biosynthesis--where are the secondary cations? Chem Soc Rev 39: 2847-2854. http://dx.doi.org/10.1039/b917107j PMid:20442917

95. Tantillo DJ (2011) Biosynthesis via carbocations: theoretical studies on terpene formation. Natural product reports 28: 1035-1053. http://dx.doi. org/10.1039/c1np00006c PMid:21541432

96. Allemann RK, Young NJ, Ma S, Truhlar DG, Gao J (2007) Synthetic efficiency in enzyme mechanisms involving carbocations: aristolochene synthase. Journal of the American Chemical Society 129: 13008-13013.   http://dx.doi.org/10.1021/ja0722067   PMid:17918834 PMCid:2528250

97. Weitman M, Major DT (2010) Challenges posed to bornyl diphosphate synthase: diverging reaction mechanisms in monoterpenes. Journal of the American Chemical Society 132: 6349-6360. http://dx.doi.org/10.1021/ ja910134x PMid:20394387

98. Hess BA, Jr., Smentek L, Noel JP, O›Maille PE (2011) Physical constraints on sesquiterpene diversity arising from cyclization of the eudesm-5-yl carbocation. Journal of the American Chemical Society 133: 12632-12641. http://dx.doi.org/10.1021/ja203342p PMid:21714557

99. Fisher JF, Meroueh SO, Mobashery S (2005) Bacterial resistance to beta-lactam antibiotics: compelling opportunism, compelling opportunity. Chem   Rev   105:   395-424.   http://dx.doi.org/10.1021/cr030102i PMid:15700950

100. Poirel L, Naas T, Nordmann P (2010) Diversity, epidemiology, and genetics of class D beta-lactamases. Antimicrobial Agents and Chemotherapy 54: 24-38. http://dx.doi.org/10.1128/AAC.01512-08 PMid:19721065 PMCid:2798486

101. Poirel L, Nordmann P (2006) Carbapenem resistance in Acinetobacter baumannii: mechanisms and epidemiology. Clin Microbiol Infect 12:   826-836.   http://dx.doi.org/10.1111/j.1469-0691.2006.01456.x PMid:16882287

102. Majiduddin FK, Palzkill T (2005) Amino acid residues that contribute to substrate specificity of class A beta-lactamase SME-1. Antimicrobial Agents and Chemotherapy 49: 3421-3427. http://dx.doi.org/10.1128/ AAC.49.8.3421-3427.2005 PMid:16048956 PMCid:1196253

103. Schneider KD, Karpen ME, Bonomo RA, Leonard DA, Powers RA (2009) The 1.4 A crystal structure of the class D beta-lactamase OXA-

1 complexed with doripenem. Biochemistry 48: 11840-11847. http://dx.doi.org/10.1021/bi901690r PMid:19919101 PMCid:2805451

104. Sun T, Nukaga M, Mayama K, Braswell EH, Knox JR (2003) Comparison of beta-lactamases of classes A and D: 1.5-A crystallographic structure of the class D OXA-1 oxacillinase. Protein Sci 12: 82-91. http://dx.doi.org/10.1110/ps.0224303 PMid:12493831 PMCid:2312410

105. Golemi D, Maveyraud L, Vakulenko S, Samama JP, Mobashery S (2001) Critical involvement of a carbamylated lysine in catalytic function of class D beta-lactamases. Proc Natl Acad Sci USA 98: 14280-14285. http://dx.doi.org/10.1073/pnas.241442898 PMid:11724923 PMCid:64673

106. Garcia-Viloca M, Poulsen TD, Truhlar DG, Gao J (2004) Sensitivity of molecular dynamics simulations to the choice of the X-ray structure used to model an enzymatic reaction. Protein Sci 13: 2341-2354. http://dx.doi.org/10.1110/ps.03504104 PMid:15322278 PMCid:2280009

107. Schnoes AM, Brown SD, Dodevski I, Babbitt PC (2009) Annotation error in public databases: misannotation of molecular function in enzyme superfamilies. PLoS Computational Biology 5: e1000605. http://dx.doi.org/10.1371/journal.pcbi.1000605 PMid:20011109 PMCid:2781113

Chapter 11

THE INTRAMOLECULAR DIELS-ALDER REACTION OF DIARYLHEPTANOIDS — QUANTUM CHEMICAL CALCULATION OF STRUCTURAL FEATURES FAVORING THE FORMATION OF PHENYLPHENALENONES

Yulia Monakhova[1] and Bernd Schneider[2]

[1]Department of Chemistry, Saratov State University, Astrakhanskaya Street 83, Saratov 410012, Russia

[2]Max Planck Institute for Chemical Ecology, Hans Knöll-Str. 8, Jena 07745, Germany

ABSTRACT

Diarylheptanoids have been reported as biosynthetic precursors of phenylphenalenones in plants. Quantum chemical calculations of molecular geometry and orbitals were used to elaborate which structural features are required to determine if diarylheptanoids can undergo an intramolecular Diel-Alder reaction to form phenylphenalenones. The computational data showed that an ortho-quinone- or a hydoxyketone-bearing ring A, containing the dienophile moiety, and a heptadiene chain with conjugated cisoid double bonds at C-4/C-6 and a saturated segment consisting of two sp³-carbon atoms, are required. Only four diarylheptanoids out of eighteen studied compounds proved to be suitable candidates. Among them are two 3,5-dideoxy compounds and two other compounds oxygenated only at C-3, suggesting that lachnanthocarpone, a representative of the 6-oxygenated phenylphenalenones, and anigorufone, a representative of the 6-deoxy phenylphenalenones, are not connected via a precursor-product relationship ("late reduction at C-6") but formed through partially separate pathways.

INTRODUCTION

Phenylphenalenones are a group of polycyclic plant natural products mainly occurring in the Haemodoraceae [1,2,3,4] and the Musaceae [5,6,7,8]. There is ample evidence that the biosynthesis of phenylphenalenones, first discussed by Thomas in 1961 [9], involves a diarylheptanoid, which undergoes cyclization to form the substituted tricyclic phenylphenalenone. This hypothesis was substantiated experimentally by the conversion of 1-(3,4-dihydroxyphenyl)-7-phenylhepta-4,6-dien-3-one (1a, Figure 1) to lachnanthocarpone in a one-pot chemical reaction [10] at ambient temperature. The authors proposed the electron-withdrawing *ortho*-quinone 1c (Figure 2) as an intermediate, in which the 5'-en-3',4'-dione moiety functions as a dienophile. In addition to the *ortho*-quinone ring, the structure of the linear chain is of special interest for determining which diarylheptanoids can undergo (4+2)-cycloaddition. The 4,6-diene-3-one unit and a saturated segment in the C_7-chain are characteristic features, which are hypothetically required to allow the Diels-Alder (DA) reaction to take place. If candidate diarylheptanoids would be available, then whether they could be experimentally converted to phenylphenalenones could be checked. The conversion could proceed either by chemical reaction *in vitro*[10], starting from compound 1a, or *in vivo* after administration to plant material [11]. [2-^{13}C]1-(3,4-Dihydroxyphenyl)-7-phenylhepta-4,6-dien-3-one (1a), when administered to cultured roots of*Anigozanthos preissii*, was smoothly converted to [8-^{13}C]anigorufone [11].

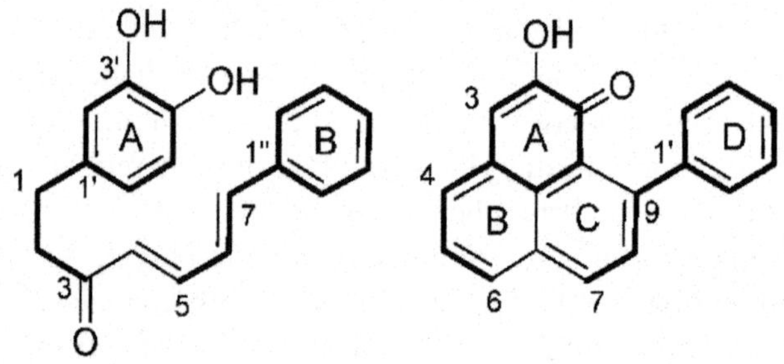

1a   **Anigorufone**

Figure 1: Structures of (4*E*,6*E*)-1-(3,4-dihydroxyphenyl)-1-phenyl-hepta-4,6-dien-3-one (1a) and anigorufone.

According to the formation of lachnanthocarpone (*in vitro* chemical reaction, [10]) and anigorufone (*in vivo* biosynthesis, [11]) from 1a, this compound seems to be a biosynthetic precursor of candidate substrates or a candidate structure itself for the DA reaction. However, other diarylheptanoids with varied substituents in ring A (catechol, hydroxyketone, *ortho*-quinone) or in the C_7-chain (e.g., 1,3-diene-, 1,3,5-triene-, 5-oxo-, 5-hydroxy-) might be able to undergo the DA reaction either under chemical conditions or in the plant. The synthesis, preferentially in an isotopically labeled form, of a complete set of diarylheptanoids possessing combinations of different structural features, followed by chemical and biosynthetic experiments for conversion into phenylphenalenones, would be very laborious. Therefore, a computational approach has been applied to calculate the conformation of plausible candidates and the highest occupied molecular orbital (HOMO)/ lowest occupied molecular orbital (LUMO) energy of the diene and the dienophile of a series of diarylheptanoid structures in order to estimate their suitability for cyclization.

RESULTS AND DISCUSSION

Diarylheptanoid Candidate Structures for Diels–Alder Cyclization

Oxygen functionalities play an important role in the formation of phenylphenalenones from their open-chain diarylheptanoid precursors [12]. The oxidation stage of the functional groups affects the electron density at the corresponding carbon atoms of the aromatic ring A and in the C_7-chain, and seems to facilitate cyclization to phenylphenalenones. The oxidation pattern of ring A, *i.e.*, the hydroxyketone or dihydroxy moiety at C-1/C-2, is a common feature of anigorufone (Figure 1) and most other phenylphenalenones. During biosynthesis, this *ortho*-dioxygenation pattern is retained from ring A of the open-chain diarylheptanoid precursor [12]. The two oxygen substituents in the *ortho* position (C-3'/C-4') of linear diarylheptanoids seem to be required or at least beneficial for the intramolecular DA cyclization because they are withdrawing electrons from the dienophile double bond (*i.e.*, C-5'/C-6') (structural criterion I). Therefore, three series of diarylheptanoids with differently 3'/4'-substituted ring A, namely catechols, hydroxyketones, and *ortho*-quinones, have been employed for computation (Figure 2).

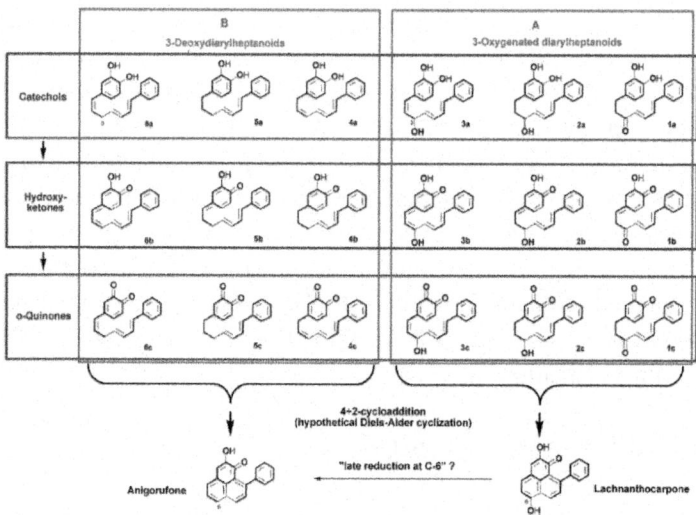

Figure 2: Diarylheptanoids potentially involved in phenylphenalenone biosynthesis.

However, an *ortho*-oxygen-substituted aryl ring is not the only requirement for cyclization. Curcumin from *Curcuma longa*, the most well-known diarylheptanoid, for example, is not converted to phenylphenalenones despite the 3-methoxy-4-hydroxyphenyl structure of the aromatic rings (Schneider, unpublished data). Thus, structural features associated with the C_7-chain are also important prerequisites for cyclization. First, the properties of the diene (position of conjugated double bonds at C-4 and C-6 within the C_7-chain and their conformation) have to be considered (structural criterion II). The conformation of the diene double bonds was *a priori* assigned cisoid in all candidate structures. Transoid conformation is inappropriate for the DA reaction and therefore was not taken into account. Moreover, to undergo (4+2)-cycloaddition, the diene and the dienophile have to be arranged in a non-planar, *i.e.*, stacked, geometry, as this arrangement facilitates the *exo*-orientation of the two moieties within the transition state. Stacked geometry requires a flexible C_7-chain, which results from the occurrence of sp³-carbon atoms [13]. Thus, the occurrence of a saturated segment in the C_7-chain next to ring A was defined as a third prerequisite (structural criterion III) for the DA reaction of diarylheptanoids.

The presence (e.g., 3-oxygenated diarylheptanoids, panel A in Figure 2) or absence of oxygen (e.g., 3,5-dideoxydiarylheptanoids, panel B in Figure 2) in the C_7-chain may also play a role in determining the suitability of a diarylheptanoid for the DA reaction. The reduction of a hydroxyl group at C-5 of diarylheptanoids (5-OH originates from a carboxyl group of a phenylpropanoid) is assumed to take place early in biosynthesis, because phenylphenalenones with oxygen at the corresponding position (C-7) have so far not been reported from natural sources. The oxygen at C-3 of diarylheptanoids (3-OH originates from the carboxyl group of the second phenylpropanoid unit) is retained at C-6 of some phenylphenalenones (e.g., lachnanthocarpone) but is lost in others (e.g., anigorufone).

The computations envisaged in this study are thought to answer the question, which out of the 18 diarylheptanoid structures shown in Figure 2 are able to be folded into a geometry that allows the diene and the dienophile to be located at the distance required to undergo the suprafacial (4+2)-cycloaddition and which therefore is the preferred candidate biosynthetic precursor of phenylphenalenones. A conclusive answer to this question would help determine whether all of the different phenylphenalenones in plants are formed from a common diarylheptanoid or if different diarylheptanoids could function as precursors of different structural types of phenylphenalenones. In this context the question arose: Are 1,2-dioxygenated and 1,2,6-trioxygenated phenylphenalenones, exemplified by anigorufone and lachnanthocarpone (Figure 2), respectively, formed through a linear pathway or a grid of pathway variants?

Molecular Geometry Calculation

As outlined above, the following structural criteria seem to be hypothetical prerequisites for the intramolecular DA reaction of diarylheptanoids and have been employed as minimal conditions for computations:

- Two oxygen substituents in ortho position (C-3'/C-4') of linear diarylheptanoids (as in all diarylheptanoids shown in Figure 2).
- Conjugated double bonds at C-4 and C-6 in the C_7-chain in cisoid conformation (as in all diarylheptanoids shown in Figure 2).
- A saturated segment (sp³-carbon atoms) in the C_7-chain next to ring A.

The compounds shown in Figure 2 were subjected to quantum chemical calculations, whether or not in each case they fulfill all of the above-mentioned structural criteria. Based on the calculated geometry of the optimized molecules (basis 6-31G*), the candidate structures can be categorized into two groups according to the "geometrical" criterion III, *i.e.*, having a saturated segment of sp³-carbon atoms next to ring A in the C_7-chain. Compounds 1a and 1c (sp³-carbon atoms C-1 and C-2) and compounds 2a, 5a,4b, 2c and 5c (three sp³-carbon atoms C-1, C-2 and C-3) fulfill this particular criterion and are marked with "+" in Table 1. These compounds are bent into a stacked geometry (see 1c in Figure 3 as an example).

Table 1: Geometrical and charge characteristics of the investigated compounds

Compound	Geometrical Criterion [a]	Distance Between Atoms [Å]		Mulliken's Charges				Difference between Atoms' Charges	
		(6'-4)	(5'-7)	6'	5'	4	7	(6'-4)	(5'-7)
1a	+	3.39	4.01	-0.11	-0.12	-0.29	-0.07	0.16	0.06
2a	+	3.48	4.07	-0.11	-0.12	-0.19	-0.11	0.08	0.02
3a	−	–b	–b	-0.08	-0.12	-0.17	-0.10	0.09	0.02
4a	−	–b	–b	-0.09	-0.12	-0.11	-0.01	0.02	0.02
5a	+	3.46	4.11	-0.11	-0.12	-0.16	-0.11	0.05	0.01
6a	−	–b	–b	-0.09	-0.12	-0.14	-0.10	0.05	0.02
1b	−	–b	–b	-0.03	-0.26	-0.21	-0.05	0.18	0.21
2b	−	–b	–b	-0.03	-0.21	-0.18	-0.09	0.15	0.12
3b	−	–b	–b	-0.01	-0.22	-0.15	-0.06	0.15	0.16
4b	+	3.47	4.10	-0.09	-0.19	-0.17	-0.10	0.08	0.09
5b	−	–b	–b	-0.04	-0.21	-0.15	-0.10	0.11	0.11
6b	−	–b	–b	-0.02	-0.21	-0.14	-0.08	0.13	0.13
1c	+	3.38	3.93	-0.05	-0.20	-0.27	-0.06	0.22	0.14
2c	+	3.50	4.02	-0.05	-0.19	-0.19	-0.10	0.14	0.09
3c	−	–b	–b	-0.05	-0.19	-0.18	-0.09	0.13	0.10
4c	−	–b	–b	-0.05	-0.19	-0.12	-0.09	0.08	0.10
5c	+	3.47	4.04	-0.05	-0.19	-0.17	-0.10	0.11	0.10
6c	−	–b	–b	-0.05	-0.19	-0.15	-0.09	0.11	0.10

[a] based on visual inspection of the optimized geometry; [b] not quoted as geometry condition was not fulfilled (the distances were more than 7 Å).

In contrast, the optimized structures of compounds 3c, 4c and 6c, marked with "–" in Table 1have only one sp³-carbon atoms in the chain at C-1 or C-3 and therefore show a stretched geometry or are only slightly bent (see 6c in Figure 3 as an example). As expected, none of the other eight candidates (3a,4a, 6a, 1b, 2b, 3b, 5b, 6b) for which structure criterion III is not fulfilled is bent into a stacked geometry (Table 1). Hence, the presence of at least two sp³-carbon atoms in the C₇-chain next to ring A seems to be essential for the intramolecular DA reaction. For structures showing bent geometry, the distances between the carbon atoms C-5›–C-7 and C-6›–C-4, which are participating in the cyclization, have been calculated (Table 1). This parameter varied between 3.93 Å and 4.11 Å for C-5› and C-7 and between 3.38 Å and 3.50 Å for C-6› and C-4. In both cases the values were the smallest for 1c. It can be concluded that, although these distances are only slightly larger for the other compounds, compound **1c** is the most privileged candidate for the DA reaction.

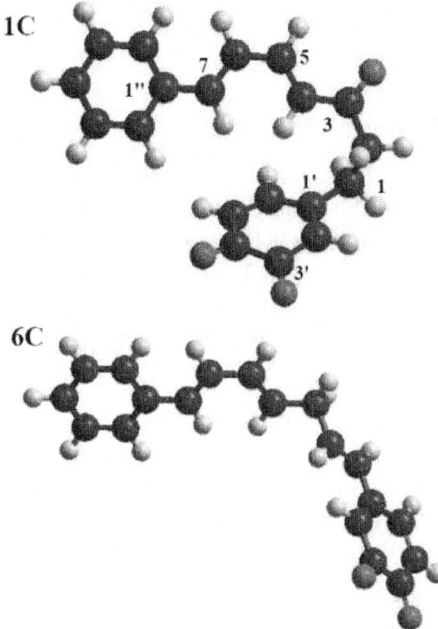

Figure 3: Diarylheptanoid 1c and 6c exemplify the two geometrical types of diaryl-heptanoids under study (the atom numbers in accordance with Figure 1 are shown for compound1c). Compound 1c is bent into a stacked geometry with a short distance between the diene carbons C-4/C-7 and the dienophile carbon atoms C-5'/C-6' and

therefore seems able to undergo the Diels-Alder cyclization. The stretched geometry of compound6c does not favor an intramolecular Diels-Alder reaction.

Mulliken's charges of the interacting atoms C-5', C-6', C-4, and C-7 are another important parameter to be considered. Table 1 contains the absolute values of Mulliken's charges on the investigated atoms. According to the theory [14,15], the dienophile carbons should have low electron density. This is consistent especially for C-5' of most compounds and more or less also for C-6', although compound **1c** is not among the best examples in this case. In contrast to the dienophile, the diene component should be electron-rich, which is not the case for the considered compounds. However, the rule of electron depletion for the dienophile and electron excess for the diene can be reversed in so-called inverse electron-demand DA reactions. Table 1 also presents charge differences for C-6'–C-4 and C-5'–C-7. Clearly, the bigger this difference is, the stronger the interaction of the two atoms and the more favored the DA reaction. Compound 1c has the largest charge differences between both positions C-6'–C-4 (0.22) and the third largest difference for the other pair of atoms C-5'–C-7 (0.135). The smallest charge differences were found for compounds from group a (catechols), namely in compound 5a for atoms C-5'–C-7 (0.01) and in compound 4a for C-6'–C-4 (0.02).

Orbital Calculation

Another aspect to be considered for the feasibility of the DA reaction is the interaction of HOMO-LUMO and the characteristics of these orbitals. Table 2 shows energy values of both HOMO and LUMO and the difference between them for all structures under consideration. There is a clear tendency: the biggest differences were found for catechols (Figure 1, group a), then three hydroxyketones (1b, 2b and5b), followed by compounds from group c (*ortho*-quinones). Interestingly, compounds 3, 4, and 6 from groups b and c have the smallest HOMO-LUMO energy differences. Figure 4 displays HOMO and LUMO of compounds 1a and 1c and illustrates their possible interactions. The DA reaction seems to occur if p-orbitals of the dienophile atoms C-6' and C-5' give the largest impact in HOMO and p-orbitals of the diene atoms (C-4 and C-7) in LUMO, or vice versa. This is the case for compounds 1c-6c as well as for some compounds from group b (2b, 4b and 6b) (Table 2). The other studied compounds do not fulfill this condition. For compound 1a, for example, p-electron density concentrates on dien chain (C-4 to C-7) and ring B (Figure 4).

Table 2: Characteristics of LUMO and HOMO energies for selected compounds

Com-pound	Orbital Crite-rion[a]	Energy of Orbitals [eV]			Com-pound	Orbital Crite-rion[a]	Energy of Orbitals [eV]		
		LUMO	HOMO	LUMO-HOMO Differ-ence			LUMO	HOMO	LUMO-HOMO Differ-ence
1a	–	-0.87	-8.94	8.07	4b	+	-0.42	-5.17	4.75
2a	–	-0.34	-8.55	8.21	5b	–	-1.21	-8.76	7.55
3a	–	-0.49	-8.66	8.17	6b	+	-1.73	-8.31	6.58
4a	–	-0.70	-8.45	7.75	1c	+	-1.80	-9.15	7.35
5a	–	-0.34	-8.53	8.19	2c	+	-1.69	-8.76	7.07
6a	–	-0.52	-8.58	8.06	3c	+	-1.77	-8.81	7.04
1b	–	-1.37	-9.10	7.73	4c	+	-1.59	-8.66	7.07
2b	+	-1.29	-8.74	7.45	5c	+	-1.63	-8.74	7.11
3b	–	-1.74	-8.16	6.42	6c	+	-1.67	-8.66	6.99

[a] based on visual inspection of LUMO and HOMO orbitals.

This finding was numerically substantiated by calculating contributions of all p-orbitals in HOMO and LUMO for C-5', C-6', C-4, and C-7 (Table 3). For example, for compound 1c the interactions C-7 (p_x)–C-6' (p_y) and C-5' (p_x)–C-4 (p_y) are possible. On the other hand, for compound 2a, for example, p-orbitals of C-6' and C-5' do not play significant roles in HOMO and LUMO. From this point of view, compounds 1c–5c, as well as 2b, 4b and 6b are suitable candidates for the intramolecular DA reaction. Although the "orbital" criterion is also fulfilled for compounds 3b and 6c, the impact of p-orbitals of C-5' in LUMO (6c) and of C-4 in LUMO (3b) is very low and, therefore, the desired HOMO-LUMO interaction is unlikely.

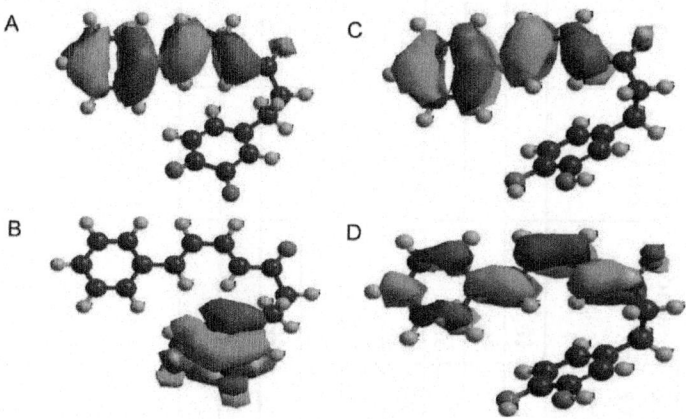

Figure 4: (A) HOMO and (B) LUMO of compound 1c and (C) HOMO and (D) LUMO of compound 1a.

This shows that the calculation of HOMO and LUMO of the positions involved in the intramolecular DA reaction alone does not allow the viability of the diarylheptanoids under study to be predicted. Instead, it is necessary to consider the combination of the geometrical and orbital parameters. From the data in Table 1, Table 2 and Table 3 it can be concluded that for some compounds (1a, 2a, 5a, 4b, 1c, 2c,5c), the "geometrical" criterion III shows that the reaction is possible, whereas for other compounds (2b,6b, 3c, 4c, 6c) orbital characteristics point to reaction possibility. Taking into account both factors, compounds 1c, 2c, 5c, and 4b seem to be suitable for the intramolecular DA reaction to form phenylphenalenones.

EXPERIMENTAL

Quantum-Chemical Calculations: We used GAMESS (US) v.7.0 [16,17,18] and HyperChem Professional v.8.0 (Hypercube, Gainesville, FL, USA) software packages for quantum chemical calculations.

Table 3: Contribution of the most important atomic orbitals of carbon atoms in LUMO and HOMO (for atom numbers, see Figure 1)

Compound	Atom number	Atomic orbital	Type of molecular orbital	Contribution	Compound	Atom number	Atomic orbital	Type of molecular orbital	Contribution
1a	7	Px	LUMO	0.23	4b	7	Px	LUMO	0.24
	4	Px	LUMO	0.19		4	Px	LUMO	0.24
	5'	Px	HOMO	0.00		5'	Px	HOMO	0.03
	6'	Py	HOMO	0.00		6'	Pz	HOMO	0.37
2a	7	Py	HOMO	0.20	5b	7	Pz	LUMO	0.03
	4	Px	HOMO	0.24		4	Pz	HOMO	0.32
	5'	Pz	LUMO	0.00		5'	Px	LUMO	0.31
	6'	Pz	HOMO	0.01		6'	Px	HOMO	0.03
3a	7	Px	HOMO	0.21	6b	7	Pz	HOMO	0.21
	4	Py	HOMO	0.27		4	Pz	LUMO	0.12
	5'	Pz	LUMO	0.03		5'	Pz	LUMO	0.20
	6'	Px	HOMO	0.07		6'	Pz	HOMO	0.16
4a	7	Py	HOMO	0.33	1c	7	Px	HOMO	0.17
	4	- [a]	- [a]	- [a]		4	Py	HOMO	0.08
	5'	Px	HOMO	0.00		5'	Px	LUMO	0.045
	6'	Pz	LUMO	0.01		6'	Py	LUMO	0.11

Compound	Position	Orbital	HOMO/LUMO	Value
5a	7	Pz	LUMO	0.24
	4	Px	HOMO	0.24
	5'	S	LUMO	0.00
	6'	Pz	HOMO	0.00
6a	7	Pz	LUMO	0.27
	4	Py	LUMO	0.26
	5'	Px	LUMO	0.02
	6'	Px	HOMO	0.26
1b	7	Py	HOMO	0.08
	4	Pz	HOMO	0.28
	5'	Px	HOMO	0.32
	6'	Px	LUMO	0.12
2b	7	Px	HOMO	0.25
	4	Py	HOMO	0.19
	5'	Pz	LUMO	0.11
	6'	Px	LUMO	0.33
3b	7	Px	HOMO	0.09
	4	Py	LUMO	0.03
	5'	Pz	LUMO	0.08
	6'	Pz	HOMO	0.21

Compound	Position	Orbital	HOMO/LUMO	Value
2c	7	Py	HOMO	0.20
	4	Px	HOMO	0.23
	5'	Px	LUMO	0.09
	6'	Pz	LUMO	0.36
3c	7	Py	HOMO	0.28
	4	Px	HOMO	0.21
	5'	Pz	LUMO	0.13
	6'	Px	LUMO	0.25
4c	7	Pz	HOMO	0.33
	4	Pz	LUMO	0.02
	5'	Py	HOMO	0.00
	6'	Py	LUMO	0.24
5c	7	Py	HOMO	0.22
	4	Px	HOMO	0.24
	5'	Px	LUMO	0.11
	6'	Pz	LUMO	0.35
6c	7	Py	HOMO	0.27
	4	Px	HOMO	0.22
	5'	Px	LUMO	0.27
	6'	Px	LUMO	0.01

a no significant contribution of p-orbitals in HOMO and LUMO.

First, we applied semi-empirical PM3 (Parametrised Model 3) method with full geometry optimization to obtain the rough geometry of molecules. The main approaches of the PM3 method include adiabatic, one-electron, MO LCAO (molecular orbital as a linear combination of atomic orbitals) and INDO (Intermediate Neglect of Differential Overlap) approximations. For details regarding the calculations, see [19]. We were particularly careful to ensure that the diene double bonds had a cisoid conformation, as this would be favorable for the DA reaction (Figure 1). All structures were then optimized at the unrestricted Hartree-Fock (UHF) [20] level of theory using the 6-31G* [21] basis set and PM3 geometry as inputs. This level of theory is sufficient for calculating the selected parameters (Mulliken's charges, distances between atoms, energy of HOMO and LUMO orbitals) for a series of similar compounds.

CONCLUSIONS

This computational study confirmed that for intramolecular Diels-Alder cyclization of diarylheptanoids to produce phenylphenalenones, both the dienophile located in ring A of the diarylheptanoid and the C_7-chain, comprising the diene and sp^3-carbon atoms, must possess optimal electronic and conformational characteristics: an *ortho*-quinone- or hydoxyketone-bearing ring A containing the dienophile moiety and a heptadiene chain with conjugated cisoid double bonds at C-4/C-6. According to the molecular geometry and orbital calculations, compounds 1c, 2c, 5c, and 4b are the best substrates for DA cyclization of diarylheptanoids to produce phenylphenalenones. The type of oxygen functionality at C-3 of the C_7-chain seems not to be crucial for the DA reaction since both compounds 1c and 2c are good candidates. This result is in agreement with previous chemical and biosynthetic studies [10,11]. Interestingly, the absence of oxygen in the C_7-chain of 4b and 5c does not prevent these diarylheptanoids from undergoing the intramolecular DA reaction. Hence, the possible DA reaction of both 3-oxygenated compounds 1c and 2cand 3,5-dideoxy compounds 4b and 5c suggests that 6-oxygenated phenylphenalenones, such as lachnanthocarpone and 6-deoxy phenylphenalenones, such as anigorufone, are not connected via a precursor-product relationship ("late reduction at C-6") but formed through early reduction at C-6 via partially separate pathways. This hypothesis remains to be confirmed experimentally by biosynthetic labeling or enzyme studies.

REFERENCES

1. Cooke, R.G.; Thomas, R.L. Colouring matters of Australian plants. XVIII. Constituents of *Anigozanthos rufus. Aust. J. Chem.* 1975, *28*, 1053–1057.

2. Hölscher, D.; Schneider, B. HPLC-NMR analysis of phenylphenalenones

and a stilbene from*Anigozanthos flavidus*. *Phytochemistry* 1999, *50*, 155–161.

3. Opitz, S.; Hölscher, D.; Oldham, N.J.; Bartram, S.; Schneider, B. Phenylphenalenone-related compounds. Chemotaxonomic markers of the Haemodoraceae from *Xiphidium caeruleum*. *J. Nat. Prod.* 2002, *65*, 1122–1130.

4. Fang, J.; Kai, M.; Schneider, B. Phytochemical profile of aerial parts and roots of *Wachendorfia thyrsiflora* L. studied by LC-DAD-SPE-NMR. *Phytochemistry* 2012, *81*, 144–152.

5. Luis, J.G.; Fletcher, W.Q.; Echeverri, F.; Grillo, T.A. Phenalenone-type phytoalexins from *Musa acuminata* synthesis of 4-phenyl-phenalenones. *Tetrahedron* 1994, *50*, 10963–10970.

6. Kamo, T.; Kato, N.; Hirai, N.; Tsuda, M.; Fujioka, D.; Ohigashi, H. Phenylphenalenone-type phytoalexins from unripe Buñgulan banana fruit. *Biosci. Biotechnol. Biochem.* 1998, *62*, 95–101.

7. Otálvaro, F.; Nanclares, J.; Vásquez, L.E.; Quiñónes, W.; Echeverri, F.; Arango, R.; Schneider, B. Phenalenone-type compounds from *Musa acuminata* var. "Yangambi km 5" (AAA) and their activity against *Mycosphaerella fijiensis*. *J. Nat. Prod.* 2007, *70*, 887–890.

8. Hölscher, D.; Dhakshinamoorthy, S.; Alexandrov, T.; Becker, M.; Bretschneider, T.; Bürkert, A.; Crecelius, A.C.; de Waele, D.; Elsen, A.; Heckel, D.G.; *et al.* Phenalenone-type phytoalexins mediate resistance of banana plants (*Musa* spp.) to the burrowing nematode *Radopholus similis.Proc. Natl. Acad. Sci. USA* 2014, *111*, 105–110.

9. Thomas, R. Studies in the biosynthesis of fungal metabolites. *Biochem. J.* 1961, *78*, 807–813.

10. Bazan, A.C.; Edwards, J.M.; Weiss, U. Synthesis of lachnanthocarpone [9-phenyl-2,6-dihydroxyphenalen-1(6)-one] by intramolecular Diels-Alder cyclization of a 1,7-diarylheptanoid orthoquinone; possible biosynthetic significance of Diels-Alder reactions. *Tetrahedron* 1978, *34*, 3005–3015.

11. Hölscher, D.; Schneider, B. A diarylheptanoid intermediate in the biosynthesis of phenylphenalenones in *Anigozanthos preissii*. *J. Chem. Soc. Chem. Comm.* 1995, 525–526.

12. Munde, T.; Brand, S.; Hidalgo, W.; Maddula, R.K.; Svatoš, A.; Schneider, B. Biosynthesis of tetraoxygenated phenylphenalenones in *Wachendorfia thyrsiflora*. *Phytochemistry* 2013, *91*, 165–176.

13. Schmitt, B.; Schneider, B. Dihydrocinnamic acids are involved in

the biosynthesis of phenylphenalenones in *Anigozanthos preissii*. *Phytochemistry* 1999, *52*, 45–53.

14. Houk, K.N.; González, J.; Li, Y. Pericyclic reaction transition states: Passions and punctilios, 1935–1995. *Acc. Chem. Res.* 1995, *28*, 81–90.

15. Brocksom, T.J.; Nakamura, J.; Ferreira, M.L.; Brocksom, U. The Diels-Alder reaction: An update.*J. Braz. Chem. Soc.* 2001, *12*, 597–622.

16. Granovsky, A.A. PC GAMESS version 7.1. Available online: http://classic.chem.msu.su/gran/gamess/index.html (accessed on 22 April 2014).

17. Nemukhin, A.V.; Grigorenko, B.L.; Granovsky, A.A. Molecular modeling by using the PC GAMESS program: From diatomic molecules to enzymes. *Moscow Univ. Chem. Bull.* 2004, *45*, 75–102.

18. Guest, M.F.; Bush, I.J.; van Dam, H.J.J.; Sherwood, P.; Thomas, J.M.H.; van Lenthe, J.H.; Havenith, R.W.A.; Kendrick, J. The GAMESS-UK electronic structure package: Algorithms, developments and applications. *Mol. Phys.* 2005, *103*, 719–747.

19. Stewart, J.J.P. *"PM3" in Encyclopedia of Computational Chemistry*; Wiley: New York, NY, USA, 1998.

20. Pople, J.A.; Nesbet, R.K. Self-consistent orbitals for radicals. *J. Chem. Phys.* 1954, *22*, 571–572.

21. Harihara, P.C.; Pople, J.A. Influence of polarization functions on molecular-orbital hydrogenation energies. *Theor. Chim. Acta* 1973, *28*, 213–222.

Chapter 12

CORRELATION FUNCTIONS IN OPEN QUANTUM-CLASSICAL SYSTEMS

Chang-Yu Hsieh and Raymond Kapral

Chemical Physics Theory Group, Department of Chemistry, University of Toronto, Toronto, ON M5S 3H6, Canada

ABSTRACT

Quantum time correlation functions are often the principal objects of interest in experimental investigations of the dynamics of quantum systems. For instance, transport properties, such as diffusion and reaction rate coefficients, can be obtained by integrating these functions. The evaluation of such correlation functions entails sampling from quantum equilibrium density operators and quantum time evolution of operators. For condensed phase and complex systems, where quantum dynamics is difficult to carry out, approximations must often be made to compute these functions. We present a general scheme for the computation of correlation functions, which preserves the full quantum equilibrium structure of the system and approximates the time evolution with quantum-classical Liouville dynamics. Several aspects of the scheme are discussed, including a practical and general approach to sample the quantum equilibrium density, the properties of the quantum-classical Liouville equation in the context of correlation function computations, simulation schemes for the approximate dynamics and their interpretation and connections to other approximate quantum dynamical methods.

INTRODUCTION

The dynamical properties of condensed-phase or complex systems are often investigated experimentally by applying external fields to weakly perturb a system and observe its relaxation back to the thermal equilibrium state. In such experiments, measurable quantities can be related to equilibrium time correlation functions via linear response theory [1,2]:

$$C_{AB}(t) \;=\; \frac{1}{Z_Q}\mathrm{Tr}\left[e^{-\beta\hat{H}}\hat{A}(0)\hat{B}(t)\right] = \frac{1}{Z_Q}\mathrm{Tr}\left[e^{-\beta\hat{H}}\hat{A}e^{\frac{i}{\hbar}\hat{H}t}\hat{B}e^{-\frac{i}{\hbar}\hat{H}t}\right]$$

where $\hat{A}$ and $\hat{B}$ are operators corresponding to some specific dynamical variables under investigation, $\hat{H}$ is the unperturbed Hamiltonian and Z_Q is the quantum canonical partition function associated with $\hat{H}$. Many experiments employing spectroscopic methods directly probe such time correlation functions.

Exact numerical evaluation of Equation (1) for real condensed phase quantum systems is prohibitive, since the computational cost scales exponentially with respect to the number of degrees of freedom (DOF). Various approaches have been developed to address this challenging problem. A common approach shared by many methods is to partition the entire system into a subsystem (whose dynamical properties are of interest) and an environment (or bath) in which the subsystem resides. Other recently developed schemes for computing quantum correlation functions do not rely on such a partition and instead utilize approximations to treat the quantum evolution of the entire system in conjunction with quantum equilibrium sampling [3–5]. In this paper, we focus on schemes based on the system-bath partition, and using this partition, the Hamiltonian reads: $\hat{H} = \hat{H}_b + \hat{h}_s + \hat{V}_c(\hat{R})$ where $\hat{H}_b = \frac{\hat{P}^2}{2M} + \hat{V}_b(\hat{R})$ and $\hat{h}_s$ represent the pure bath and subsystem Hamiltonians, respectively. The last term in $\hat{H}$ is a coupling potential that depends on the spatial coordinates of the bath wave functions. We shall always take the bath part of the Hamiltonian in the coordinate representation; however, we can represent $\hat{h}_s = \frac{\hat{p}^2}{2m} + \hat{V}_s(\hat{r})$ in some quantum basis:

$$\hat{h}_s = \sum_{ij} |i\rangle\langle i|\,\hat{h}_s\,|j\rangle\langle j|$$

Several methods based on various master equations [6–10] and path integral influence functional methods [11,12] provide approximate schemes, often in the weak coupling limit, to systematically project out the environmental DOF and yield a subsystem dynamics that incorporates dissipation and decoherence, due to coupling to the environment. However, for many applications, such as proton and electron transfer in condensed phases, it is desirable to explicitly simulate, even approximately, the bath dynamics, since specific local bath DOF may be crucial for a description of the dynamics of the quantum subsystem. For this purpose, several semiclassical [13–15] and mixed quantum classical [16,17] (MQC) methods, which either treat the entire dynamics semiclassically or simulate the dynamics of the bath and subsystem with different levels of rigor (e.g., classical versus quantum mechanical), have been formulated. Many semiclassical and mixed quantum-classical approaches, adopting powerful classical simulation techniques, evaluate Equation (1) by combined Monte Carlo-molecular dynamics (MC-MD) techniques.

In this paper, we formulate MC-MD schemes to evaluate Equation (1) within the framework of the quantum-classical Liouville equation (QCLE) [18]. The QCLE employs a partial Wigner representation of the environmental (bath) DOF and may be derived from full quantum dynamics by truncating the quantum evolution operator to the first order in a small parameter related to the ratio of the characteristic masses of quantum and bath DOF [18]. In particular, we suppose that the quantum subsystem has a finite-dimension Hilbert space. Under this assumption, Equation (1) is cast in the following form [19,20]:

$$
C_{AB}(t) = \frac{1}{Z_Q} \sum_{n_1,n_2} \int dX \left[\left(e^{-\beta \hat{H}} A \right)_W^{n_1 n_2} (X) B_W^{n_2 n_1}(X,t) \right]
\tag{2}
$$

where the n_j indices label the basis states (in some chosen quantum basis), X = (R, P) represents the Wigner-transformed phase space point for the bath, N_B is the number of bath DOF and the subscript, W, on an operator indicates a partial Wigner transform on the bath DOF; e.g., an operator is partially Wigner transformed as $\hat{B}_W(X) = \int dZ \langle R - \frac{Z}{2} | \hat{B} | R + \frac{Z}{2} \rangle e^{\frac{i}{\hbar} P \cdot Z}$.

Two main tasks are involved in evaluating Equation (2) with an MC-MD algorithm. First, one needs to sample initial conditions (for an ensemble of trajectories) from the partially Wigner-transformed quantum density, $\left(\hat{\rho}_{eq} \hat{A} \right)_W (X)$ with $\rho_{eq} = e^{-\beta H} / Z_Q$. There exist numerical algorithms to accomplish such a task [21,22]. Second, one needs to propagate the initial points in the phase space. These time-evolved trajectories may then be used to construct the matrix elements, $B_W^{nm}(X,t)$, needed to compute the correlation function. Various simulation methods, whose structure depends on the basis chosen to represent the quantum degrees of freedom in the QCLE, have been devised to simulate the mixed quantum-classical dynamics [23–31]. Simulation methods that utilize an adiabatic basis can be cast into the form of surface-hopping dynamics, but in a way that includes coherent evolution segments that account for creation and destruction of coherence in a proper manner. More recently, as in some semiclassical approaches [32], the mapping basis [33] was used to describe the quantum degrees of freedom in the QCLE in a continuous classical-like manner, leading to a trajectory description in the full system phase space [30,31,34–36].

The goals and outline of the paper are as follows: We first consider how the two ingredients, quantum equilibrium sampling and evolution of quantum operators, which are needed to compute quantum correlation functions, may be carried out. In Section 2, we describe a path-integral scheme to perform

MC sampling from the partially Wigner transformed quantum density. In the Appendix, we also discuss a simplified, but approximate sampling scheme that is useful in the high-temperature limit. Another aim of this paper is to demonstrate how a recently-developed simulation method for the QCLE, the forward-backward trajectory solution (FBTS), can be used to efficiently obtain quantum correlation functions. To place these results in proper context, in Section 3, we sketch the important features and properties of the QCLE and discuss both the adiabatic Trotter-based surface-hopping (TBSH) algorithm and the FBTS, which is formulated in the mapping basis. In this section, we also present the explicit form of the N-level generalization of the TBSH algorithm. Comparisons of the trajectories that underlie these algorithms allow us to investigate how completely different ensembles of trajectories can be used to simulate the same observable correlation function. The implementation and utility of the simulation algorithms are illustrated on the dynamics in a two-level system coupled to a quartic oscillator embedded in a bath of independent harmonic oscillators, described in Section 4. Finally, in Section 5, we comment on the advantages, challenges and potential problems in adopting an approximate mixed quantum-classical dynamics for the computation of quantum time-correlation functions.

SAMPLING FROM THE PARTIALLY WIGNER-TRANS-FORMED DENSITY

In general, analytical expressions for the Wigner transform of the density matrix cannot be determined easily. In this section, we present a path-integral-based scheme to perform MC sampling from the Wigner-transformed density, $\left(\hat{\rho}_{eq}\hat{A}\right)_W^{n_1 n_2}$, in Equation (2).

First, we recall the definition of partial Wigner transform:

$$\left(\hat{\rho}_{eq}\hat{A}\right)_W^{n_1 n_2}(X) = \frac{1}{Z_Q}\int dZ \left\langle n_1, R - \frac{Z}{2}\left| e^{-\beta\hat{H}}\hat{A} \right| n_2, R + \frac{Z}{2}\right\rangle e^{\frac{i}{\hbar}P\cdot Z}$$

(3)

where R represents the vector of bath coordinates, n denotes a basis state for the subsystem and $\hat{H} = \frac{\hat{P}^2}{2M} + \hat{V}_b(\hat{R}) + \hat{h}(\hat{R})$ with $\hat{h}(\hat{R}) \equiv \hat{h}_s + \hat{V}_c(\hat{R})$. One way to compute the integral on the right side of Equation (3) is to first factorize $e^{-\beta\hat{H}} = \prod e^{-\beta_L\hat{H}}$ into L − 1 pieces with $\beta_L = \beta/(L - 1)$. Following the standard procedures for path integral calculations, we then insert resolutions of the identity, $\mathcal{I} = \int dR_i \sum_{m_i} |m_i, R_i\rangle\langle m_i, R_i|$ between every pair of factorized operators and apply the approximation,

$$e^{-\beta_L \hat{H}} \approx e^{-\beta_L \frac{P^2}{2M}} e^{-\beta_L \left(\hat{V}_b(\hat{R}) + \hat{h}(\hat{R})\right)}$$. The integrand on the right side of Equation (3) can then be written as follows:

$$\left\langle n_1, R - \frac{Z}{2} \middle| e^{-\beta \hat{H}} \hat{A} \middle| n_2, R + \frac{Z}{2} \right\rangle$$

$$= \int \prod_{i=1}^{L-1} dR_i \sum_{\{m_i\}} \left\langle n_1, R - \frac{Z}{2} \middle| e^{-\beta_L \hat{H}} \middle| m_1, R_1 \right\rangle \left\langle m_1, R_1 \middle| e^{-\beta_L \hat{H}} \middle| m_2, R_2 \right\rangle \cdots$$

$$\times \left\langle m_{L-1}, R_{L-1} \middle| \hat{A} \middle| n_2, R + \frac{Z}{2} \right\rangle,$$

$$= \left(\frac{M}{2\pi\beta_L \hbar^2}\right)^{\frac{N_B(L-1)}{2}} \int \prod_{i=1}^{L-1} dR_i \sum_{\{m_i\}} \left\{ \prod_{i=1}^{L-2} \mathcal{M}_{i,i+1}(R_i) e^{-\beta_L V_b(R_i)} e^{-\frac{M}{2\beta_L \hbar^2}(R_i - R_{i+1})^2} \right\}$$

$$\times \left\langle n_1, R - \frac{Z}{2} \middle| \left(e^{-\beta_L \hat{H}} \middle| m_1, R_1 \right\rangle \left\langle m_{L-1}, R_{L-1} \middle| \hat{A}\right) \middle| n_2, R + \frac{Z}{2} \right\rangle$$

$$\tag{4}$$

Where

$$\mathcal{M}_{i,j} = \left\langle m_i \middle| e^{-\beta_L \hat{h}(R_i)} \middle| m_j \right\rangle = \begin{cases} e^{-\beta_L h^{ij}(R_i)}, & i = j \\ -\beta_L h^{ij}(R_i) e^{-\beta_L h^{ij}(R)}, & i \neq j \end{cases} \tag{5}$$

which is correct to order $\mathcal{O}(\beta_L^2)$ Substituting Equation (4) into Equation (3), the new integrand of the Wigner transform becomes $\hat{A} = \left(e^{-\beta_L \hat{H}} \middle| m_1, R_1 \right\rangle \left\langle m_{L-1}, R_{L-1} \middle| \hat{A}\right)$ An analytical approximation for the Wigner transform of $\mathcal{A}$ can be obtained easily in most cases when $\hat{A}$ is a pure observable subsystem or if it depends on just one of the conjugate variables: R or P. Since $\beta_L \ll 1$, it is possible to replace the term, $e^{-\beta_L} \hat{H}$, inside $\mathcal{A}$ with its high-temperature approximation (discussed in the Appendix). Letting $\mathcal{A}_W(X)$ be the partial Wigner transform of $\mathcal{A}$, Equation (3) reads:

$$\left(\hat{\rho}_{eq} \hat{A}\right)_W^{n_1 n_2}(X) = \frac{\mathcal{G}^{N_B(L-1)/2}}{Z_Q} \int \prod_{i=1}^{L-1} dR_i \sum_{\{m_i\}} \left\{ \prod_{i=1}^{L-2} \mathcal{M}_{i,i+1}(R_i) e^{-\beta_L V_b(R_i)} e^{-\pi \mathcal{G}(R_i - R_{i+1})^2} \right\} A_W^{n_1 n_2}(X) \tag{6}$$

Where $\mathcal{G} = \left(\frac{M}{2\pi\beta_L \hbar^2}\right)$. Substituting Equation (6) into Equation (2), the time correlation function becomes:

$$C_{AB}(t) = \frac{\mathcal{G}^{N_B(L-1)/2}}{Z_Q} \sum_{n_1,n_2} \sum_{\{m_i\}} \int \prod_{i=1}^{L-1} dR_i \left\{ \prod_{i=1}^{L-2} \mathcal{M}_{i,i+1}(R_i) e^{-\beta_L V_b(R_i)} e^{-\pi \mathcal{G}(R_i - R_{i+1})^2} \right\}$$

$$\times \int dX \left(\hat{A}\right)_W^{n_1 n_2}(X) B_W^{n_2 n_1}(X, t) \tag{7}$$

Following [37], we remark that the initial phase space coordinate X = (R, P) and auxiliary variables, $\{R_i\}$, can be sampled from probability densities constructed from $\mathcal{A}_w(X)$ and $|M_{i,i+1}(R_i)|e^{-\beta}_L V_b(R_i)e^{-\pi G(R}_i-R_{i+1})^2$, respectively.

QUANTUM-CLASSICAL LIOUVILLE EQUATION

In this section, we discuss how one can simulate the time-evolved matrix elements, $B_W^{n_2 n_1}(X, t)$ in Equation (2) using the QCLE:

$$\frac{\partial \hat{B}_W(X,t)}{\partial t} = \frac{i}{\hbar}[\hat{H}_W, \hat{B}_W] - \frac{1}{2}(\{\hat{H}_W, \hat{B}_W\} - \{\hat{B}_W, \hat{H}_W\})$$

$$= i\hat{\mathcal{L}}\hat{B}_W(X,t) = \frac{i}{\hbar}\left(\overrightarrow{\mathcal{H}}_\Lambda \hat{B}_W - \hat{B}_W \overleftarrow{\mathcal{H}}_\Lambda\right) \qquad (8)$$

where $\Lambda = \overleftarrow{\nabla}_P \overrightarrow{\nabla}_R - \overleftarrow{\nabla}_R \overrightarrow{\nabla}_P$. The arrow on top of a differential operator indicates the direction in which it acts. In the first line, the square bracket and the curly brackets denote the quantum commutator and classical Poisson brackets, respectively. The two kinds of Lie bracket act together as the generator of the mixed quantum-classical dynamics. Due to the fact that $\hat{H}_w(X)$ and $\hat{B}_w(X, t)$ are quantum operators with respect to the subsystem DOF, two differently ordered Poisson brackets are needed to properly account for the mixed dynamics. However, in general, the dynamics described by the QCLE does not have a Lie algebraic structure, a feature that is common to mixed quantum-classical approaches [38]. In the second line, we introduce the abstract, quantum-classical Liouville (QCL) superoperator, $\hat{L}$. Finally, the third equality is another equivalent representation of QCLE in terms of the forward and backward mixed quantum-classical Hamiltonians:

$$\overrightarrow{\mathcal{H}}_\Lambda = \hat{H}_W\left(1 + \frac{\hbar\Lambda}{2i}\right), \quad \overleftarrow{\mathcal{H}}_\Lambda = \left(1 + \frac{\hbar\Lambda}{2i}\right)\hat{H}_W \qquad (9)$$

The QCLE has many desirable features, such as the conservation of energy, momentum and phase space volumes. Furthermore, the QCLE is equivalent to full quantum dynamics for arbitrary quantum subsystems, which are bilinearly coupled to a harmonic bath. For instance, commonly used spin boson models are of this type. In this circumstance, the combination of quantum and classical brackets in the QCLE does have a Lie algebraic structure. For the more general bath and coupling potentials, the QCLE provides an approximate description of the quantum dynamics. In this case, comparisons of simulations of QCL dynamics with exact quantum results have indicated that it is quantitatively accurate for a wide range of systems [36,39–48] The QCLE equation can be simulated using ensembles of trajectories, which, in combination with the quantum initial condition sampling discussed above, provides a way to compute quantum correlation functions. As we shall see, the nature of the trajectories that enter in the simulations depends on the algorithm and should not be ascribed physical significance. It is only the observable, in this case, the correlation function, that has physical meaning and is independent of the manner in which it is simulated, provided the simulation algorithm is capable of exactly solving the QCLE, which is not always the case. One of the goals of this paper is to illustrate how a recently-developed FBTS [31] can be used to easily compute quantum correlation functions. For this purpose, it is interesting to contrast the solution using this scheme, and the trajectory description that underlies it, with the previously-developed and frequently-used TBSH algorithm [26]. Taking the adiabatic representation of the QCL superoperator is the key step in implementing the TBSH algorithm. The last representation of QCLE in Equation (8) resembles the quantum Liouville equation and forms the starting point of the FBTS.

Adiabatic Trotter-Based Surface Hopping

In order to discuss the nature of the trajectory description involved in the TBSH algorithm, we briefly describe how it is implemented and, in particular, present the explicit generalization to an N-level quantum subsystem, which was only outlined in [26]. We first consider the adiabatic representation of the QCLE, since the TBSH algorithm is cast in this basis. The adiabatic basis is defined by $\hat{h}_W(R) \mid \alpha; R\rangle = E_a(R) \mid \alpha; R\rangle$, where $\hat{h}_W(R) = \hat{H}_W(R) - P^2/2M$ is taken to be the adiabatic Hamiltonian for a static configuration of R in this section. In the adiabatic basis, the QCLE reads:

$$\frac{\partial B_W^{\alpha\alpha'}}{\partial t} = i\mathcal{L}_{\alpha\alpha',\beta\beta'} B_W^{\beta\beta'}(X,t)$$

(10)

where the matrix elements of the QCL superoperator are given by:

$$i\mathcal{L}_{\alpha\alpha',\beta\beta'} = (i\omega_{\alpha\alpha'} + iL_{\alpha\alpha'})\delta_{\alpha\beta}\delta_{\alpha'\beta'} - \mathcal{J}_{\alpha\alpha',\beta\beta'} = i\mathcal{L}^0_{\alpha\alpha'}\delta_{\alpha\beta}\delta_{\alpha'\beta'} - \mathcal{J}_{\alpha\alpha',\beta\beta'}$$

(11)

with $\omega_{\alpha\alpha'} = (E_\alpha - E_{\alpha'})/\hbar$. (The Einstein summation convention will be used throughout the following sections, although sometimes, sums will be explicitly written if there is the possibility of confusion.) The Liouville operator, iL, may be separated into two contributions: The classical propagator is defined as:

$$iL_{\alpha\alpha'} = \frac{P}{M} \cdot \frac{\partial}{\partial P} + \frac{1}{2}(F_\alpha + F_{\alpha'}) \cdot \frac{\partial}{\partial R}$$

(12)

Where $F^\alpha = \langle \alpha; R \mid \frac{\partial \hat{h}_W(R)}{\partial R} \mid \alpha; R\rangle$ is the Hellmann-Feynman force. The superoperator, $J_{\alpha\alpha',\beta\beta'}$ is responsible for nonadiabatic transitions and associated momentum changes in the bath. For an N-level system, there exist $N(N-1)/2$ unique transitions. In the following, we define J as a sum of $J_{\lambda\lambda'}$, which introduces transitions only between the specific pair of λ and λ' adiabatic states:

$$
\begin{aligned}
\mathcal{J}_{\alpha\alpha',\beta\beta'} &= \sum_{\lambda>\lambda'} (\mathcal{J}_{\lambda\lambda'})_{\alpha\alpha',\beta\beta'} \\
&= \sum_{\lambda>\lambda'} \left\{ -d_{\lambda\lambda'} \cdot \frac{P}{M} ((\delta_{\lambda\alpha}\delta_{\lambda'\beta} - \delta_{\lambda'\alpha}\delta_{\lambda\beta})\delta_{\alpha'\beta'} + ((\delta_{\lambda\alpha'}\delta_{\lambda'\beta'} - \delta_{\lambda'\alpha'}\delta_{\lambda\beta'})\delta_{\alpha\beta}) \right. \\
&\qquad \left. -\frac{1}{2}\hbar\omega_{\lambda\lambda'}d_{\lambda\lambda'} \cdot \frac{\partial}{\partial P}((\delta_{\lambda\alpha}\delta_{\lambda'\beta} + \delta_{\lambda'\alpha}\delta_{\lambda\beta})\delta_{\alpha'\beta'} + (\delta_{\lambda\alpha'}\delta_{\lambda'\beta'} + \delta_{\lambda'\alpha'}\delta_{\lambda\beta'})\delta_{\alpha\beta}) \right\} \\
&= -\frac{P}{M} \cdot d_{\alpha\beta} \left(1 + \frac{1}{2}S_{\alpha\beta} \cdot \frac{\partial}{\partial P}\right)\delta_{\alpha'\beta'} + \frac{P}{M} \cdot d_{\beta'\alpha'}\left(1 - \frac{1}{2}S_{\beta'\alpha'} \cdot \frac{\partial}{\partial P}\right)\delta_{\alpha\beta}
\end{aligned}
\tag{13}
$$

Where $d_{\alpha\beta} = \langle \alpha; R| \partial/\partial R |\beta; R\rangle$ and $S_{\alpha\beta} = \hbar\omega_{\alpha\beta}d_{\alpha\beta}\left(\frac{P}{M} \cdot d_{\alpha\beta}\right)^{-1}$. The second equality gives the adiabatic representation of $J_{\lambda\lambda'}$. We remark that it is difficult to exactly simulate the term, J, involving bath momentum derivatives within the context of a trajectory-based algorithm. Using the identity that $\frac{1}{2}S_{\alpha\beta} \cdot \frac{\partial}{\partial P} = \hbar\omega_{\alpha\beta}M \cdot \partial/\partial(\hat{d}_{\alpha\beta} \cdot P)^2$, where M is a diagonal matrix of the masses of the bath particles and $\hat{d}_{\alpha\beta}$ is the unit vector along $d_{\alpha\beta}$, allows us to employ the momentum-jump approximation:

$$
\left(1 + \frac{c}{2}S_{\alpha\beta} \cdot \frac{\partial}{\partial P}\right)f(P) \approx e^{\frac{c}{2}S_{\alpha\beta} \cdot \frac{\partial}{\partial P}}f(P) = e^{c\hbar\omega_{\alpha\beta}M \cdot \partial/\partial(\hat{d}_{\alpha\beta} \cdot P)^2}f(P) = f(P + \Delta P_c)
\tag{14}
$$

where c = 1, 2 corresponding to single and double hops, respectively, and $\Delta P_c = \hat{d}_{\alpha\beta}\text{sgn}\left(\hat{d} \cdot P\right)\sqrt{(\hat{d}_{\alpha\beta} \cdot P)^2 + c\hbar\omega_{\alpha\beta}M} - \hat{d}\left(\hat{d} \cdot P\right)$. We have a translation operator with respect to the variable, $(\hat{d}_{\alpha\beta} \cdot P)^2$, in the above equation. Decomposing $P = P_\perp + P_\parallel = P_\perp + \hat{d}_{\alpha\beta}\text{sgn}(\hat{d}_{\alpha\beta} \cdot P)\sqrt{(\hat{d}_{\alpha\beta} \cdot P)^2}$, it becomes obvious that the translation operator updates $P\hat{d}$ components by ΔP_c, as presented in Equation (14). This momentum update conserves the energy of surface-hopping trajectories. Apart from technical issues associated with sampling when the algorithm is implemented, this is the only approximation made to QCL evolution. In fact, it is this approximation that gives this algorithm a surface-hopping structure that has some features in common with Tully's surface-hopping method; however, coherence and decoherence are automatically incorporated in the evolution. The QCLE does not have such sudden momentum changes, and its evolution is described by continuous momentum changes in the course of the evolution. Comparisons of results using this algorithm with exact quantum solutions indicate that the momentum-jump is rarely the source of problems.

Equation (10) admits a formal solution:

$$\hat{B}_W^{\alpha\alpha'}(X,t) = \left(e^{i\mathcal{L}t}\right)_{\alpha\alpha',\beta\beta'} B_W^{\beta\beta'}(X) \tag{15}$$

Thus, our following discussion focuses on evaluating:

$$\left(e^{i\mathcal{L}t}\right)_{\alpha\alpha',\alpha_K\alpha'_K} = \sum_{(\alpha_1\alpha'_1)\dots(\alpha_K\alpha'_K)} \prod_{j=1}^{K} \left(e^{i\mathcal{L}\Delta t_j}\right)_{\alpha_{j-1}\alpha'_{j-1},\alpha_j\alpha'_j} \tag{16}$$

In the above equation, we simply factorize the propagator into K pieces with $\Delta t_j = t_j - t_{j-1} = \Delta t$. In each small time slice, we perform the symmetric Trotter decomposition:

$$\left(e^{i\mathcal{L}\nabla t}\right)^{\alpha\alpha',\beta\beta'} \approx \mathcal{M}^{\beta\beta'}\left(e^{\jmath-1}\cdot e^{\jmath-1} + \frac{\jmath}{\nabla t}\right)e^{\Gamma\beta\beta',\nabla t\backslash\jmath} \mathcal{Q}^{\beta\beta','\sigma\sigma'}\mathcal{M}^{\sigma\sigma'}\left(e^{\jmath-1} + \frac{\jmath}{\nabla t}\cdot e^{\jmath}\right)e^{\Gamma^{\sigma\sigma'},\nabla t\backslash\jmath} \tag{17}$$

where: $W_{\alpha\alpha'}(t_1, t_2) = e^{i\omega_{\alpha\alpha'}(t_2-t_1)}$, and:

$$\mathcal{Q}_{\alpha\alpha',\beta\beta'} = \left(e^{\mathcal{J}\Delta t}\right)_{\alpha\alpha',\beta\beta'} = \left(e^{\sum_{\lambda>\lambda'}\mathcal{J}_{\lambda\lambda'}\Delta t}\right)_{\alpha\alpha',\beta\beta'} \approx \left(\prod_{\lambda>\lambda'}e^{\mathcal{J}_{\lambda\lambda'}\Delta t}\right)_{\alpha\alpha',\beta\beta'} \tag{18}$$

We observe that it is possible to express $e^{\mathcal{J}_{\lambda\lambda'}}\Delta t$ in the following block-diagonal matrix form:

$$e^{\mathcal{J}_{\lambda\lambda'}\Delta t} = \mathcal{M}^{\lambda\lambda'} \oplus \mathcal{K}^{\lambda\lambda'}_{\xi_1} \cdots \oplus \mathcal{K}^{\lambda\lambda'}_{\xi_{N-2}} \oplus \mathcal{N}^{\lambda\lambda'} \tag{19}$$

where ξ_i is one of the $N-2$ adiabatic states other than λ and λ'. In the above equation, M is a four by four matrix, defined with respect to the basis, $\{(\lambda, \lambda), (\lambda, \lambda'), (\lambda', \lambda), (\lambda',\lambda')\}$:

$$\mathcal{M}^{\lambda\lambda'} = \begin{pmatrix} \cos^2(a) & -\cos(a)\sin(a)\hat{\jmath}_{\lambda\lambda'} & -\cos(a)\sin(a)\hat{\jmath}_{\lambda\lambda'} & \sin^2(a)\hat{\jmath}_{\lambda\to\lambda'} \\ \cos(a)\sin(a)\hat{\jmath}_{\lambda\lambda'} & \cos^2(a) & -\sin^2(a) & -\sin(a)\cos(a)\hat{\jmath}_{\lambda\lambda'} \\ \cos(a)\sin(a)\hat{\jmath}_{\lambda\lambda'} & -\sin^2(a) & \cos^2(a) & -\sin(a)\cos(a)\hat{\jmath}_{\lambda\lambda'} \\ \sin^2(a)\hat{\jmath}_{\lambda\to\lambda'} & \cos(a)\sin(a)\hat{\jmath}_{\lambda\lambda'} & \cos(a)\sin(a)\hat{\jmath}_{\lambda\lambda'} & \cos^2(a) \end{pmatrix} \tag{20}$$

with $a = (P/M) \cdot d_{\lambda\lambda'} \Delta t$, and $\hat{\jmath}_{\lambda\lambda'}$ and $\hat{\jmath}_{\lambda\to\lambda'}$ are the momentum-jump operators, $e^{\frac{1}{2}S_{\lambda\lambda'}\frac{\partial}{\partial P}}$ and $e^{S_{\lambda\lambda'}\frac{\partial}{\partial P}}$, defined in Equation (14) with c = 1, 2, respectively.

In Equation (19), there exists another set of four by four matrices, $\mathcal{K}^{\lambda\lambda'}_{\xi_i}$, with

$i = 1, ..., N - 2$. Each of these matrices is defined with respect to a basis of the form, $\{(\lambda, \xi_i), (\lambda', \xi_i)\} \oplus \{(\xi_i, \lambda), (\xi_i, \lambda')\}$:

$$
\mathcal{K}_\xi^{\lambda\lambda'} = \begin{pmatrix} \cos(a) & -\sin(a)\hat{j}_{\lambda\lambda'} \\ \sin(a)\hat{j}_{\lambda\lambda'} & \cos(a) \end{pmatrix} \oplus \begin{pmatrix} \cos(a) & -\sin(a)\hat{j}_{\lambda\lambda'} \\ \sin(a)\hat{j}_{\lambda\lambda'} & \cos(a) \end{pmatrix}
$$

$$(21)$$

Finally, there is a null matrix, $N^{\lambda\lambda'}$, of a size of $(N - 2)^2$, and the associated null space is spanned by basis vectors, (ξ_1, ξ_2), where $\xi_i \neq \lambda^{()}$. We remark that one has to permute the basis vectors in order to construct these block-diagonal matrices [26].

At this point, we have specified all the necessary details in order to simulate the QCL dynamics in the adiabatic basis:

$$
B_W^{\alpha\alpha'}(X,t) = \sum_{\substack{(\alpha_1\alpha_1'),...,\\(\alpha_K\alpha_K')}} \left[\prod_{j=1}^{K} \mathcal{W}_{\alpha_{j-1}\alpha_{j-1}'} e^{iL_{\alpha_{j-1}\alpha_{j-1}'}\Delta t} \mathcal{Q}_{\alpha_{j-1}\alpha_{j-1}',\alpha_j\alpha_j'} \mathcal{W}_{\alpha_j\alpha_j'} e^{iL_{\alpha_j\alpha_j'}\Delta t} \right] B_W^{\alpha_K\alpha_K'}(X)
$$

$$(22)$$

Where $\alpha_0^{(')} = \alpha^{(')}$, The explicit summation over all quantum indices, $(\alpha_1\alpha_1')\ldots(\alpha_K\alpha_K')$, can also be evaluated stochastically. For instance, given a pair of indices, $(\alpha_j\alpha_{j-1})$, one can determine the next pair at the time slice, $j + 1$, by drawing an MC sample from the transition probability:

$$
P(\alpha_{j+1}, \alpha_{j+1}'|\alpha_j, \alpha_j') = \frac{|\mathcal{Q}_{\alpha_j\alpha_j',\alpha_{j+1}\alpha_{j+1}'}|}{\sum_{\beta_{j+1},\beta_{j+1}'} |\mathcal{Q}_{\alpha_j\alpha_j',\beta_{j+1}\beta_{j+1}'}|} \quad (23)
$$

If the sampled new pair of indices differs from the starting pair, then the sampled Q matrix element must contain the proper momentum-jump operators to update the energy of the trajectory after the jump. In any actual implementation of this algorithm, it is desirable to restrict to nonadiabatic transitions between one pair of states in every time slice. Under this assumption, one can then approximate:

$$
\mathcal{Q}_{\alpha\alpha',\beta\beta'} \approx \begin{cases} \delta_{\alpha\beta}\delta_{\alpha'\beta'} & \text{if no hop happens,} \\ \left(e^{\mathcal{J}_{\mu\gamma}}\right)_{\alpha\alpha',\beta\beta'} & \text{if } (\alpha,\alpha') \to (\beta,\beta') \text{ involves transition between } (\mu,\gamma) \text{ states,} \\ 0 & \text{if } (\alpha,\alpha') \to (\beta,\beta') \text{ involves transitions between two or more pairs of states.} \end{cases}
$$

$$(24)$$

In this algorithm, we see that the trajectories in the ensemble that are used to simulate the time evolution are non-Newtonian in character, consisting of

Newtonian segments where the system evolves on adiabatic surfaces, or the mean of two adiabatic surfaces, interspersed with quantum transitions and momentum changes.

Forward-Backward Trajectory Solution

This scheme is motivated by another way of writing the formally exact solution [38] of the QCLE using the last line of Equation (8):

$$\hat{B}_W(X,t) = \mathcal{S}\left(e^{i\overrightarrow{\mathcal{H}}_\Lambda t/\hbar}\,\hat{B}_W(X)e^{-i\overleftarrow{\mathcal{H}}_\Lambda t/\hbar}\right) \tag{25}$$

The S operator [31,38] specifies the order in which the forward and backward evolution operators act on B_W (X). The ordering of evolution operators is critical because of the lack of an underlying Lie algebraic structure [38] of the QCLE.

One approach to solve Equation (25) is to apply the mapping transformation in which N discrete quantum states of the subsystem are represented by the continuous position and momenta of Nfictitious harmonic oscillators. The properties of the original subsystem are then obtained via an ensemble average involving trajectories in the phase space of the fictitious oscillators. More precisely, in the mapping representation, a subsystem state, $|\lambda\rangle$, is replaced by $|m_\lambda\rangle = |0_1, \ldots, 1_\lambda, \ldots 0_N\rangle$, a product state specifying the occupation numbers (limited to zero or one) of N fictitious harmonic oscillators [33,49].

Creation and annihilation operators, $\hat{a}_\lambda^\dagger$ and $\hat{a}_\lambda$ satisfy the commutation relation $[\hat{a}_\lambda, \hat{a}_{\lambda'}^\dagger] = \delta_{\lambda,\lambda'}$ for harmonic oscillators. The actions of these operators on the single-excitation mapping states are $\hat{a}_\lambda^\dagger |0\rangle = |m_\lambda\rangle$ and $\hat{a}_\lambda |m_\lambda\rangle = |0\rangle$, where $|0\rangle = ||0_1 \ldots 0_N\rangle$ is the ground state of the mapping basis.

Next, we define the mapping version of operators, $\hat{B}_m(X) = B_W^{\lambda\lambda'}(X)\hat{a}_\lambda^\dagger\hat{a}_{\lambda'}$, such that matrix elements of $\hat{B}_w$ in the subsystem basis are equal to the matrix elements of the corresponding mapping operator: $B_W^{\lambda\lambda'}(X) = \langle\lambda|\hat{B}_W(X)|\lambda'\rangle = \langle m_\lambda|\hat{B}_m(X)|m_{\lambda'}\rangle$. In particular, the mapping Hamiltonian is:

$$\hat{H}_m = H_b(X) + h^{\lambda\lambda'}(R)\hat{a}_\lambda^\dagger\hat{a}_{\lambda'} \equiv H_b(X) + \hat{h}_m \tag{26}$$

where we applied the mapping transformation only on the part of the Hamiltonian that involves the subsystem DOF in Equation (26). The mapping Hamiltonian, $\hat{h}_m$, is always a quadratic Hamiltonian with respect to the quantum DOF. The pure bath term, $\hat{H}_b(X)$, acts as an identity operator in the subsystem basis and is mapped onto the identity operator of the mapping space directly. The mapped formal solution of QCLE now reads:

$$\hat{B}_m(X, t) = \mathcal{S}\left(e^{i\overrightarrow{\mathcal{H}}_\Lambda^m t/\hbar} \hat{B}_m(X) e^{-i\overleftarrow{\mathcal{H}}_\Lambda^m t/\hbar} \right)$$

(27)

Where $\overrightarrow{\mathcal{H}}_\Lambda^m$ is given by $\overrightarrow{\mathcal{H}}_\Lambda^m = \hat{H}_m(1 + \hbar\Lambda/2i)$, with an analogous definition for $\overleftarrow{\mathcal{H}}_\Lambda^m$.

We now introduce the coherent states, $|z\rangle$, in the mapping space, $\hat{a}_\lambda|z\rangle = z_\lambda|z\rangle$ and $\langle z|\hat{a}_\lambda^\dagger = z_\lambda^*\langle z|$, where $|z\rangle = |z_1, \ldots, z_N\rangle$, and the eigenvalue is $z_\lambda = (q_\lambda + ip_\lambda)/\sqrt{\hbar}$. The variables $q = (q_1, \ldots, q_N)$ and $p = (p_1, \ldots, p_N)$ are mean coordinates and momenta of the harmonic oscillators encoded in the coherent state, $|z\rangle$, respectively. The coherent states form an overcomplete basis with the inner product between any two such states, $\langle z | z' \rangle = e^{-(|z-z'|^2)-i(z \cdot z'^* - z^* \cdot z')}$. Finally, we remark that the coherent states provide the resolution of identity:

$$\mathcal{I} = \int \frac{d^2 z}{\pi^N} |z\rangle \langle z|$$

(28)

where $d^2 z = d(\Re(z))d(\Im(z)) = dqdp/(2\hbar)^N$.

Similar to the path integral approach for solving the quantum dynamics, we decompose the forward and backward evolution operators in Equation (27) into a concatenation of M short-time evolutions with $\Delta t_i = \tau$ and $M\tau = t$. In each short-time interval, Δt_i, we introduce two sets of coherent states, $|z_i\rangle$ and $|z_i'\rangle$, via Equation (28) to expand the forward and backward time evolution operators, respectively. The time evolution (generated by a quadratic Hamiltonian) of coherent states can be represented by trajectory evolution in the phase space of (q, p). After some algebra, the matrix elements of Equation (27) can be approximated by:

$$B_W^{\lambda\lambda'}(X,t) = \sum_{\mu\mu'} \int dx dx' \phi(x)\phi(x')\frac{1}{\hbar}(q_\lambda + ip_\lambda)(q'_{\lambda'} - ip'_{\lambda'})B_W^{\mu\mu'}(X_t)$$

$$\times \frac{1}{\hbar}(q_\mu(t) - ip_\mu(t))(q'_{\mu'}(t) + ip'_{\mu'}(t)) \tag{29}$$

where x = (q, p) gives the real and imaginary parts of z, dx = dqdp and $\phi(x) = (\hbar)^{-N} e^{-\sum_\nu(q_\nu^2 + p_\nu^2)/\hbar}$ is the normalized Gaussian distribution function. In deriving Equation (29), we have invoked an orthogonality approximation on the inner product between subsequent coherent state variables, $\langle z_i | e^{\frac{i}{\hbar}\hat{h}t} | z_{i+1} \rangle = \langle z_i(t) | z_{i+1} \rangle \approx \pi^N \delta(z_{i+1} - z_i(t_i))$, with i being the time step index. This approximation is necessary to construct a continuous trajectory of z(t). In the extended phase space of (X(t), z(t), z'(t)), the trajectories follow Hamiltonian dynamics:

$$\frac{d\chi_\mu}{dt} = \frac{\partial H_e(\chi, \pi)}{\partial \pi_\mu}, \qquad \frac{d\pi_\mu}{dt} = -\frac{\partial H_e(\chi, \pi)}{\partial \chi_\mu} \tag{30}$$

Where

$$H_e(\chi, \pi) = P^2/2M + V_0(R) + \frac{1}{2\hbar}h^{\lambda\lambda'}(R)(q_\lambda q_{\lambda'} + p_\lambda p_{\lambda'} + q'_\lambda q'_{\lambda'} + p'_\lambda p'_{\lambda'})$$

with $V_0(R) = V_b(R) - T$ rĥ(R), $\chi = (R, q, q')$ and $\pi = (P, p, p')$. We remark that the FBTS trajectories manifestly conserve energy. Furthermore, simulating the dynamics with a standard velocity Verlet type of symplectic integrator has a stationary solution proportional to $H_{pseudo} = H_e(\chi, \pi) + \Delta t^2 \delta H$, as discussed in [35].

The main approximation introduced in the derivation of the FBTS, Equation (29), is the orthogonality approximation. The simplest improvement to the algorithm is to refrain from applying this approximation at every time step. In [36], we outlined a practical approach to evaluate the set of selected integrals of z_i and z'_i (which could be evaluated analytically if the orthogonality approximation were applied). We termed this extension of FBTS as the jump FBTS (JFBTS). Since the computational cost grows quickly with respect to the number of jumps inserted, one needs to make a trade-off between numerical efficiency and accuracy.

In the simplest approach, one selects every (M/K) time step from a total of M steps to fully evaluate the coherent state integrals:

$$B_W^{\lambda\lambda'}(X,t) = \sum_{\mu\mu'} \sum_{\substack{s_0 s_0' \dots \\ s_{K-1} s_{K-1}'}} \int \prod_{v=0}^{K} dx dx' \phi(x_v)\phi(x_v')$$

$$\times \frac{1}{\hbar}(q_{0\lambda} + ip_{0\lambda})(q_{0\lambda'}' - ip_{0\lambda'}')B_W^{\mu\mu'}(X_t)$$

$$\times \frac{1}{\hbar}\left\{ \prod_{v=1}^{K} \left(q_{(v-1)s_{v-1}}(\tau_v) - ip_{(v-1)s_{v-1}}(\tau_v) \right) \left(q_{vs_v} + ip_{vs_v} \right) \right\}$$

$$\times \frac{1}{\hbar}\left\{ \prod_{v=1}^{K} \left(q_{(v-1)s_{v-1}}'(\tau_v) + ip_{(v-1)s_{v-1}}'(\tau_v) \right) \left(q_{vs_v}' - ip_{vs_v}' \right) \right\}$$

$$\times \frac{1}{\hbar}(q_{K\mu}(\tau_{K+1}) - ip_{K\mu}(\tau_{K+1}))(q_{K\mu'}'(\tau_{K+1}) + ip_{K\mu'}'(\tau_{K+1}))$$

$$\tag{31}$$

where the subscripts, v and s, refer to the v-th time step and the s-th component of the q and pvectors, respectively, and $\tau_v = t_{i_v} - t_{i_{v-1}}$ with $t_{i_0} = 0$ and $t_{i_{K+1}} = t$. According to this prescription, the continuous FB trajectories experience K discontinuous jumps in the (x, x') phase space. Between subsequent jumps, the evolution of the FB trajectory is governed by Equation (30). Simulations show that with a sufficient number of jumps, numerically exact solutions of the QCLE can be obtained [36].

Comparisons between Algorithms

The differences between the two QCLE simulation algorithms can be traced to the quantum basis that is used and the way that feedback between quantum and classical systems is treated. In the case of the TBSH algorithm, the trajectories are propagated through a Hellmann-Feynman force, or the mean of two Hellmann-Feynman forces [Equation (12)], with intermittent surface hops that switch the adiabatic surfaces on which the trajectories propagate. In the case of FBTS, one not only propagates the bath dynamical variables as trajectories, but also the quantum dynamical variables, which are associated with fictitious harmonic oscillators. In this extended phase space, we have exact Hamiltonian dynamics. In particular, the force acting on the bath particles simultaneously involves all N adiabatic surfaces, which is similar to, but different from, the Ehrenfest mean-field approach. The very different characteristics of the trajectories in two algorithms manifest the artificial character of the trajectory dynamics. Thus, one should not attach physical significance to single trajectories in the computation. All physical properties of the system can only be extracted from a proper ensemble average of a large set of trajectories, as implied in Equation (2). Nevertheless, insight into the trajectory dynamics of each algorithm will help to judge the simulation efficiency for various classes of models.

For certain problems, such as proton transfer reactions, where the time scales of the bath and subsystem are well-separated, even during nonadiabatic transitions, the TBSH algorithm can yield quantitatively accurate results with a few hops. There are also dynamical problems in which distinct bath motions can be explicitly correlated with the subsystem's quantum states. For instance, in the simple Tully I model [35,50], trajectories populated on the excited state will cross the avoided crossing point, while the ground state trajectories will eventually be reflected and retrace their paths in the opposite direction. This kind of behavior is, however, completely missed when one propagates trajectories in a single effective mean field. Again, the inherent multi-configuration nature of surface-hopping-like algorithms is a more appropriate choice for this case. However, a recent study [51] has indicated that the "jump" version of mean-field-like algorithms can improve the simulation results in cases of this type.

Alternatively, there are also many examples where one would expect FBTS to be the preferred simulation method. In general, the TBSH algorithm has con-vergence issues, as the MC weights associated with nonadiabatic hops grows rapidly. Even for the simple spin boson model, one can identify parameter regimes where this numerical instability is clearly observed. In these cases, the FBTS and JFBTS are certainly the alternatives that one should adopt for efficient simulations.

AN EXAMPLE: QUARTIC OSCILLATOR IN A HARMONIC BATH

As a specific example to illustrate the formalism outlined above, we consider a two-level system coupled to a quartic bistable oscillator with a single pair of phase space coordinates $X_0 = (R_0, P_0)$. The quartic oscillator is, in turn, coupled to an Ohmic heat bath of N_b independent harmonic oscillators with phase space coordinates $X_i = (R_i, P_i)$ and $i = 1 \ldots N_b$. The partially Wigner transformed Hamiltonian, expressed in the diabatic basis, $\{|R\rangle, |L\rangle\}$, reads:

$$\hat{H}_W = \begin{pmatrix} \hbar\gamma_0 R_0 & -\hbar\Omega \\ -\hbar\Omega & -\hbar\gamma_0 R_0 \end{pmatrix} + \left(\frac{P_0^2}{2M_0} + V_n(R_0) + \sum_{j=1}^{N_b} \frac{P_j^2}{2M_j} + \frac{M_j\omega_j^2}{2}\left(R_j - \frac{\gamma_b c_j}{M_j\omega_j^2}R_0\right)^2 \right) \mathbf{I}$$

(32)

Where $V_n(R_0) = -M_0\omega_0^2 R_0^2/2 + AR_0^4/4$ and $\mathbf{I}$ is an identity matrix. We take $N_b = 40$ harmonic oscillators for the discretization of the Ohmic heat bath. Following the discretization scheme introduced in [52], we set $\omega_j = \omega_c \ln(1-j\omega_c/\delta\omega)$ and $c_j = (\xi\hbar\omega M_j)^{1/2}\omega_j$ with $\delta\omega = (1-\exp(\omega_{max}/\omega_c))/N_b$. The parameters, ω_c and ω_{max}, are the characteristic and cut-off frequencies for the Ohmic bath, respectively. The Kondo parameter is ξ.

The adiabatic states for the subsystem are:

$$|+; R_0\rangle = \frac{1}{\mathcal{N}(R_0)}[(1-G)|R\rangle - (1+G)|L\rangle]$$

$$|-; R_0\rangle = \frac{1}{\mathcal{N}(R_0)}[(1+G)|R\rangle + (1-G)|L\rangle] \tag{33}$$

where $\quad \mathcal{N}(R_0) = \sqrt{2(1+G^2(R_0))} \quad$ and

$G(R_0) = (\gamma_0 R_0)^{-1}\left[-\Omega + \sqrt{\Omega^2 + \gamma_0^2 R_0^2}\right]$. The adiabatic energies are given by

$$E_\pm = V_n(R_0) \pm \hbar\sqrt{\Omega^2 + \gamma_0^2 R_0^2} = V_n(R_0) \pm \epsilon_\pm(R_0).$$

We shall study the autocorrelation functions, C_{LL}, with $\hat{A} = B = |L\rangle\langle L|$. The entire system is assumed to be in thermal equilibrium initially. Using the high-temperature approximation presented in the Appendix, the correlation function of interest can be given in a compact form:

$$C_{LL}(t) = \int dX_0 dX_b W(R_0)\mathcal{G}\left(P_0; \frac{M_0}{\beta}\right)\prod_{j=1}^{N_b}\mathcal{G}\left(P_j; \frac{M_j}{\beta}\right)\mathcal{G}\left(R_j - \frac{\gamma_b c_j}{M_j\omega_j^2}R_0; \frac{1}{\beta M_j\omega_j^2}\right)$$

$$\times \sum_{n=L,R}\sum_{\alpha,\alpha'} F_{\alpha\alpha'}(X_0)\langle n|\alpha; R_0\rangle\langle \alpha'; R_0|L\rangle B_W^{Ln}(X_t) \tag{34}$$

Where $\mathcal{G}(x; \sigma^2) = (2\pi\sigma^2)^{-1/2}e^{-x^2/2\sigma^2}$ and:

$$W(R_0) = \frac{e^{-\beta\left(\frac{A}{4}R_0^4 - \frac{1}{2}M_0\omega_0^2 R_0^2\right)}\left(e^{-\beta\epsilon_+(R_0)} + e^{\beta\epsilon_-(R_0)}\right)}{\int dR_0 e^{-\beta\left(\frac{A}{4}R_0^4 - \frac{1}{2}M_0\omega_0^2 R_0^2\right)}\left(e^{-\beta\epsilon_+(R_0)} + e^{\beta\epsilon_-(R_0)}\right)}$$

(35)

An MC evaluation of the integrals can be done by sampling P_0, R_b, P_b from the Gaussian distributions and sampling R_0 from $W(R_0)$, respectively. The time-evolved matrix element, $B_W^{nm}(X_t)$, will be computed using both the TBSH and the FBTS algorithms. Finally, we note that the path-integral-based sampling scheme introduced in Section 2 should be adopted to sample phase-space points from $(\rho_{eq})_W(X)$ for more generalized situations, including cases of low-temperature, arbitrary subsystem-bath divisions of a composite system, strong subsystem-bath couplings and an arbitrary potential energy profile.

In this study, we report numerical results in the energy unit, $\hbar\omega_c$, and distance unit, $\sqrt{\hbar/M_j\omega_c}$, for each environmental DOF. We consider two sets of parameters. In the first case, we use the following parameter values, a = 1.0, ω_0 = 1.2, γ_0 = 0.05 γ_b = 1.0, Ω = 0.3, ξ = 0.1, ω_{max} = 3 and β = 0.2, in the dimensionless units. Figure 1a presents the potential surface profiles [53], $W_\alpha(R_0)$. The two diabatic surfaces, $W_{L,R}(R_0)$, remain close to each other, and the two adiabatic surfaces, $W_\pm(R_0)$, share essentially the same characteristics. In this case, a mean-field-based algorithm, like FBTS, should be accurate and efficient. This problem can also be handled easily in the adiabatic basis, since the surface-hopping trajectories will be initialized in both the adiabatic ground and excited states, because the system is in a thermal equilibrium state at t = 0. Furthermore, the coupling parameter, γ_b, was purposely chosen to be small in order to minimize the number of nonadiabatic transitions (or hops) encountered in the TBSH algorithm. In panel (b), $C_{LL}(t)$ is computed using both algorithms. The agreement between these results is good.

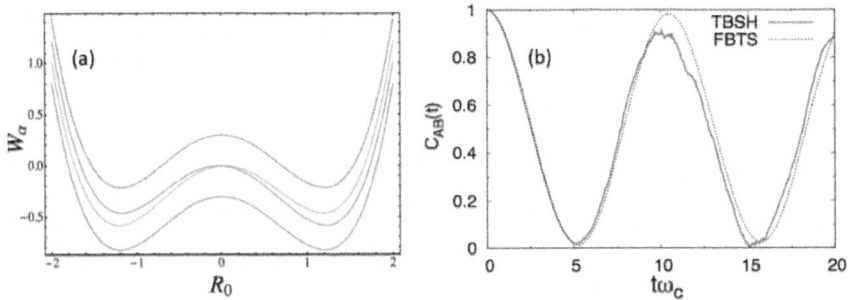

Figure 1: (a) Potential surface profiles, $W_\alpha(R_0)$, for the ground adiabatic state (black, dotted), excited adiabatic state (black, dotted) and for the diabatic states, L (green) and R (red). (b) $C_{LL}(t)$ correlation function. These results are associated with the first set of parameters.

Next, we consider the following parameter set, a = 0.8, ω_0 = 0.6, γ_0 = 0.3 γ_b = 1.0, Ω = 0.1, ξ = 0.1, ω_{max} = 3, and β = 0.2 in the dimensionless units. Figure 2a shows the potential surface profiles, $W_\alpha(R_0)$, obtained from this set of parameters. In this case, the adiabatic, $W_\pm(R_0)$, and diabatic surfaces, $W_{L,R}(R_0)$, only differ markedly near the region of the barrier top, where an avoided crossing point indicates significant mixing of the two diabatic states. Nonadiabatic effects should be most prominent near this barrier top. A stronger coupling, γ_0, is also chosen in this case. Figure 2bpresents the autocorrelation functions. In the main figure of panel (b), the blue curves ($C_{LL}(t)$ computed by the FBTS) start with the full correlation at one, then gradually reduce to 1/2, which implies that the subsystem is in an equal admixture of the two diabatic states in the asymptotic limit. The TBSH simulation results are only valid for very short times (as shown in the inset of the Figure 2b), due to instability arising from the accumulation of weights, even with filtering [54]. The thermal equilibrium distribution, $W(R_0)$, has a bimodal distribution profile, as illustrated in Figure 2a; however, for the (inverse) temperature, $\beta = 0.2$, the double-peaked structure is very broad. The $W(R_0)$ distribution profile (blue curve in Figure 2a) suggests that the thermal equilibrium state has a non-trivial contribution from the excited surface. Sampling from $W(R_0)$ yields many R_0 values near the barrier top, where several hops immediately take place for this strong-coupling case, and the instability sets in early in the simulation. Lowering β will produce a more pronounced double-peak structure for $W(R_0)$, but the quartic oscillator's momentum, P_0, will fluctuate with a larger variance in the presence of the heat bath in this case. Since nonadiabatic transitions depend non-trivially on a= $P_0 \cdot d_{12}(R_0)\Delta t$ in the TBSH algorithm, large momentum fluctuations will eventually affect the long-time result. This case shows some of the practical limitations of the TBSH algorithm for the computation of this correlation function.

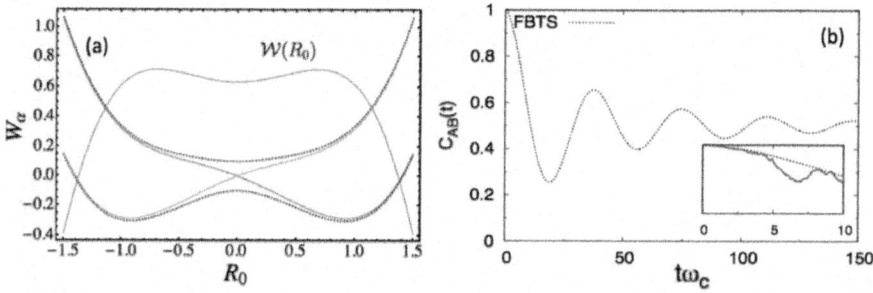

Figure 2: (a) Potential surface profiles, $W_\alpha(R_0)$, for the ground adiabatic state (black, dotted), excited adiabatic state (black, dotted) and for the diabatic states, L (green) and R (red). The blue curve is a plot of the un-normalized distribution function, $W(R_0)$, Equation (35). (b) $C_{LL}(t)$ correlation functions. (Inset) Short-time $C_{LL}(t)$ computed by the FBTS (blue) and TBSH (red) algorithms. These results are associated with the second set of parameters.

CONCLUSIONS

The scheme for computing the quantum correlation function in Equation (2) combines a numerically exact quantum initial sampling method with dynamics described by the QCLE; thus, the approximations in the simulation method reside in the dynamics. It is easier to compute the equilibrium properties of a quantum system, for instance, by using the imaginary-time Feynman path integral method, than to obtain dynamical properties by using similar real-time Feynman path integrals without adopting further approximations. Since we approximate the quantum dynamics of the entire system, quantum subsystem plus bath, by QCL dynamics, it is appropriate to comment on some of its features.

It is known that the quantum-classical bracket, defined in terms of the commutator and Poisson brackets in Equation (8), does not possess a Lie algebraic structure, since it fails to satisfy the Jacobi identity [2,38]. This lack of a proper algebraic structure is shared by all known MQC methods and simply reflects the inconsistency in mixing classical and quantum mechanical dynamics. One consequence of this inconsistency is that the partial Wigner transform, $\hat{\rho}W_e(R, P)$, of the full canonical equilibrium density function, $\hat{\rho}_{eq} = e^{-\beta\hat{H}}/Z_Q$, is not stationary under the QCLE; however, $\rho W_e(R, P)$ can be written as an expansion in the mass ratio (or $\hbar$), and it has been shown that the full quantum equilibrium density is conserved under the QCL dynamics up to $O(\hbar)$. Therefore, the detailed balance relation is also satisfied to this order. The violation of a detailed balance is a common problem that affects all major MQC methods, including the two most popular approaches, Ehrenfest mean-field [55] and Tully's fewest switching surface hopping [56] (FSSH), to various degrees. Of course, as noted earlier, for the class of models where an arbitrary quantum system is bilinearly coupled to a harmonic bath, the dynamics is exact, and a detailed balance is exactly satisfied.

The dynamics described by the QCLE can be related to that prescribed by other methods. In [57], it was shown that one could derive both Ehrenfest mean-field dynamics and a version of surface-hopping dynamics starting from the QCLE. In the former case, one simply drops all the "correlations" (including entanglement) between the subsystem and bath densities in the QCLE [58]. In the later case, one projects out all the off-diagonal matrix elements of the density in the QCLE to obtain a generalized master equation for the subsystem alone. Then, one considers decoherence to suppress the coherences in order to recover a simple "surface hopping" dynamics [59] similar to that prescribed in the FSSH algorithm. Furthermore, it had been proven [60] that the QCLE and the partially linearized path integral (PLPI) method [61–64] share the same starting mathematical foundation. In particular, the most recent PLPI algorithm, called PLDM (Partially Linearized Density Matrix) method [64], is very similar to the FBTS presented in this paper [31]. One can also draw comparisons between methods based on the QCLE and semiclassical initial value representations. For instance, numerical schemes based on the Poisson bracket mapping equation (PBME) [30], an approximate equation derived from the mapping-transformed QCLE, and the linearized semiclassical initial value representations [65] share the same set of equations of motion for the trajectories.

Mixed quantum-classical methods are often the only feasible approach to explore the dynamics of large complex systems, such as condensed phase or biochemical systems, where only a few light-mass DOF need be treated quantum mechanically. In many rate processes of interest, such as electron transfer or proton transfer, the local polar solvent motions are responsible for important features of the reaction mechanism. As a result, it is essential that the dynamics of these environmental degrees of freedom be treated in detail. Open quantum system methods that trace out all bath details cannot capture important aspects of such dynamics.

Some recent work [48,66] has suggested interesting ways to combine the QCLE and the generalized master equation [67–69] approach. Simulation tests on spin boson models [48] and a two-level system coupled to an anharmonic bath [68] indicate that accurate, long-time dynamical properties of such systems can be efficiently calculated with an improved memory kernel (which takes the short-time QCLE computation of some bath correlation functions as the input) for the general master equation. This type of hybrid approach may eventually prove to be useful for studies of more complex systems.

Finally, we provide comments that may help in choosing between the two algorithms for simulations. The TBSH algorithm, without filtering, provides a very accurate QCL dynamics before the onset of the sign problem associated with its heavy reliance on Monte Carlo sampling. While filtering can be used to extend simulations to much longer times, the problems related to Monte Carlo sampling limits its usefulness in performing long-time simulations, as vividly illustrated in Section 4. However, the TBSH is found to be the preferred simulation method (in comparison to the FBTS) when one investigates bath dynamical properties of systems in the vicinity of conical intersections and avoided crossings. For instance, the TBSH results accurately capture the intricate geometric phases [46] and the bimodal structure in the momentum distribution [35] in the Tully 1 model (a single avoided crossing model), while the FBTS fails to reproduce these delicate features, even though it provides fairly accurate population dynamics, as reported in [36]. Since the FBTS trajectory dynamics is based on a mean-field description, one finds that the results are usually very accurate (even in the long-time limit) when the energy gap between diabatic energy surfaces is small in comparison to the typical subsystem-bath coupling strength. Another advantage of the FBTS is the availability of the JFBTS [36] algorithm, which implements systematic correction of FBTS results towards the exact QCL dynamics and provides a simple method to gauge the sufficiency of the FBTS results.

Research was supported in part by a grant from the Natural Sciences and Engineering Research Council of Canada.

APPENDIX HIGH TEMPERATURE LIMIT

Many realistic chemical and biological processes take place at room temperature, in which case, it is often justified to apply a classical approximation to the bath. In this Appendix, we make two assumptions: As in most condensed phase models, we consider a pure subsystem observable, $\hat{A}$, such that $\left(\hat{\rho}_{eq}\hat{A}\right)_W = (\hat{\rho}_{eq})_W$. We also assume that the environment is further partitioned into an immediate part that can couple nonlinearly to the quantum subsystem and shield the subsystem from the larger set of environmental DOF, often modeled as a heat bath of independent harmonic oscillators. Furthermore, we write $X = \{X_b, X_n\}$, where n refers to the few DOF that couple directly to the quantum subsystem and b refers to the remainder of the large number of coordinates that only couple to the n-labeled coordinates. Similarly, we re-label different parts of the Hamiltonian as follows: $\hat{H} = \hat{H}_s + \hat{H}_n + \hat{V}_{sn} + \hat{H}_b + \hat{V}_{bn}$ with $\hat{H}_i = \hat{K}_i + \hat{V}_i$ and i = s, b, n. The quantities, $\hat{K}_i$ and $\hat{V}_i$, are the total kinetic energy and isolated potential, respectively, of the i-th system. Potential energy terms with a subscript of two letters imply a coupling potential between two components of the composite system. In addition, we introduce $\hat{h}_W(R_n) = \hat{H}_s + V_{sn}(R_n) + V_n(R_n), \hat{H}_{bn}(R_n) = \hat{H}_b + \hat{V}_{bn}(R_n)$ and $\hat{H}_{en} = \hat{h}_W(R_n) + \hat{K}_n$. In the following, we express the distance in units of $\lambda_j = \sqrt{\hbar/M_j\omega_c}$ and energy in units of $\hbar\omega_c$, where ω_c is the cut-off frequency of the heat bath.

Under these assumptions, one needs to evaluate the partial Wigner transform of $e^{-\beta\hat{H}}$ alone. In the high temperature limit, we factorize the un-normalized equilibrium density matrix operator, $\hat{\rho} = e^{-\beta\hat{H}_{sn}}e^{-\beta\hat{H}_{bn}}$. The partial Wigner transform of this approximate density operator reads:

$$\hat{\rho}_W(X) = \int dZ e^{-iP_n \cdot Z}\left\langle R_n + \frac{Z}{2}\left| e^{-\beta\hat{H}_{sn}}\right| R_n - \frac{Z}{2}\right\rangle \rho_b(X_b; R_n)$$

(36)

Where $\rho_b(X_b; R_n) = \int dZ_b e^{-iP_b \cdot Z_b}\left\langle R_b + \frac{Z_b}{2}\left| e^{-\beta\hat{H}_{bn}(R_n)}\right| R_b - \frac{Z_b}{2}\right\rangle$ is the Wigner transform of the un-normalized equilibrium density matrix for the heat bath.

We next apply a symmetric Trotter decomposition to the matrix element of Equation (36):

$$\left\langle R_n + \frac{Z}{2} \middle| e^{-\beta \hat{H}_{sn}} \middle| R_n - \frac{Z}{2} \right\rangle \approx \left\langle R_n + \frac{Z}{2} \middle| e^{-\frac{\beta}{2}\Delta \hat{H}_W \left(R_n + \frac{Z}{2}\right)} e^{-\beta \hat{H}_{ho}} e^{-\frac{\beta}{2}\Delta \hat{H}_W \left(R_n - \frac{Z}{2}\right)} \middle| R_n - \frac{Z}{2} \right\rangle$$

$$= \left(\frac{\omega}{2\pi \sinh(\omega\beta)}\right)^{N_n/2} \exp\left(-\coth\left(\frac{\omega\beta}{2}\right)\frac{\omega Z^2}{4}\right) \exp\left(-\tanh\left(\frac{\omega\beta}{2}\right)\omega R_n^2\right)$$

$$\times e^{-\frac{\beta}{2}\Delta \hat{H}_W \left(R_n + \frac{Z}{2}\right)} e^{-\frac{\beta}{2}\Delta \hat{H}_W \left(R - \frac{Z}{2}\right)} \tag{37}$$

In this equation, the symmetric Trotter decomposition separates the subsystem potential in $\hat{h}_W(R_n)$ into harmonic, $\hat{V}_{ho} = \frac{1}{2}\omega^2 R_n^2$ n, and anharmonic, $\Delta\hat{H}(R_n) = \hat{h}_W(R_n) - V_{ho}(R_n)$, contributions; furthermore, we define $\hat{H}_{ho} = \hat{K}_n + \hat{V}_{ho}$.

The anharmonic term in Equation (37) can be approximated as follows:

$$e^{-\beta \Delta \hat{h}_W \left(R_n + \frac{Z}{2}\right)} e^{-\beta \Delta \hat{h}_W \left(R_n - \frac{Z}{2}\right)}$$

$$= \sum_{\alpha,\alpha'} e^{-\beta \tilde{E}_\alpha(R_n)} \left[\delta_{\alpha\alpha'} + \frac{Z}{2}O_{\alpha\alpha'}(R_n)d_{\alpha\alpha'}(R_n)\right] |\alpha; R_n\rangle \langle\alpha'; R_n \tag{38}$$

where $|n\rangle$ is the subsystem basis and $|\alpha; R_n\rangle$ is the real-valued adiabatic state with adiabatic energy $E_\alpha(R_n)$ with respect to the Hamiltonian, $\hat{h}_W(R_n)$. The adjusted energy is $\tilde{E}_\alpha(R_n) = E_\alpha(R_n) - V_{ho}(R_n)$. The O function in Equation (38) reads:

$$O_{\alpha\alpha'}(R_n) = \left[1 - e^{-\frac{\beta}{2}(\tilde{E}_{\alpha'}(R_n) - \tilde{E}_\alpha(R_n))}\right]^2 \tag{39}$$

and $d_{\alpha\alpha'} = \langle\alpha; R_n|\nabla_{R_n}|\alpha'; R_n\rangle$. Details of a similar derivation for Equations (37) and (38) may be found in [70].

Substituting Equation (37) into Equation (36) and integrating out the Z variable, Equation (36) simplifies to:

$$\hat{\rho}_W(X) = \left(\frac{1}{2\pi\hbar \cosh\left(\frac{\omega\beta}{2}\right)}\right)^{N_n} \rho_b(X_b; R_n) e^{-\frac{P_n^2}{\omega}\tanh\left(\frac{\omega\beta}{2}\right)} \sum_\lambda e^{-\beta E_\lambda(R_n)}$$

$$\sum_{\alpha,\alpha'} |\alpha; R_n\rangle \langle\alpha'; R_n| F_{\alpha\alpha'}(X_n) \tag{40}$$

Where

$$F_{\alpha\alpha'}(X_n) = \frac{e^{-\beta E_\alpha(R_n)}}{\sum_\lambda e^{-\beta E_\lambda(R_n)}} \left[\delta_{\alpha\alpha'} - i\frac{P_n}{\omega} \tanh(\omega\beta/2)O_{\alpha\alpha'}(R_n)d_{\alpha\alpha'}(R_n) \right]$$

$$(41)$$

Now, the canonical partition function is determined by:

$$
\begin{aligned}
Z_Q &= \sum_\alpha \int dX_n dX_b \rho_W^{\alpha\alpha}(X) \\
&= \left(\frac{1}{\cosh(\frac{\omega\beta}{2})} \right)^{N_n} \left(\prod_{j=1}^{N_b} \frac{\pi}{\sinh(\omega_j\beta/2)} \right) \sqrt{\frac{\pi\omega}{\tanh(\omega\beta/2)}} \int dR_n \sum_\lambda e^{-\beta E_\lambda(R_n)} \\
&= \left(\frac{1}{\cosh(\frac{\omega\beta}{2})} \right)^{N_n} Z_b Z_{sn}
\end{aligned}
$$

$$(42)$$

where Z_b is defined by the expression in the second bracket on the second line and Z_{sn} is defined by the expression behind the second bracket on the second line. Z_b and Z_{sn} are the bath and subsystem (with its immediate environment) canonical partition functions, respectively. In summary, the time correlation function takes the following simple form:

$$C^{AB}(t) = \frac{1}{Z_Q} \sum \int dX \langle \nu_1 | \hat{\rho}_W(X) | \nu_3 \rangle \langle \nu_3 | \hat{A} | \nu_5 \rangle \langle \nu_5 | \hat{B}_W(X,t) | \nu_1 \rangle$$

$$(43)$$

where $\hat{\rho}_W(X)$ and Z_Q are given by Equations (40) and (42), respectively.

REFERENCES

1. Zwanzig, R. Time-correlation functions and transport coefficients in statistical mechanics. Annu. Rev. Phys. Chem 1965, 16, 67–102.

2. Kapral, R.; Ciccotti, G. A Statistical Mechanical Theory of Quantum Dynamics in Classical Environments. In Bridging Time Scales: Molecular Simulations for the Next Decade; Nielaba, P., Mareschal, M., Ciccotti, G., Eds.; Springer: Berlin, Germany, 2002; pp. 445–472.

3. Bonella, S.; Monteferrante, M.; Pierleoni, C.; Ciccotti, G. Path integral based calculations of symmetrized time correlation functions. I. J. Chem. Phys 2010, 133, 164104.

4. Bonella, S.; Monteferrante, M.; Pierleoni, C.; Ciccotti, G. Path integral based calculations of symmetrized time correlation functions. II. J. Chem. Phys 2010, 133, 164105.

5. Monteferrante, M.; Bonella, S.; Ciccotti, G. Linearized symmetrized

quantum time correlation functions calculation via phase pre-averaging. Mol. Phys 2011, 109, 3015–3027.

6. Redfield, A. The Theory of Relaxation Processes. In Advances in Magnetic Resonance; Waugh, J., Ed.; Academic Press: New York, NY, USA, 1965; Volume 1.

7. Lindblad, G. On the generators of quantum dynamical semigroups. Commun. Math. Phys1976, 48, 119–130.

8. Weiss, U. Quantum Dissipative Systems; World Scientific: Singapore, Singapore, 1999.

9. Blum, K. Density Matrix Theory and Applications; Plenum: New York, NY, USA, 1981.

10. Breuer, H.P.; Petruccione, F. The Theory of Open Quantum Systems; Oxford University Press: Oxford, UK, 2007.

11. Feynman, R.P.; Vernon, J.F.L.. The theory of a general quantum mechanical system interacting with a linear dissipative system. Ann. Phys 1963, 24, 118–173.

12. Feynman, R.P.; Hibbs, A.R.. Quantum Mechanics and Path Integrals; McGraw-Hill: New York, NY, USA, 1965.

13. Herman, M.F. Dynamics by semiclassical methods. Annu. Rev. Phys. Chem 1994, 45, 83–111.

14. Thoss, M.; Wang, H.B.. Semiclassical description of molecular dynamics based on initial-value representation methods. Annu. Rev. Phys. Chem 2004, 55, 299–332.

15. Kay, K.G. Semiclassical initial value treatments of atoms and molecules. Annu. Rev. Phys. Chem 2005, 56, 255–280.

16. Tully, J.C. Nonadiabatic Dynamics. In Modern Methods for Multidimensional Dynamics Computations in Chemistry; Thompson, D.L., Ed.; World Scientific: New York, NY, USA, 1998; p. 34.

17. Kapral, R. Progress in the theory of mixed quantum-classical dynamics. Ann. Rev. Phys. Chem 2006, 57, 129–157.

18. Kapral, R.; Ciccotti, G. Mixed quantum-classical dynamics. J. Chem. Phys 1999, 110, 8919–8929.

19. Sergi, A.; Kapral, R. Quantum-classical limit of quantum correlation functions. J. Chem. Phys 2004, 121, 7565–7576.

20. Nassimi, A.; Kapral, R. Mapping approach for quantum-classical time correlation functions 1. Can. J. Chem 2009, 87, 880–890.

21. Filinov, V.; Bonella, S.; Lozovik, Y.; Filinov, A., Zacharov, I. Quantum

Molecular Dynamics Using Wigner Representation. In Classical Dynamics in Condensed Phase Simulations; Berne, B.J., Ciccotti, G., Coker, D.C., Eds.; World Scientic: Singapore, Singapore, 1998.

22. Basire, M.; Borgis, D.; Vuilleumier, R. Computing Wigner distributions and time correlation functions using the quantum thermal bath method: Application to proton transfer spectroscopy. Phys. Chem. Chem. Phys 2013, 15, 12591–12601.

23. Martens, C.C.; Fang, J.Y.. Semiclassical-limit molecular dynamics on multiple electronic surfaces. J. Chem. Phys 1996, 106, 4918–4930.

24. Donoso, A.; Martens, C.C.. Simulation of coherent nonadiabatic dynamics using classical trajectories. J. Phys. Chem. A 1998, 102, 4291–4300.

25. MacKernan, D.; Kapral, R.; Ciccotti, G. Sequential short-time propagation of quantum-classical dynamics. J. Phys. Condens. Matter 2002, 14, 9069–9076.

26. MacKernan, D.; Ciccotti, G.; Kapral, R. Trotter-based simulation of quantum-classical dynamics. J. Phys. Chem. B 2008, 112, 424–432.

27. Horenko, I.; Salzmann, C.; Schmidt, B.; Schutte, C. Quantum-classical Liouville approach to molecular dynamics: Surface hopping Gaussian phase-space packets. J. Chem. Phys 2002,117, 11075–11088.

28. Wan, C.; Schofield, J. Mixed quantum-classical molecular dynamics: Aspects of the multithreads algorithm. J. Chem. Phys 2000, 113, 7047–7054.

29. Wan, C.; Schofield, J. Solutions of mixed quantum-classical dynamics in multiple dimensions using classical trajectories. J. Chem. Phys 2002, 116, 494–506.

30. Kim, H.; Nassimi, A.; Kapral, R. Quantum-classical Liouville dynamics in the mapping basis. J. Chem. Phys 2008, 129, 084102.

31. Hsieh, C.Y.; Kapral, R. Nonadiabatic dynamics in open quantum-classical systems: Forward-backward trajectory solution. J. Chem. Phys 2012, 137, 22A507.

32. Stock, G.; Thoss, M. Classical description of nonadiabatic quantum dynamics. Adv. Chem. Phys 2005, 131, 243–375.

33. Schwinger, J. On Angular Momentum. In Quantum Theory of Angular Momentum; Biedenharn, L.C., Dam, H.V., Eds.; Academic Press: New York, NY, USA, 1965; p. 229.

34. Nassimi, A.; Bonella, S.; Kapral, R. Analysis of the quantum-classical Liouville equation in the mapping basis. J. Chem. Phys 2010, 133, 134115.

35. Kelly, A.; van Zon, R.; Schofield, J.; Kapral, R. Mapping quantum-classical Liouville equation: Projectors and trajectories. J. Chem. Phys 2012, 136, 084101.

36. Hsieh, C.Y.; Kapral, R. Analysis of the forward-backward trajectory solution for the mixed quantum-classical Liouville equation. J. Chem. Phys 2013, 138, 134110.

37. Ananth, N.; Miller, T.F.. Exact quantum statistics for electronically nonadiabatic systems using continuous path variables. J. Chem. Phys 2010, 133, 234103.

38. Nielsen, S.; Kapral, R.; Ciccotti, G. Statistical mechanics of quantum-classical systems. J. Chem. Phys 2001, 115, 5805–5815.

39. Kim, H.; Hanna, G.; Kapral, R. Analysis of kinetic isotope effects for nonadiabatic reactions.J. Chem. Phys 2006, 125, 084509.

40. Hanna, G.; Kapral, R. Quantum-classical Liouville dynamics of nonadiabatic proton transfer.J. Chem. Phys 2005, 122, 244505.

41. Kim, H.; Kapral, R. Solvation and proton transfer in polar molecule nanoclusters. J. Chem. Phys 2005, 125, 234309.

42. Kim, H.; Kapral, R. Proton and deuteron transfer reactions in molecular nanoclusters.ChemPhysChem 2008, 9, 470–474.

43. Hanna, G.; Geva, E. Vibrational energy relaxation of a hydrogen-bonded complex dissolved in a polar liquid via the mixed quantum-classical lionville method. J. Phys. Chem. B 2008,112, 4048–4058.

44. Horenko, I.; Schmidt, B.; Schutte, C. A theoretical model for molecules interacting with intense laser pulses: The floquet-based quantum-classical Liouville equation. J. Chem. Phys2001, 115, 5733–5743.

45. Morales, C.M.; Thompson, W.H.. Mixed quantum-classical molecular dynamics analysis of the molecular-level mechanisms of vibrational frequency shifts. J. Phys. Chem. A 2007, 111, 5422–5433.

46. Kelly, A.; Kapral, R. Quantum-classical description of environmental effects on electronic dynamics at conical intersections. J. Chem. Phys 2010, 133, 084502.

47. Kim, H.W.; Kelly, A.; Park, J.W.; Rhee, Y.M.. All-atom semiclassical dynamics study of quantum coherence in photosynthetic fennamatthewsolson complex. J. Am. Chem. Soc 2012,134, 11640–11651.

48. Kelly, A.; Markland, T.E.. Efficient and accurate surface hopping for long time nonadiabatic quantum dynamics. J. Chem. Phys 2013, 139, 014104.

49. Stock, G.; Thoss, M. Semiclassical description of nonadiabatic quantum

dynamics. Phys. Rev. Lett 1997, 78, 578–581.

50. Ananth, N.; Venkataraman, C.; Miller, W.H.. Semiclassical description of electronically nonadiabatic dynamics via the initial value representation. J. Chem. Phys 2007, 127, 084114.

51. Huo, P.; Coker, D.F.. Consistent schemes for non-adiabatic dynamics derived from partial linearized density matrix propagation. J. Chem. Phys 2012, 137, 22A535.

52. Makri, N.; Thompson, K. Semiclassical influence functionals for quantum systems in anharmonic environments. Chem. Phys. Lett 1998, 291, 101–109.

53. Sergi, A.; Kapral, R. Quantum-classical dynamics of nonadiabatic chemical reactions. J. Chem. Phys 2003, 118, 8566–8575.

54. Bonella, S.; Coker, D.F.; Kernan, D.M.; Kapral, R.; Ciccotti, G. Trajectory Based Simulations of Quantum-Classical Systems. In Energy Transfer Dynamics in Biomaterial Systems; Burghardt, I., May, V., Micha, D.A., Bittner, E.R., Eds.; Springer: Berlin/Heidelberg, Germany, 2009; Volume 93.

55. Parandekar, P.V.; Tully, J.C.. Mixed quantum-classical equilibrium. J. Chem. Phys 2005,122, 094102.

56. Schmidt, J.R.; Parandekar, P.V.; Tully, J.C.. Mixed quantum-classical equilibrium: Surface hopping. J. Chem. Phys 2008, 129, 044104.

57. Grunwald, R.; Kelly, A.; Kapral, R. Quantum Dynamics in Almost Classical Environments. In Energy Transfer Dynamics in Biomaterial Systems; Springer: Berlin/Heidelberg, Germany, 2009; pp. 383–413.

58. Gerasimenko, V.I. Dynamical equations of quantum-classical systems. Theor. Math. Phys1982, 50, 49–55.

59. Grunwald, R.; Kapral, R. Decoherence and quantum-classical master equation dynamics. J. Chem. Phys 2007, 126, 114109.

60. Bonella, S.; Ciccotti, G.; Kapral, R. Linearization approximation and Liouville Quantum-classical dynamics. Chem. Phys. Lett 2010, 484, 399–404.

61. Bonella, S.; Coker, D.F.. Semiclassical implementation of the mapping Hamiltonian approach for nonadiabatic dynamics using focused initial distribution sampling. J. Chem. Phys 2003, 118, 4370–4385.

62. Bonella, S.; Coker, D.F.. LAND-map, a linearized approach to nonadiabatic dynamics using the mapping formalism. J. Chem. Phys 2005, 122, 194102.

63. Dunkel, E.; Bonella, S.; Coker, D.F.. Iterative linearized approach to

nonadiabatic dynamics.J. Chem. Phys 2008, 129, 114106.

64. Huo, P.; Coker, D.F.. Partial linearized density matrix dynamics for dissipative, non-adiabatic quantum evolution. J. Chem. Phys 2011, 135, 201101.

65. Liu, J.; Miller, W.H.. Real time correlation function in a single phase space integral beyond the linearized semiclassical initial value representation. J. Chem. Phys 2007, 126, 234110.

66. Shi, Q.; Geva, E. A derivation of the mixed quantum-classical Liouville equation from the influence functional formalism. J. Chem. Phys 2004, 121, 3393–3404.

67. Shi, Q.; Geva, E. A new approach to calculating the memory kernel of the generalized quantum master equation for an arbitrary systembath coupling. J. Chem. Phys 2003, 119, 12063–12076.

68. Shi, Q.; Geva, E. A semiclassical generalized quantum master equation for an arbitrary system-bath coupling. J. Chem. Phys 2004, 120, 10647–10658.

69. Zhang, M.L.; Ka, B.J.; Geva, E. Nonequilibrium quantum dynamics in the condensed phase via the generalized quantum master equation. J. Chem. Phys 2006, 125, 044106.

70. Kim, H.; Kapral, R. Nonadiabatic quantum-classical reaction rates with quantum equilibrium structure. J. Chem. Phys 2005, 123, 194108.

Chapter 13

ADSORPTION AND QUANTUM CHEMICAL STUDIES ON THE INHIBITION POTENTIALS OF SOME THIOSEMICARBAZIDES FOR THE COR-ROSION OF MILD STEEL IN ACIDIC MEDIUM

Eno E. Ebenso [1], David A. Isabirye [1] and Nnabuk O. Eddy [2]

[1]Department of Chemistry, North West University (Mafikeng Campus), Mmabatho 2735, South Africa

[2]Department of Chemistry, Ahmadu Bello University, Zaria, Nigeria

ABSTRACT

Three thiosemicarbazides, namely 2-(2-aminophenyl)-N phenylhydrazineca-rbothioamide (AP4PT), N,2-diphenylhydrazinecarbothioamide (D4PT) and 2-(2-hydroxyphenyl)-N-phenyl hydrazinecarbothioamide (HP4PT), were investigated as corrosion inhibitors for mild steel in H_2SO_4 solution using gravimetric and gasometric methods. The results revealed that they all inhibit corrosion and their % inhibition efficiencies (%IE) follow the order: AP4PT > HP4PT > D4PT. The %IE obtained from the gravimetric and gasometric experiments were in good agreement. The thermodynamic parameters obtained support a physical adsorption mechanism and the adsorption followed the Langmuir adsorption isotherm. Some quantum chemical parameters were calculated using different methods and correlated with the experimental %IE. Quantitative structure activity relationship (QSAR) approach was used on a composite index of some quantum chemical parameters to characterize the inhibition performance of the studied molecules. The results showed that the %IE were closely related to some of the quantum chemical parameters, but with varying degrees. The calculated/theoretical %IE of the molecules were found to be close to their experimental %IE. The local reactivity has been studied through the Fukui and condensed softness indices in order to predict both the reactive centers and to know the possible sites of nucleophilic and electrophilic attacks.

INTRODUCTION

There has been a growing interest in the use of organic compounds as inhibitors for the aqueous corrosion of metals. The protection of metal surfaces against corrosion is an important industrial and scientific topic. Inhibitors are one of the practical means of preventing corrosion, particularly in acidic media. Inhibitors can adhere to a metal surface to form a protective barrier against corrosive agents in contact with metal. The effectiveness of an inhibitor to provide corrosion protection depends to a large extent on the interaction between the inhibitor and the metal surface. The adsorbed inhibitors can affect the corrosion reaction, either by the blocking effect of the adsorbed inhibitor on the metal surface or by the effects attributed to the change in the activation barriers of the anodic and cathodic reactions of the corrosion process. Organic compounds, which can donate electrons to unoccupied d orbitals of metal surface to form coordinate covalent bonds and can also accept free electrons from the metal surface by using their antibonding orbitals to form feedback bonds, constitute excellent corrosion inhibitors. The most effective inhibitors are those compounds containing heteroatoms like nitrogen, oxygen, sulfur and phosphorus, as well as aromatic rings. The inhibitory activity of these molecules is accompanied by their adsorption to the metal surface. Free electron pairs on heteroatoms or π electrons are readily available for sharing to form a bond and act as nucleophile centers of inhibitor molecules and greatly facilitate the adsorption process over the metal surface, whose atoms act as electrophiles. Recently, the effectiveness of an inhibitor molecule has been related to its spatial as well as electronic structure [1–4]. Quantum chemical methods are ideal tools for investigating these parameters and are able to provide insight into the inhibitor–surface interaction.

Thiosemicarbazides, thiosemicarbazones and their derivatives have continued to be the subject of extensive investigation in chemistry and biology owing to their broad spectrum of antitumor [5], antibacterial [6,7], antiviral [8–10], antifungal [11], antimalarial [12] and antineoplastic [13] activities, and recently reported corrosion inhibiting properties [14–23]. Recently, Kandemirli and Sagdinc [24] reported on the theoretical studies of corrosion inhibition of some amides and thiosemicarbazones using some quantum chemical calculations. The data available so far are largely incomplete and it is not yet possible to draw very good conclusions about the characteristics of this set of compounds and their derivatives. Therefore, the objective of this study is to present an experimental and theoretical study on the adsorption, electronic and molecular structures of three thiosemicarbazides, namely 2-(2-aminophenyl)-N-phenylhydrazinecarbothioamide (AP4PT), N,2-diphenylhydrazinecarbothioamide (D4PT) and 2-(2-hydroxyphenyl)-N-phenyl hydrazinecarbothioamide (HP4PT), used as inhibitors, and to determine the relationship between some quantum chemical parameters/descriptors from the structure of the compounds and the inhibition efficiencies obtained using different methods. Our aim is to also find good theoretical parameters to characterize the inhibition property of the inhibitors, to establish correlations between inhibition efficiencies and some of the electronic properties of the studied molecules using different quantum chemical/theoretical methods, quantitative structure activity relationship (QSAR) approach and local reactivity indices.

RESULTS AND DISCUSSION

Effect of AP4PT, HP4PT and D4PT and Temperature

Figure 1 shows the chemical and optimized structures of the inhibitors (AP4PT, HP4PT and D4PT). Table 1 shows the corrosion rates and the % inhibition efficiencies of AP4PT, HP4PT and D4PT in acid media. The corrosion rate of mild steel for the blank solution (1 M H_2SO_4) is higher than those obtained for solutions containing various concentrations of AP4PT, HP4PT and D4PT. This indicates that the corrosion of mild steel in H_2SO_4 solution is inhibited by various concentrations of AP4PT, HP4PT and D4PT. It was also found that the corrosion rate of mild steel decreases with increase in the concentration of the inhibitor, but decreases with increasing temperature, which indicates that the inhibitory potentials of AP4PT, HP4PT and D4PT for mild steel corrosion increase with increasing concentration but decrease with increase in temperature. The values of % inhibition efficiencies obtained from the hydrogen evolution method are close to that obtained using the weight loss method (Table 1). From the calculated values of the inhibition efficiencies of AP4PT, HP4PT and D4PT, it is indicative that these inhibitors are adsorption inhibitors and that their inhibition efficiencies decrease in the following trend, AP4PT > HP4PT > D4PT. Furthermore, from the observed trend for the variation of inhibition efficiency with temperature, it is evident that the mechanism of adsorption of the inhibitors on mild steel surface is by a physical adsorption mechanism. For physical adsorption, the inhibition efficiency is expected to decrease with increasing temperature, but for chemical adsorption, the inhibition efficiency is expected to increase with increasing temperature [25].

Optimised structure	Chemical structures and IUPAC names

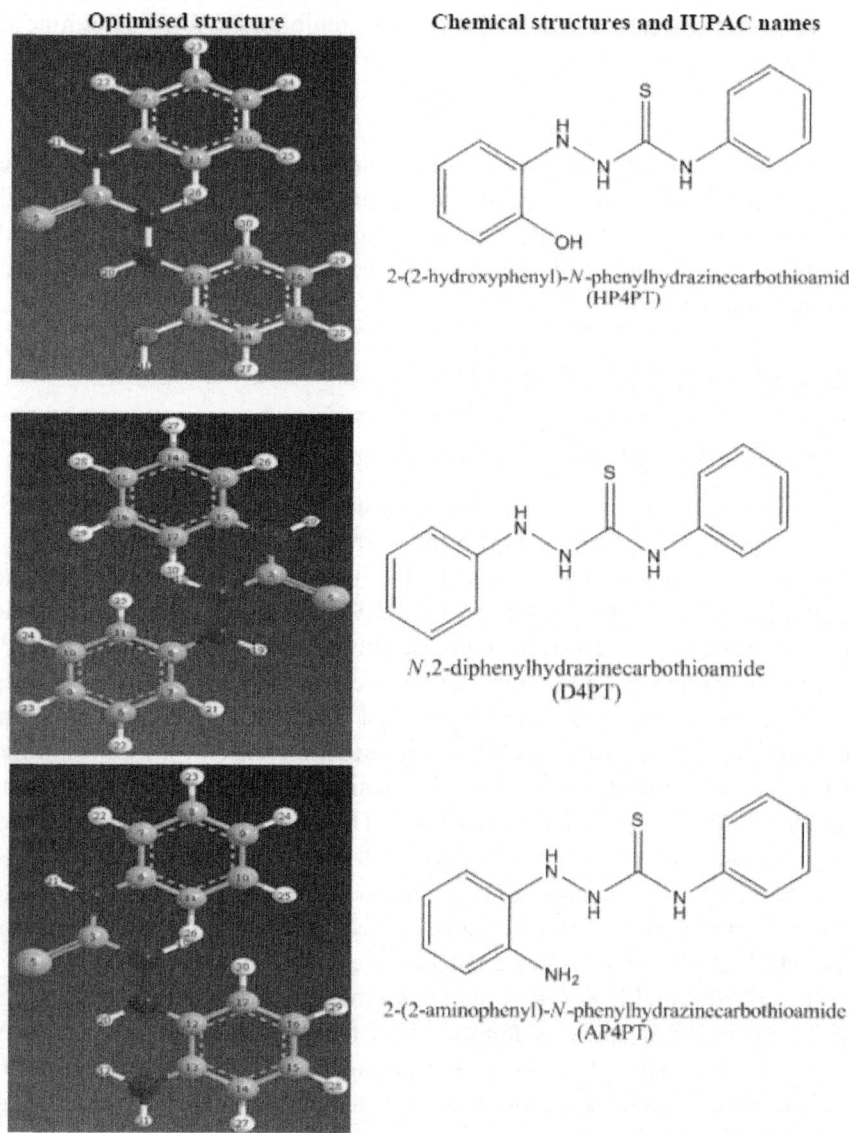

2-(2-hydroxyphenyl)-*N*-phenylhydrazinecarbothioamide
(HP4PT)

N,2-diphenylhydrazinecarbothioamide
(D4PT)

2-(2-aminophenyl)-*N*-phenylhydrazinecarbothioamide
(AP4PT)

Figure 1: Chemical and optimized structures of the studied thiosemicarbazides.

Table 1: Inhibition efficiencies (%IE) and corrosion rates (CR) of the studied thiosemicarbazides for the corrosion of mild steel in H_2SO_4 solutions using both the weight loss (at 303 and 333 K) and hydrogen evolution techniques at 303 K only

Systems	%IE (CR)	
	303 K	333 K
1 M H_2SO_4 (Blank)	(34.69)	(96.23)
4×10^{-4} M AP4PT + 1 M H_2SO_4	94.0 [a](2.08) [92.2][b]	78.92 (20.29)
8×10^{-4} M AP4PT + 1 M H_2SO_4	96.0 (1.37) [94.6]	90.80 (8.85)
12×10^{-4} M AP4PT + 1 M H_2SO_4	97.0(1.04) [95.9]	91.90 (7.79)
16×10^{-4} M AP4PT + 1 M H_2SO_4	97.6 (0.82) [96.1]	92.80 (6.93)
20×10^{-4} M AP4PT + 1 M H_2SO_4	98.7 (0.44) [97.2]	93.40 (6.35)
4×10^{-4} M HP4PT + 1 M H_2SO_4	90.7 (3.23) [88.6]	52.80 (45.42)
8×10^{-4} M HP4PT + 1 M H_2SO_4	93.1 (2.38) [91.8]	75.90 (23.19)
12×10^{-4} M HP4PT + 1 M H_2SO_4	94.4 (1.93) [92.9]	79.00 (20.21)
16×10^{-4} M HP4PT + 1 M H_2SO_4	95.1 (1.70) [94.2]	89.84 (9.78)
20×10^{-4} M HP4PT + 1 M H_2SO_4	96.2 (1.30) [95.0]	91.21 (8.46)
4×10^{-4} M D44PT + 1 M H_2SO_4	91.2 (3.05) [90.1]	54.42 (43.86)
8×10^{-4} M D4PT + 1 M H_2SO_4	91.9 (2.79) [88.4]	66.40 (32.33)
12×10^{-4} M D4PT + 1 M H_2SO_4	93.8 (2.15) [91.5]	70.50 (28.39)
16×10^{-4} M D4PT + 1 M H_2SO_4	94.4 (1.93) [93.2]	79.20 (20.02)
20×10^{-4} M D4PT + 1 M H_2SO_4	95.1 (1.70) [93.9]	89.98 (9.64)

[a] %IE obtained from the weight loss technique.
[b] %IE obtained from the hydrogen evolution technique.

The activation energies for the corrosion of mild steel in the absence and presence of the inhibitors were calculated using the logarithmic form of the Arrhenius Equation shown below [25]:

$$\log \frac{CR_2}{CR_1} = \frac{E_a}{2.303R}\left(\frac{1}{T_1} - \frac{1}{T_2}\right) \tag{1}$$

where CR_1 and CR_2 are the corrosion rates of mild steel at the temperatures T_1 (303 K) and T_2 (333 K), respectively. E_a is the activation energy for the reaction and R is the molar gas constant. Values of the E_a calculated from Equation 1 are presented inTable 2. The activation energies obtained for the inhibited corrosion of mild steel are within the limit expected (<80 KJ/mol) for the mechanism of physical adsorption, hence the adsorption of AP4PT, HP4PT and D4PT on mild steel surface is consistent with the mechanism of charge transfer from the inhibitor to the metal surface [26].

Table 2: Some thermodynamics parameters for the adsorption of the studied thiosemi-carbazides using the weight loss technique

Systems	Activation Energy, E_a (kJ mol^{-1})	Heat of adsorption, Q_{ads} (kJ mol^{-1})
1 M H$_2$SO$_4$ (Blank)	28.28	-
4×10^{-4} M AP4PT + 1 M H$_2$SO$_4$	63.13	-30.03
8×10^{-4} M AP4PT + 1 M H$_2$SO$_4$	51.72	-18.64
12×10^{-4} M AP4PT + 1 M H$_2$SO$_4$	55.83	-21.97
16×10^{-4} M AP4PT + 1 M H$_2$SO$_4$	59.15	-24.10
20×10^{-4} M AP4PT + 1 M H$_2$SO$_4$	74.00	-35.24
4×10^{-4} M HP4PT + 1 M H$_2$SO$_4$	73.27	-45.42
8×10^{-4} M HP4PT + 1 M H$_2$SO$_4$	63.11	-30.52
12×10^{-4} M HP4PT + 1 M H$_2$SO$_4$	65.10	-31.46
16×10^{-4} M HP4PT + 1 M H$_2$SO$_4$	48.49	-16.49
20×10^{-4} M HP4PT + 1 M H$_2$SO$_4$	51.91	-18.71
4×10^{-4} M D44PT + 1 M H$_2$SO$_4$	73.89	-45.33
8×10^{-4} M D4PT + 1 M H$_2$SO$_4$	67.91	-36.66
12×10^{-4} M D4PT + 1 M H$_2$SO$_4$	71.53	-38.71
16×10^{-4} M D4PT + 1 M H$_2$SO$_4$	64.83	-31.21
20×10^{-4} M D4PT + 1 M H$_2$SO$_4$	48.11	-16.17

Thermodynamics/Adsorption Considerations

The heats of adsorption of AP4PT, HP4PT and D4PT on mild steel surface were calculated using the following Equation [25]:

$$Q_{ads} = 2.303R \left[\log\left(\frac{\theta_2}{1-\theta_2}\right) - \log\left(\frac{\theta_1}{1-\theta_1}\right) \right] x \left(\frac{T_{1x}T_2}{T_2 - T_1}\right) kJmol^{-1} \tag{2}$$

where Q_{ads} is the heat of adsorption, R is the gas constant, θ_1 and θ_2 are the degrees of surface coverage of the inhibitors at the temperatures T_1 (303 K) and T_2 (333 K), respectively. Calculated values of Q_{ads} are negative, indicating that the adsorption of the inhibitors on the mild steel surface is exothermic (see Table 2).

The adsorption characteristics of the inhibitors were investigated by fitting the experimental data obtained for the degrees of surface coverage into different adsorption isotherms. The tests revealed that the adsorption of AP4PT, HP4PT and D4PT can best be described by the Langmuir adsorption isotherm. The Equation for the Langmuir adsorption isotherm can be written as follows [27,28]:

$$\theta = KC \times 1/(1 + KC) \tag{3}$$

where K designates the adsorption equilibrium constant and C is the concentration of the inhibitor in the bulk electrolyte. From the rearrangement of Equation 3, Equations 4 and 5 are obtained.

$$1/K + C = C/\theta \tag{4}$$

$$\log(C/\theta) = \log C - \log K \qquad (5)$$

Figure 2 shows the plots of values of $\log(C/\theta)$ versus $\log C$. The plots were found to be linear indicating the application of the Langmuir isotherm to the adsorption of AP4PT, HP4PT and D4PT on mild steel surface. Values of adsorption parameters deduced from the Langmuir adsorption isotherms are presented in Table 3. The results obtained indicated that the slopes and R^2values were very close to unity, which signifies strong adherence of AP4PT, HP4PT and D4PT to the adsorption of the Langmuir model.

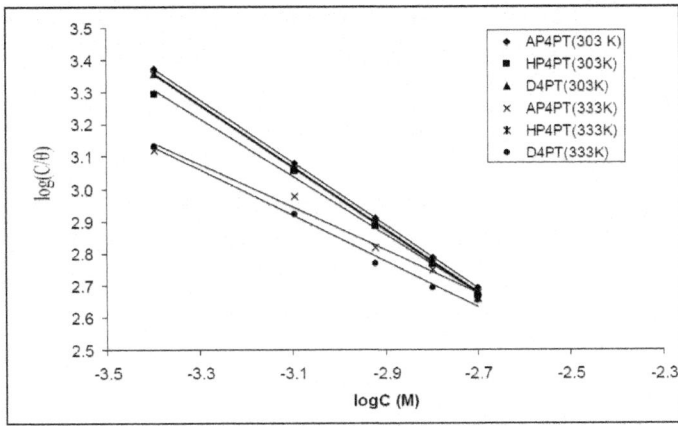

Figure 2: Langmuir isotherm for the adsorption of AP4PT, HP4PT and D4PT on mild steel surface.

Table 3: Langmuir parameters for the adsorption for AP4PT, HP4PT and D4PT on mild steel surface

Inhibitor	Temperature (K)	Slope	$\log K$	ΔG^0(kJ mol^{-1})	R^2
AP4PT	303	0.971	0.0716	-10.51	1.0000
	333	0.6625	0.8893	-16.76	0.9843
HP4PT	303	0.8987	0.9975	-15.88	0.9975
	333	0.9727	0.051	-11.42	0.9999
D4PT	303	0.9644	0.0786	-10.55	1.0000
	333	0.7074	0.7243	-15.71	0.9938

The values of the adsorption equilibrium constant (K) obtained from the intercept of the Langmuir adsorption isotherms are related to the free energy of adsorption according to Equation 6 [28,29];

$$\Delta G^0{}_{ads} = -2.303RT \log(55.5\ K) \qquad (6)$$

where $\Delta G^0{}_{ads}$ is the free energy of adsorption, R is the gas constant and T is the temperature of the system. Calculated values of the free energies are also presented in Table 3. The free energies ranged from −10.51 to −16.76 kJ/mol and are within the range expected for the transfer of charge from the inhibitor to the metal surface. Therefore, the adsorption of AP4PT, HP4PT and D4PT is spontaneous. Generally, values of ΔG^0_{ads} up to −20 kJ/mol signify physisorption, the inhibition acts due to electrostatic interactions between the charged molecules and the charged metal, while values around −40 kJ/mol or less are associated with chemisorption as a result of sharing or transfer of electrons from the organic molecules to the metal surface to form a coordinate type of bond (chemisorption). The values obtained from this study ranged from −10.51 to −16.76 kJ/mol, which support the mechanism of physical adsorption [30].

Quantum Chemical Studies

Quantum chemical calculations have been widely used to study reaction mechanisms [31]. They have also been proved to be a very powerful tool for studying inhibition of the corrosion of metals [32–34]. It has been found that the effectiveness of a corrosion inhibitor can be related to its electronic and spatial molecular structure [35–39]. In this study, the relationship between quantum chemical parameters and inhibition efficiency was investigated.

Table 4 shows the values of some quantum chemical parameters, namely the energy of the highest occupied molecular orbital (E_{HOMO}), energy of the lowest unoccupied molecular orbital (E_{LUMO}), the energy gap ($E_{LUMO-HOMO}$), the total electronic energy of the molecules (EE), core core repulsion (CC), dipole moment (μ), logP (substituent constant - measure of the differential solubility of a compound in two solvents and characterizes the hydrophobicity/hydrophilicity of a molecule), molecular polarizability (pol), cosmo volume (molecular volume) (cosVol) and cosmo area (molecular surface area or solvent accessible molecular surface area) (cosAr). The quantum chemical parameters were computed for five different Hamiltonians, namely, parametric method 6 (PM6), parametric method 3 (PM3), Austin model 1 (AM1), Recife model 1 (RM1) and modified neglect of diatomic overlap (MNDO) [40]. The results obtained from semi empirical computations, are presented in Table 4. As can be seen from Table 4, the E_{HOMO}, E_{LUMO}, ΔE and dipole moment values calculated for AP4PT by PM3 method heavily deviate from the data obtained from other Hamiltonians. This can be explained as follows. All the semi empirical methods contain sets of parameters. Atomic and diatomic parameters exist in PM6, while MNDO, AM1, PM3, and MNDO-d use only single-atom parameters. Not all parameters are optimized for all methods; for example, in MNDO and AM1 the two electron one center integrals are normally taken from atomic spectra. Therefore, in AP4PT, atomic and diatomic parameters are very significant. The frontier molecular orbital energies (i.e., E_{HOMO} and E_{LUMO}) are significant parameters for the prediction of the reactivity of a chemical species. The E_{HOMO} is often associated with the electron donating ability of a molecule [36–38]. Therefore, increasing values of E_{HOMO} indicates higher tendency for the donation of electron(s) to the appropriate acceptor molecule with low energy and empty molecular orbital. According to Eddy and Ebenso [40], increasing values of E_{HOMO} facilitate the adsorption of the inhibitor. Consequently, the inhibition efficiency of the inhibitor would be enhanced by improving the transport process through the adsorbed layer. From Table 4, it is evident that the E_{HOMO} for the inhibitors decreases in the order; (AP4PT > HP4PT > D4PT), which is consistent with the experimental % inhibition efficiency results. However, the E_{LUMO} decreases in a similar order. This can be explained as follows. The E_{LUMO} indicates the ability of the molecule to accept electrons. Therefore, the lower the value of E_{LUMO} the more apparent it is that the molecule would accept electrons. Also, the $E_{LUMO-HOMO}$ (energy gap) was also found to decrease in the order similar to that of the E_{LUMO}. Literature reveals that a larger value of the energy gap indicates low reactivity to a chemical species because the energy gap is related to the softness or hardness of a molecule. A soft molecule is more reactive than a hard molecule because a hard molecule has a larger energy gap [37,38].

Table 4: Quantum chemical parameters for the studied thiosemicarbazides

Inhibitor	Models	E_{HOMO} (eV)	E_{LUMO} (eV)	ΔE (eV)	EE (eV)	C-C (eV)	cosAr (Å^2)	cosVol (Å^3)	μ (Debye)	logP	Pol (Å^3)
AP4PT	PM6	-7.168	-0.846	6.322	-19082.68	16440.61	264.16	283.67	2.874	2.64	30.58
	PM3	-4.986	-2.219	2.767	-18634.24	16068.10	264.16	283.67	6.549	2.64	30.58
	AMI	-7.390	-0.522	6.868	-19175.65	16316.84	264.16	283.67	3.050	2.64	30.58
	RMI	-7.197	-0.450	6.747	-19276.05	16439.86	264.16	283.67	3.473	2.64	30.58
	MNDO	-7.441	-0.531	6.910	-19245.14	16350.65	264.16	283.67	3.076	2.64	30.58
HP4PT	PM6	-7.658	-0.896	6.762	-19094.75	16353.37	263.66	282.72	4.517	3.33	29.86
	PM3	-7.769	-1.058	6.711	-18716.62	16028.44	263.66	282.72	4.642	3.33	29.86
	AMI	-7.854	-0.538	7.316	-19220.30	16260.58	263.66	282.72	4.550	3.33	29.86
	RMI	-7.678	-0.481	7.197	-19304.37	16375.01	263.66	282.72	5.067	3.33	29.86
	MNDO	-7.889	-0.546	7.343	-19290.79	16293.87	263.66	282.72	4.460	3.33	29.86
D4PT	PM6	-7.927	-0.940	6.987	-17084.54	14634.46	254.24	270.42	4.870	3.50	29.23
	PM3	-8.014	-1.131	6.883	-16675.34	14280.63	254.24	270.42	5.098	3.50	29.23
	AMI	-8.103	-0.598	7.505	-17145.89	14506.42	254.24	270.42	4.854	3.50	29.23
	RMI	-7.914	-0.537	7.377	-17240.09	14627.01	254.24	270.42	5.307	3.50	29.23
	MNDO	-8.112	-0.605	7.507	-17213.07	14538.32	254.24	270.42	4.806	3.50	29.23

Values of logP (substituent constant) were also found to have a good relationship with the corrosion inhibition efficiencies of the studied inhibitors. Substituent constants are empirical quantities, which account for the variation of the structure and do not depend on the parent structure but vary with the substituent [39]. According to Eddy and Ebenso [40], logP accounts for the hydrophobicity of an actual molecule. Hydrophobicity of an organic molecule increases with decreasing water solubility. In corrosion studies, hydrophobicity is related to the mechanism of formation of the oxide/hydroxide layer on the metal surface (which reduces the corrosion process drastically). From the results obtained, based on the increasing value of logP, the inhibition efficiencies of the studied thiosemicarbazides increase in the following order, AP4PT > HP4PT > D4PT, which is consistent with experimental obtained % inhibition efficiency results.

The dipole moment (μ) is an index that can also be used for the prediction of the direction of a corrosion inhibition process. Dipole moment is the measure of polarity in a bond and is related to the distribution of electrons in a molecule [41]. Although literature is inconsistent on the use of 'μ' as a predictor for the direction of a corrosion inhibition reaction, it is generally agreed that the adsorption of polar compounds possessing high dipole moments on the metal surface should lead to better inhibition efficiency. Comparison of the results obtained from quantum chemical calculations with experimental inhibition efficiencies indicated that the % inhibition efficiencies of the inhibitors increase with increasing value of the dipole moment.

El Ashry et al. [42] noted that core core repulsion energy is a quantum chemical parameter that may have excellent correlation with inhibition efficiency. They reported that the inhibition efficiency of some Schiff bases decreased with increasing value of core core repulsion energy. Similarly, the inhibition efficiencies of the studied thiosemicarbazides were found to decrease with increasing values of core core repulsion. This study on quantum chemical descriptors has been extended to include the total and the electronic energies of the molecules. From the results, it is evident that based on the decreasing values of the total energy (TE) and electronic energy (EE), as well as decreasing value of cosmo area and cosmo volume, the trend for the variation of the inhibition efficiency follows the order similar to experimental %inhibition efficiencies (AP4PT > HP4PT > D4PT).

Polarizability is the ratio of induced dipole moment to the intensity of the electric field. The induced dipole moment is proportional to polarizability [43]. Some attempts have been made to relate the polarizability of some corrosion inhibitors to their inhibition efficiency. According to Arslan et al. [38], the minimum polarizability principle (MPP) expects that the natural direction of evolution of any system is towards a state of minimum polarizability. From the results obtained from quantum chemical calculations, the trend for the increase in the inhibition efficiencies of the inhibitors with respect to increasing polarizability correlates well with the order of the experimental % inhibition efficiencies results (AP4PT > HP4PT > D4PT). Correlations between the calculated quantum chemical parameters were also carried out. Figure 3 shows plots for the variation of the experimental inhibition efficiencies with some quantum chemical parameters. The figure reveals that the degree of linearity (R^2) between the plotted quantum chemical parameters and the experimental inhibition efficiencies were very close to unity, which indicated a high degree of linearity. However, the plots were developed from parameters obtained from PM6 Hamiltonians. R^2 values for other Hamiltonians are presented in Table 5. From the results obtained, the highest degree of linearity between the experimental inhibition efficiencies and the E_{HOMO}, E_{LUMO}, $E_{LUMO-HOMO}$ and μ were obtained from the AM1, PM6, AM1 and MNDO Hamiltonians, respectively. However, R^2 values with respect to EE, CC, cosVol and cosAr were relatively low.

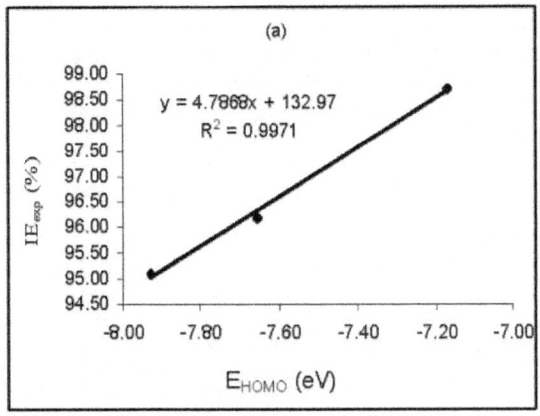

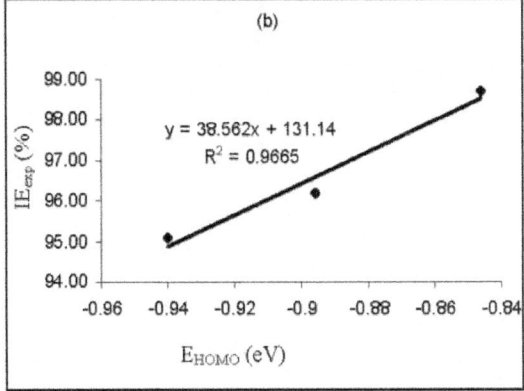

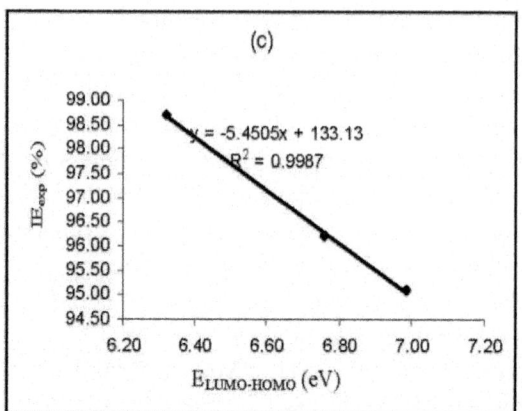

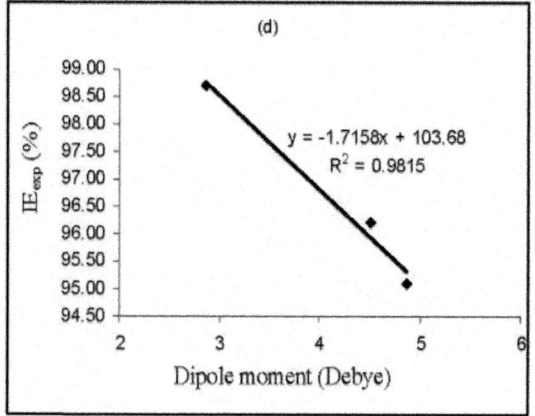

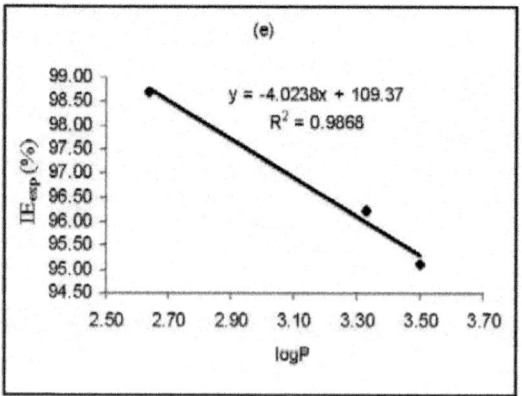

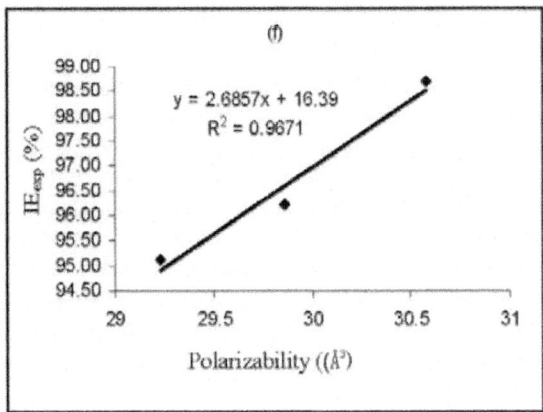

Figure 3: Variation of experimental inhibition efficiency (IEexp) of the studied thios-emicarbazides with (a) E_{HOMO} (b) E_{LUMO} (c) $E_{LUMO-HOMO}$ (d)Dipole moment; (e) logP and (f) Polarizability obtained from PM6 calculations.

Table 5: Correlation coefficients, r (degree of linearity, R^2) between quantum chemical parameters and experimental inhibition efficiencies of the studied thiosemicarbazides

Quantum parameters	PM6	PM3	AM1	RM1	MNDO
E_{HOMO} (eV)	0.9916	0.9917	0.8800	0.9362	0.8669
	(0.9971)	(0.9481)	(0.9977)	(0.993)	(0.9910)
E_{LUMO} (eV)	-0.998	-0.994	-0.9983	-0.9989	-0.9980
	(0.9665)	(0.8766)	(0.7326)	(0.8538)	(0.7254)
$E_{LUMO-HOMO}$ (eV)	-0.7678	-0.5319	-0.7552	-0.7410	-0.7420
	(0.9987)	(0.931)	(0.9999)	(0.9995)	(0.9989)
EE (eV)	0.7977	0.5777	0.7847	0.7733	0.7730
	(0.5357)	(0.5053)	(0.5221)	(0.5290)	(0.5217)
CC (eV)	0.7989	0.5982	0.7953	0.7816	0.7838
	(0.5834)	(0.5602)	(0.5681)	(0.5722)	(0.5683)
cosAr ($Å^2$)	0.8104	0.6138	0.8071	0.7936	0.7958
	(0.5853)	(0.5853)	(0.5853)	(0.5853)	(0.5853)
cosVol ($Å^3$)	-0.9819	0.9542	-0.9814	-0.9784	-0.9899
	(0.6045)	(0.6045)	(0.6045)	(0.6045)	(0.6048)
μ (Debye)	-0.9857	-0.9934	-0.9867	-0.9901	-0.9896
	(0.9815)	(0.7412)	(0.9792)	(0.9672)	(0.9874)
logP	0.9918	0.9162	0.9910	0.9878	0.9884
	(0.9868)	(0.9868)	(0.9668)	(0.9668)	(0.9868)
Polarizability	0.9985	0.9737	0.9989	0.9997	0.9960
($Å^2$)	(0.9671)	(0.9671)	(0.9671)	(0.9671)	(0.9671)

Quantitative Structure Activity Relationship (QSAR)

According to Karelson and Lobanov [44], quantitative structure-activity and structure property relationship studies are unquestionably of great importance in modern chemistry and biochemistry. The concept of QSAR/QSPR is to transform searches for compounds with desired properties using chemical intuition and experience into mathematically quantified and computed form. Once a correlation between structure and activity/property is found, any number of compounds, including those not yet synthesized, can readily be screened on the computer [45,46].

Most recent studies on the use of QSPR/QSAR for corrosion employ quantum chemical calculations as an attractive source of new molecular descriptor. According to Vera et al. [47], QSAR/QSPR can be used to relate the inhibition efficiency of most inhibitors to structural parameters (quantum and topological), which can be theoretically calculated with the ultimate aim of obtaining a molecular design of new corrosion inhibitors. El Ashry et al. [48] also stated that although the QSAR is a useful tool for the development of new corrosion inhibitors, the development of Equations for calculating the corrosion inhibition efficiency may lead to a prediction of the efficiency of some inhibitors. Attempts were made to establish the relationship between corrosion inhibition efficiencies and the calculated quantum chemical parameters using linear regression analysis. The linear model approximated the inhibition efficiency (IE_{Theor}) according to the following Equation [49]:

$$IE_{Theor} = Ax_i C_i + B \qquad (7)$$

where A and B are the regression coefficients determined by regression analysis, x_i is a quantum chemical index characteristic of the molecule i, C_i is the experimental concentration of the inhibitor. Equation 7 did not give a good correlation between the experimental and theoretical inhibition efficiencies therefore, a non linear model, which was first proposed by Lukovits et al.[50] for the study of interaction of corrosion inhibitors with metal surface in acidic solutions, was used. This model is based on the Langmuir adsorption isotherm (which assumes that the coverage of the metal surface by the inhibitor's molecule is the primary cause of corrosion inhibition) and can be written as follows [38]:

$$IE_{Theor}(\%) = \frac{(Ax_i + B)C_i}{1 + (Ax_i + B)C_i} \times 100 \qquad (8)$$

Using the non linear model, multiple regressions were performed between the inhibition efficiencies of the inhibitors and some quantum chemical parameters/descriptors. The solutions of the above non linear Equation are given by Equations 9 to 13 for PM6, PM3, AM1, RM1 and MNDO, respectively. The corresponding correlation coefficients (r) were 0.821, 0.8589, 0.7500, 0.8155 and 0.8068, respectively. Values of inhibition efficiencies calculated from Equations 9 to 13 are presented in Table 6.

Table 6: Theoretical inhibition efficiencies of the studied thiosemicarbazides obtained from various models

Inhibitor	C (M)	PM6 (%)	PM3 (%)	AM1 (%)	RM1 (%)	MNDO (%)
AP4PT	4×10^{-4}	90.98	91.18	90.37	91.20	91.31
	8×10^{-4}	95.27	95.39	94.94	95.40	95.46
	12×10^{-4}	96.80	96.88	96.57	96.88	96.92
	16×10^{-4}	97.58	97.64	97.40	97.64	97.68
	20×10^{-4}	98.05	98.10	97.91	98.11	98.13
HP4PT	4×10^{-4}	91.30	90.40	91.30	91.11	91.35
	8×10^{-4}	95.45	94.96	95.45	95.34	95.48
	12×10^{-4}	96.92	96.58	96.92	96.84	96.94
	16×10^{-4}	97.67	97.41	97.67	97.61	97.68
	20×10^{-4}	98.13	97.92	98.13	98.08	98.14
D4PT	4×10^{-4}	91.28	90.39	91.28	91.07	91.32
	8×10^{-4}	95.44	94.95	95.44	95.33	95.47
	12×10^{-4}	96.91	96.58	96.92	96.83	96.93
	16×10^{-4}	97.67	97.41	97.67	97.61	97.68
	20×10^{-4}	98.13	97.92	98.13	98.08	98.14

$$IE_{Theor} = \frac{(1.0127E_{HOMO} + E_{LUMO} + 0.99E_{LUMo\text{-}HOMO} + \mu + LogP + Pol + 69.12)*C_i \times 100}{(1 + (1.0127E_{HOMO} + E_{LUMO} + 0.99E_{LUMo\text{-}HOMO} + \mu + LogP + Pol + 69.12)*C_i}$$

$$(9)$$

$$IE_{Theor} = \frac{(1.0175E_{HOMO} + E_{LUMO} + 0.997E_{LUMo\text{-}HOMO} + \mu + LogP + Pol + 58.61)*C_i \times 100}{(1 + (1.0175E_{HOMO} + E_{LUMO} + 0.997E_{LUMo\text{-}HOMO} + \mu + LogP + Pol + 58.61)*C_i}$$

$$(10)$$

$$IE_{Theor} = \frac{(1.0148E_{HOMO} + E_{LUMO} + 0.991E_{LUMo\text{-}HOMO} + \mu + LogP + Pol + 65.19)*C_i \times 100}{(1 + (1.0148E_{HOMO} + E_{LUMO} + 0.991E_{LUMo\text{-}HOMO} + \mu + LogP + Pol + 65.19)*C_i}$$

$$(11)$$

$$IE_{Theor} = \frac{(1.0148E_{HOMO} + E_{LUMO} + 0.991E_{LUMo\text{-}HOMO} + \mu + LogP + Pol + 65.19)*C_i \times 100}{(1 + (1.0148E_{HOMO} + E_{LUMO} + 0.991E_{LUMo\text{-}HOMO} + \mu + LogP + Pol + 65.19)*C_i}$$

$$(12)$$

$$IE_{Theor} = \frac{(1.0131E_{HOMO} + E_{LUMO} + 0.987E_{LUMo\text{-}HOMO} + \mu + LogP + Pol + 72.63)*C_i \times 100}{(1 + (1.0131E_{HOMO} + E_{LUMO} + 0.987E_{LUMo\text{-}HOMO} + \mu + LogP + Pol + 72.63))*C_i}$$

$$(13)$$

Figures 4–6 are plots showing the variation of experimental inhibition efficiencies with theoretical inhibition efficiencies for AP4PT, HP4PT and D4PT, respectively. From the plots, it is evident that there is a strong relationship between the theoretical and experimental inhibition efficiencies, indicating that these models can be used to predict the inhibition efficiencies of new corrosion inhibitors that are structurally related to the studied thiosemicarbazides.

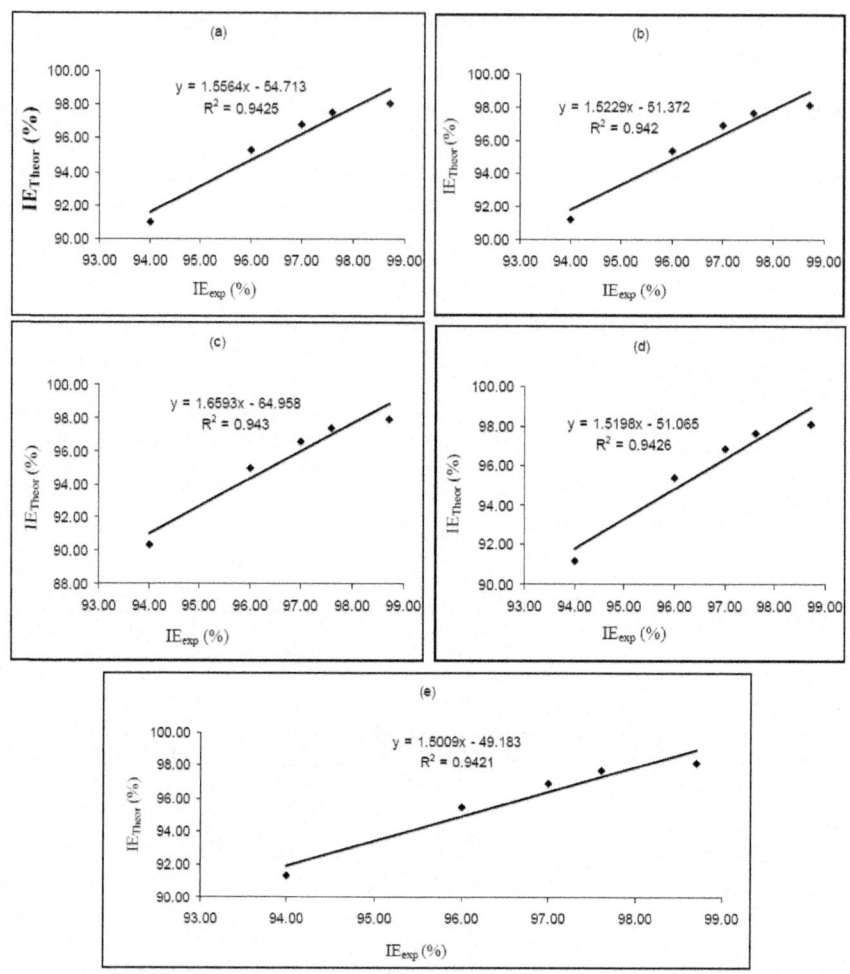

Figure 4: Variation of experimental inhibition efficiency (IE_{exp}) of AP4PT with the theoretical inhibition efficiencies (IE_{Theor}) calculated for (a) PM6 (b)PM3 (c) AM1 (d) RM1 and (e) MNDO Hamiltonians.

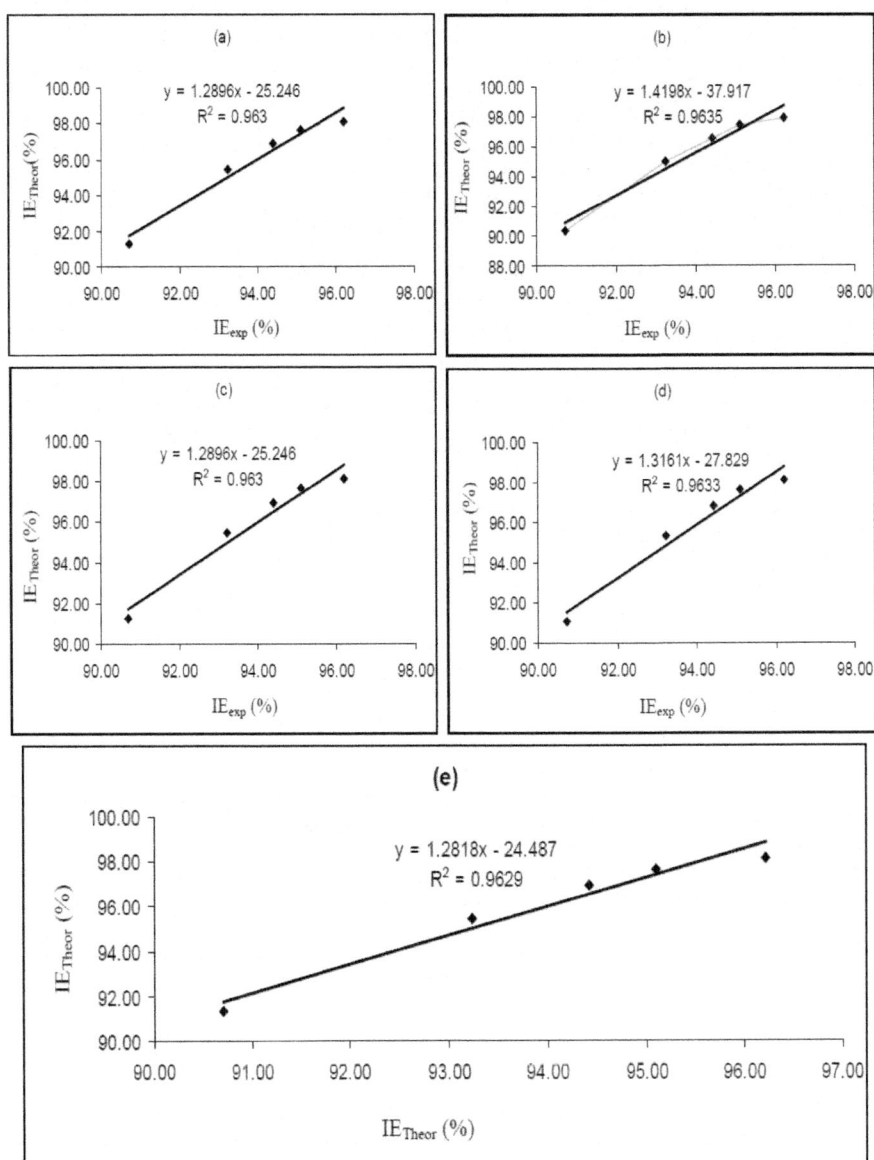

Figure 5: Variation of experimental inhibition efficiency (IE_{exp}) of HP4PT with the theoretical inhibition efficiencies (IE_{Theor}) calculated for (a) PM6 (b)PM3 (c) AM1 (d) RM1 and (e) MNDO Hamiltonians.

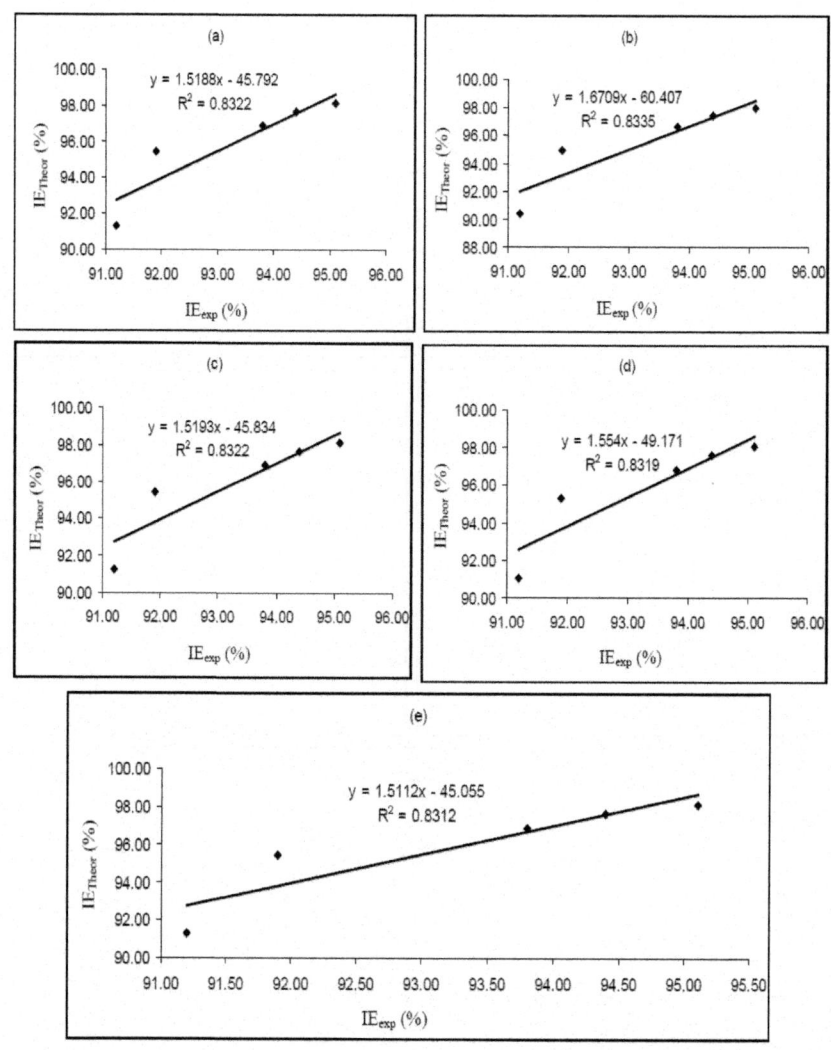

Figure 6: Variation of experimental inhibition efficiency (IE$_{exp}$) of D4PT with the theoretical inhibition efficiencies (IE$_{Theor}$) calculated for **(a)** PM6 **(b)**PM3 **(c)** AM1 **(d)** RM1 and **(e)** MNDO Hamiltonians.

Density Functional Theory (DFT)

DFT is based on solving the time independent Schrodinger Equation for the electrons of molecular systems as a function of the positions of the nuclei [51]. The premise behind the density functional theory is that the energy of a molecule can be determined from the election density instead of a wave function

[52]. However, in this study, MP2 method (Moller-Plesset perturbation theory method at level 2) was adopted for computation because of the bulkiness of the studied inhibitors.

Ionization Energy and Electron Affinity

Ionization potential (IP) and the electron affinity (EA) were calculated using the finite difference approximation as follows [53]:

$$IP=E_{(N-1)}-E_{(N)} \qquad (14)$$

$$EA=E_{(N)}-E_{(N+1)} \qquad (15)$$

where $E_{(N-1)}$, $E_{(N)}$ and $E_{(N+1)}$ are the ground state energies of the system with N - 1, N and N + 1 electrons, respectively. Values of IE and EA calculated from Equations 14 and 15 are presented in Table 7. The results obtained indicate that the inhibition efficiencies of the inhibitors increase with increasing ionization energy but decrease with decreasing value of electron affinity. This is because IP is directly related with the E_{HOMO}, while EA is related to the E_{LUMO}. This explains why the trend for the variation of inhibition efficiencies of the inhibitors with IP and EA are similar to those obtained for E_{HOMO} and E_{LUMO} data.

Table 7: Calculated quantum chemical descriptors for the studied thiosemicarbazides

Inhibitor	Model	E_N (eV)	E_{N-1} (eV)	E_{N-1} (eV)	IP (eV)	EA (eV)	η (eV)	S (/eV)	χ (eV)	δ
AP4PT	PM6	-2642.06	-2635.75	-2643.14	6.31	1.08	5.23	0.19	3.69	0.3160
	PM3	-2566.14	-2562.45	-2570.28	5.69	2.14	3.55	0.28	3.92	0.4345
	AM1	-2858.81	-2852.21	-2858.91	6.60	0.10	6.50	0.15	3.35	0.2808
	RM1	-2836.25	-2829.91	-2837.33	6.34	1.08	5.26	0.19	3.71	0.3127
	MNDO	-2894.50	-2887.83	-2895.06	6.67	0.56	6.11	0.16	3.62	0.2770
HP4PT	PM6	-2741.38	-2734.67	-2743.09	6.71	1.71	5.00	0.20	4.21	0.2790
	PM3	-2688.18	-2681.28	-2687.19	6.90	-0.99	7.89	0.13	2.95	0.2563
	AM1	-2959.71	-2952.78	-2959.76	6.93	0.05	6.88	0.15	3.49	0.2551
	RM1	-2929.36	-2922.64	-2930.42	6.72	1.06	5.66	0.18	3.89	0.2747
	MNDO	-2996.92	-2989.95	-2997.48	6.97	0.56	6.41	0.16	3.77	0.2523
D4PT	PM6	-2450.08	-2443.10	-2451.81	6.98	1.73	5.25	0.19	4.36	0.2519
	PM3	-2394.72	-2387.57	-2395.94	7.15	1.22	5.93	0.17	4.18	0.2374
	AM1	-2639.48	-2632.31	-2639.63	7.17	0.15	7.02	0.14	3.66	0.2379
	RM1	-2613.08	-2606.13	-2614.17	6.95	1.09	5.86	0.17	4.02	0.2543
	MNDO	-2674.75	-2667.56	-2675.36	7.19	0.61	6.58	0.15	3.90	0.2356

Global Softness and Hardness

In DFT, the ground state energy $E(\rho)$ of an atom/molecule can be expressed in terms of its electron density, $\rho(r)$. The first and second derivatives of $E(\rho)$ with respect to the number of electrons (N) defines the chemical potential (σ) and the global hardness (η) of a molecule as follows [46]:

$$\sigma=(\delta E/\delta N)_{v(r)} \qquad\qquad (16)$$

$$\eta=(\delta^2 E/\delta^2 N)_{v(r)} \qquad\qquad (17)$$

where $v(r)$ indicates that the differentiation is carried out under constant external potentials. Using the finite difference approximation, the global softness can be evaluated as $S = 1/(IP - EA)$. The global hardness, η, which is the inverse of the global softness can be evaluated using Equation 18 below:

$$S=1/[(E_{(N-1)}-E_{(N)})-(E_{(N)}-E_{(N+1)})] \qquad\qquad (18)$$

Calculated values of S and η are also presented in Table 7. From Table 7, it is evident that the inhibitor with the least value of global hardness (hence the highest value of global softness) is the best and vice versa. This is because a soft molecule is more reactive than a hard molecule. This observation is consistent with the results obtained from experimental %inhibition efficiencies.

The fraction of electron transferred, δ, was calculated using the Equation 19 below [36]:

$$\delta=(\chi_{Fe}-\chi_{inh})/2(\eta_{Fe}+\eta_{inh}) \qquad\qquad (19)$$

where χ_{Fe} and χ_{inh} are the electronegativity of Fe and the inhibitor, respectively, and can be evaluated as $\chi = (IP + EA)/2$. η_{Fe} and η_{inh} are the global hardness of Fe and the inhibitor, respectively (calculated using Equation 21). In order to apply Equation 19 to the present study, the theoretical values of $\chi_{Fe} = 7$ ev and $\eta_{Fe} = 0$ were used for the computation of δ values for the various Hamiltonians. Calculated values of δ are presented in Table 7. The results indicate that δ values correlates strongly with experimental inhibition efficiencies. Thus, the highest fraction of electrons is associated with the best inhibitor (AP4PT), while the least fraction is associated with the inhibitor that has the least inhibition efficiency (D4PT).

Local Selectivity

The local selectivity of a corrosion inhibitor is best analyzed using the Fukui function. The Fukui indices permit the distinction of each part of a molecule on the basis of its chemical behavior due to different substituent functional groups.

According to Fuentealba et al. [52], the Fukui function can be formally defined as

$$f(r)=(\delta\sigma/\delta v)_N \qquad\qquad (20)$$

where the functional derivative must be taken at constant number of electron, N. If it is assumed that the total energy, E, is a function of $v(r)$ and is an exact differential, then the Maxwell relations between derivatives can be applied to derive the following Equation:

$$f(r)=(\delta\rho(r)/\delta N)_v \tag{21}$$

Equation 21 is the most standard presentation of the Fukui function. The Fukui function is provoked by the fact that if an electron δ is transferred to an N electron molecule, it will tend to distribute so as to minimize the energy of the resulting $N + \delta$ electron system. The resulting change in electron density is the nucleophilic (f^+) and electrophilic (f^-) Fukui functions, which can be calculated using the finite difference approximation as follows [52]:

$$f^+ = (\delta\rho(r)/\delta N)^+_v = q_{(N+1)} - q_{(N)} \tag{22}$$

$$f^- = (\delta\rho(r)/\delta N)^-_v = q_{(N)} - q_{(N-1)} \tag{23}$$

where ρ, $q_{(N+1)}$, $q_{(N)}$ and $q_{(N-1)}$ are the density of electron and the Mulliken charge of the atom with $N + 1$, N and N-1 electrons.

Calculated values of $q_{(N+1)}$, $q_{(N)}$, $q_{(N-1)}$, f^+ and f^- for AP4PT, HP4PT and D4PT are presented in Tables 7–10. The site for nucleophilic attack is the site where the value of f^+ is maximum, while the site for electrophilic attack is controlled by the values of f^-. Assuming that the protonated forms of the inhibitors' molecules have a net positive charge, it can be deduced that the sites for nucleophilic attack are the nitrogen atom (N 4) for the three inhibitors. However, the sites for electrophilic attack are in the carbon atoms (C12, C6 and C6) for AP4PT, HP4PT and D4PT, respectively.

The HOMO and LUMO orbitals of AP4PT, HP4PT and D4PT are presented in Figure 7. The figure clearly reveals the information that governs the nucleophilic and electrophilic attacks on the studied inhibitors. The information obtained from the HOMO and LUMO orbitals are consistent with the findings obtained from the Fukui function.

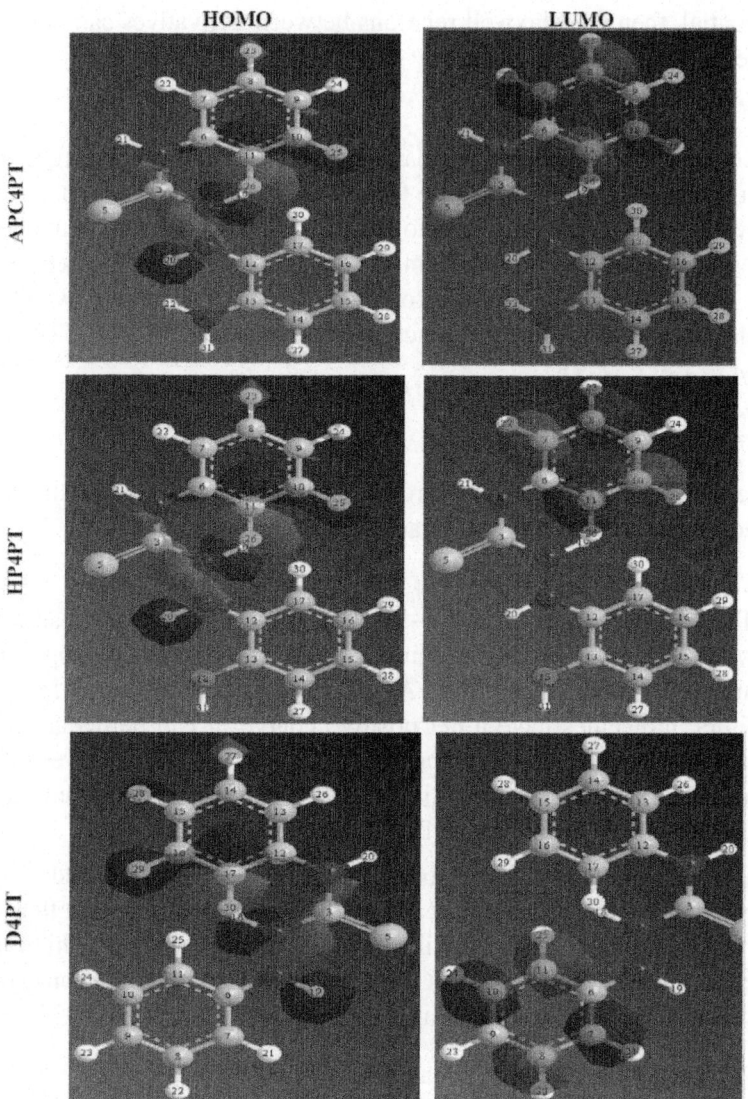

Figure 7: Molecular orbital of the studied thiosemicarbazides showing the HOMO and the LUMO.

The local softness, S, for an atom can be expressed as the product of the condensed Fukui function (f) and the global softness (S), as follows [52–54];

$$S^+ = (f^+)S \qquad (24)$$

$$S^- = (f^-)S \qquad (25)$$

The local softness contains information similar to those obtained from the condensed Fukui function plus additional information about the total molecular softness, which is related to the global reactivity with respect to a reaction partner. Calculated values of S^+ and S^- are presented in Tables 8–10. From the results obtained, the sites for electrophilic and nucleophilic attack in the studied thiosemicarbazides are slightly similar. Other indices that can be used to predict the reactive sites of a corrosion inhibitor are the relative nucleophilicity and electrophilicity, which is defined as (S^+/S^-) and (S^-/S^+), respectively. These functions have been successfully applied for the prediction of reactivity sequences of carbonyl compounds toward nucleophilic attack. It was observed that the atoms with highest value of relative nucleophilicity and electrophilicity are similar to those obtained for the Fukui and global softness functions (though the results are not presented here).

Table 8: Fukui and global softness indices for nucleophilic and electrophilic attacks in HP4PT calculated from Mulliken (Lowdin) charges

Atom (No)	f_x^+ (|e|)	f_x^- (|e|)	S_x^+ (eV|e|)	S_x^- (eV|e|)
1 N	-0.0245(-0.0346)	0.5191(0.3439)	-0.0059(-0.0083)	0.0825(0.1246)
2 N	-0.0023(0.0003)	2.7382(2.8610)	-0.0006(0.0001)	0.6866(0.6572)
3 C	**0.0009**(0.0042)	-3.5149(-3.1289)	0.0002(0.0010)	-0.7509(-0.8436)
4 N	**0.0011(0.0071)**	-5.1867(-4.9549)	0.0003(0.0017)	-1.1892(-1.2448)
5 S	-0.1030(-0.0974)	1.6665(1.3058)	-0.0247(-0.0234)	0.3134(0.4000)
6 C	-0.0198(-0.0132)	-3.8634(-3.9023)	-0.0048(-0.0032)	-0.9366(-0.9272)
7 C	-0.0841(-0.1110)	-4.0753(-4.0663)	-0.0202(-0.0266)	-0.9759(-0.9781)
8 C	-0.1022(-0.1425)	-4.0387(-4.0031)	-0.0245(-0.0342)	-0.9607(-0.9693)
9 C	0.0000(0.0131)	-4.0731(-4.0514)	0.0000(0.0031)	-0.9723(-0.9775)
10 C	-0.0928(-0.1269)	-4.0352(-4.0019)	-0.0223(-0.0305)	-0.9605(-0.9684)
11 C	-0.1004(-0.1281)	-4.0622(-4.1127)	-0.0241(-0.0307)	-0.9870(-0.9749)
12 C	-0.0070(-0.0075)	**4.1171(4.0717)**	-0.0017(-0.0018)	0.9772(0.9881)
13 C	-0.0086(-0.0082)	**4.1127(4.0689)**	-0.0021(-0.0020)	0.9765(0.9870)
14 C	-0.0084(-0.0103)	3.9200(3.9422)	-0.0020(-0.0025)	0.9461(0.9408)
15 C	-0.0070(-0.0085)	3.9152(3.9360)	-0.0017(-0.0020)	0.9446(0.9396)
16 C	-0.0048(-0.0045)	3.9426(3.9744)	-0.0012(-0.0011)	0.9539(0.9462)
17 C	0.0012(0.0057)	3.9019(3.9011)	0.0003(0.0014)	0.9363(0.9365)
18 O	-0.0010(-0.0014)	1.6743(1.7849)	-0.0002(-0.0003)	0.4284(0.4018)

Table 9: Fukui and global softness indices for nucleophilic and electrophilic attacks in D4PT calculated from Mulliken (Lowdin) charges

Atom (No)	f_x^+ (|e|)	f_x^- (|e|)	S_x^+ (eV|e|)	S_x^- (eV|e|)
1 N	-0.0460(-0.0547)	-0.0527(-0.0617)	-0.0017(-0.0021)	-0.0020(-0.0023)
2 N	0.0059(0.0104)	-0.0063(-0.0101)	0.0002(0.0004)	-0.0002(-0.0004)
3 C	-0.1280(-0.1782)	**0.0106(0.0241)**	-0.0049(-0.0068)	0.0004(0.0009)
4 N	**0.0241(0.0366)**	-0.0263(-0.0361)	0.0009(0.0014)	-0.0010(-0.0014)
5 S	-0.2715(-0.2551)	-0.5881(-0.6199)	-0.0103(-0.0097)	-0.0223(-0.0236)
6 C	-0.0031(-0.0016)	**0.0143(0.0227)**	-0.0001(-0.0001)	0.0005(0.0009)
7 C	-0.0158(-0.0174)	-0.0167(-0.0201)	-0.0006(-0.0007)	-0.0006(-0.0008)
8 C	-0.0097(-0.0110)	-0.0057(-0.0049)	-0.0004(-0.0004)	-0.0002(-0.0002)
9 C	-0.0213(-0.0275)	-0.0245(-0.0317)	-0.0008(-0.0010)	-0.0009(-0.0012)
10 C	-0.0070(-0.0059)	-0.0048(-0.0030)	-0.0003(-0.0002)	-0.0002(-0.0001)
11 C	-0.0069(-0.0075)	-0.0151(-0.0183)	-0.0003(-0.0003)	-0.0006(-0.0007)
12 C	-0.0653(-0.0862)	**0.0082(0.0217)**	-0.0025(-0.0033)	0.0003(0.0008)
13 C	-0.0207(-0.0232)	-0.0251(-0.0298)	-0.0008(-0.0009)	-0.0010(-0.0011)
14 C	-0.0256(-0.0293)	-0.0052(-0.0026)	-0.0010(-0.0011)	-0.0002(-0.0001)
15 C	-0.0359(-0.0483)	-0.0268(-0.0342)	-0.0014(-0.0018)	-0.0010(-0.0013)
16 C	-0.0263(-0.0291)	-0.0047(0.0006)	-0.0010(-0.0011)	-0.0002(0.0000)
17 C	-0.0251(-0.0269)	-0.0456(-0.0579)	-0.0010(-0.0010)	-0.0017(-0.0022)

Table 10: Fukui and global softness indices for nucleophilic and electrophilic attacks in AP4PT calculated from Mulliken (Lowdin) charges

Atom (No)	f_x^+ (|e|)	f_x^- (|e|)	S_x^+ (eV|e|)	S_x^- (eV|e|)
1 N	-0.0239(-0.0340)	-0.0113(-0.0051)	-0.0045(-0.0065)	-0.0022(-0.0010)
2 N	-0.0034(-0.0010)	-0.1666(-0.2265)	-0.0007(-0.0002)	-0.0317(-0.0430)
3 C	**0.0011(0.0044)**	-0.0009(0.0002)	0.0002(0.0008)	-0.0002(0.0000)
4 N	**0.0009(0.0070)**	-0.0148(-0.0176)	0.0002(0.0013)	-0.0028(-0.0033)
5 S	-0.1005(-0.0948)	-0.1067(-0.1053)	-0.0191(-0.0180)	-0.0203(-0.0200)
6 C	-0.0178(-0.0102)	**0.0055(0.0130)**	-0.0034(-0.0019)	0.00105(0.0025)
7 C	-0.0879(-0.1168)	-0.0190(-0.0230)	-0.0167(-0.0222)	-0.0036(-0.0044)
8 C	-0.1014(-0.1408)	-0.0069(-0.0064)	-0.0193(-0.0268)	-0.0013(-0.0012)
9 C	0.0007(0.0142)	-0.0194(-0.0237)	0.0001(0.00270)	-0.0037(-0.0045)
10 C	-0.0947(-0.1304)	-0.0013(0.0032)	-0.0180(-0.0248)	-0.0003(0.0006)
11 C	-0.0990(-0.1257)	-0.0272(-0.0362)	-0.0188(-0.0239)	-0.0052(-0.0069)
12 C	**0.0020(0.0051)**	-0.0074(-0.0025)	0.0004(0.0010)	-0.0014(-0.0005)
13 C	-0.0106(-0.0116)	-0.0261(-0.0257)	-0.0020(-0.0022)	-0.0050(-0.0049)
14 C	-0.0075(-0.0088)	-0.0221(-0.0227)	-0.0014(-0.0017)	-0.0042(-0.0043)
15 C	-0.0123(-0.0155)	-0.0487(-0.0639)	-0.0023(-0.0030)	-0.0093(-0.0121)
16 C	-0.0015(0.0004)	-0.0255(-0.0282)	-0.0003(0.0001)	-0.0049(-0.0054)
17 C	-0.0041(-0.0018)	-0.0553(-0.0682)	-0.0008(-0.0003)	-0.0105(-0.0130)
18 N	-0.0036(-0.0042)	-0.6434(-0.4673)	-0.0007(-0.0008)	-0.1223(-0.0888)

EXPERIMENTAL TECHNIQUES

Materials

Materials used for the study were mild steel sheets of composition (wt %); Mn (0.6), P (0.36), C (0.15) and Si (0.03) and the rest Fe. Each sheet was mechanically pressed cut to form different coupons, each of dimension 5 × 4 × 0.11 cm. Each coupon was degreased by washing with ethanol, dipped in acetone and allowed to air dry before they were preserved in a desiccator. All reagents used for the study were Analar grade and double distilled water was used for their preparation. The inhibitors 2-(2-aminophenyl) Nphenylh ydrazinecarbothioamide(AP4PT),N,2diphenylhydrazinecarbothioamide(D 4PT) and 2-(2-hydroxyphenyl)- phenylhydrazinecarbothioamide(HP4PT) were synthesized as described earlier by Kittur and Mahajan Shetti [55]. The concentrations of inhibitor used for the study was 4×10^{-4} to 20×10^{-4} M in 1L solution of 1M H_2SO_4.

Gravimetric (Weight Loss) Method

In the gravimetric experiment, a previously weighed mild steel coupon was completely immersed in 250 mL of the test solution in an open beaker. The beaker was inserted into a water bath maintained at 303 K. After every 24 h, each sample was withdrawn from the test solution, washed in a solution containing 50% NaOH and 100 g/L of zinc dust. The washed coupons were dipped in acetone and allowed to air dry before re-weighing. The difference in weight for a period of 168 h (7 days) was taken as total weight loss. The experiments were repeated at 333 K. From the weight loss results, the inhibition efficiency (%I) of the inhibitor, degree of surface coverage and corrosion rates were calculated using Equations 26, 27, and 28. respectively [56];

$$\%IE=(1-W_1/W_2)\times100 \qquad (26)$$
$$\theta=1-W_1/W_2 \qquad (27)$$
$$CR(gh^{-1}cm^{-2})=W/At \qquad (28)$$

where W_1 and W_2 are the weight losses (g) for mild steel in the presence and absence of the inhibitor in H_2SO_4 solution, θ is the degree of surface coverage of the inhibitor, A is the area of the mild steel coupon (in cm^2), t is the period of immersion (in hours) and W is the weight loss of mild steel after time, t. All the measurements were performed in triplicate and the mean value recorded.

Gasometric (Hydrogen Evolution) Method

The hydrogen evolution technique (gasometric) experiment was carried out at

303 K as described in literature [56]. From the volume of hydrogen evolved per minute, inhibition efficiencies were calculated using Equation 29 below.

$$\%IE = \left(1 - \frac{V_{Ht}^{1}}{V_{Ht}^{o}}\right) x\ 100$$

(29)

where V_{Ht}^{1} and V_{Ht}^{o} are the volumes of H_2 gas evolved at time 't' for inhibited and uninhibited solutions, respectively.

Quantum Chemical Calculations

Single point energy calculations were carried out using AM1, PM6, PM3, MNDO and RM1 Hamiltonian in the MOPAC 2008 software for Windows [57]. Calculations were performed on an IBM compatible Intel Pentium IV (2.8 GHz, 4 GB RAM) computer. The following quantum chemical parameters were calculated: the energy of the highest occupied molecular orbital (E_{HOMO}), the energy of the lowest unoccupied molecular orbital (E_{LUMO}), the dipole moment (μ), the total energy (TE), the electronic energy (EE), the ionization potential, the cosmo area (cosAr) and the cosmo volume (cosVol). The polarizability (Pol) and logP were also calculated using Hyperchem release 8.0.3 for windows [58]. The Mulliken and Lowdin charges (q) for nucleophilic and electrophilic attacks were computed using GAMES computational software [59]. The correlation type and method used for the calculation was MP2 while the basis set was set to STO3G*.

Statistical analyses were performed using SPSS program version 15.0 for Windows. Non-linear regression analyses were performed by unconstrained sum of squared residuals for loss function and estimation methods of Levenberg-Marquardt using SPSS program version 15.0 for Windows [60].

CONCLUSIONS

All the methods used showed that the three thiosemicarbazides possess good inhibition properties for the corrosion of mild steel in H_2SO_4 at the temperatures studied and their % inhibition efficiencies increased with increasing concentration of the inhibitors and decreasing temperature. The % inhibition efficiencies obtained from the gravimetric and gasometric experiments were in good agreement. The thermodynamic parameters obtained support a physical adsorption mechanism. Adsorption of the inhibitors on the mild steel surface followed the Langmuir adsorption isotherm. The calculated/theoretical % inhibition efficiencies of the molecules were found to be close to their experimental % inhibition efficiencies. From the local reactivity indices, it was found that the sites for electrophilic attack are in the carbon atoms (C12, C6 and C6) for AP4PT, HP4PT and D4PT respectively and that for nucleophilic attack are the nitrogen atom (N 4) for the three inhibitors.

REFERENCES AND NOTES

1.   Valdez, LMR; Villafane, AM; Mitnik, DG. CHIH-DFT theoretical study of isomeric thiatriazoles and their potential activity as corrosion inhibitors. J. Mol. Struct.: Theochem 2005, 716, 61–65.

2.   Stoyanova, AE; Peyerimhoff, SD. On the relationship between corrosion inhibiting effect and molecular structure.Electrochim. Acta 2002, 47, 1365–1371.

3.   Gomez, B; Likhanova, NV; Aguilar, MAD; Palou, RM; Vela, A; Gasquez, JL. Quantum chemical study of the inhibitive properties of 2-Pyridyl-Azoles. J. Phys. Chem 2006, B110, 8928–8934.

4.   Finsgar, M; Lesar, A; Kokaij, A; Milosev, I. A comparative electrochemical and quantum chemical calculation study of BTAH and BTAOH as copper corrosion inhibitors in near neutral chloride solution. Electrochim. Acta 2008, 53, 8287–8297.

5.   Dobek, AS; Klayman, DL; Dickson, ET, Jr; Scovill, JP; Tramont, EC. Inhibition of clinically significant bacterial organisms in vitro by 2-acetylpyridine thiosemicarbazones. Antimocrob. Agents Chemother 1980, 18, 27–36.

6.   Offiong, OE; Martelli, S. Antifungal and antibacterial activity of 2-acetylpyridine-(4-phenylthiosemicarbazone) and its metal (II) complexes. IL Farmaco 1992, 42, 1543–1554.

7.   Offiong, OE; Martelli, S. Stereochemistry and antitumor activity of platinum metal complexes of 2-acetylpyridine thiosemicarbazones. Transit. Met. Chem 1997, 22, 263–269.

8.   Easmon, J; Heinisch, G; Holzer, W; Rosenwirth, B. Synthesis and antiviral activity of thiosemicarbazone derivatives of pyridazinecarbaldehydes and alkyl pyridazinyl ketones. Arnein-Forsch/Drug Res 1989, 39, 1196–1201.

9.   Easmon, J; Heinisch, G; Holzer, W; Rosenwirth, B. Pyridazines. 63. Novel thiosemicarbazones derived from formyl- and acyldiazines: synthesis, effects on cell proliferation, and synergism with antiviral agents. J. Med. Chem 1992, 35, 3288–3296.

10.   West, DX; El-Sawaf, AK; Bain, GA. Metal complexes of N (4)-substituted

analogues of the antiviral drug methisazone {1-methylisatin thiosemicarbazone}. Transit. Met. Chem 1997, 23, 1–6.

11. Offiong, OE; Martelli, S. Synthesis, biological activity of novel metal complexes of 2-acetylpyridine thiosemicarbazones.IL Farmaco 1995, 50, 625–632.

12. Klayman, DL; Bartosevich, JE; Griffin, TS; Mason, CJ; Scovil, JP. 2-Acetylpyridine thiosemicarbazones. 1. A new class of potential antimalarial agents. J. Med. Chem 1979, 22, 855–862.

13. Brockman, RW; Thompson, JR; Bell, MJ; Skipper, HE. Observations on the antileukemic activity of pyridine-2-carboxaldehyde thiosemicarbazone and thiocarbohydrazone. Cancer Res 1956, 16, 167–170.

14. Ekpe, UJ; Ibok, UJ; Ita, BI; Offiong, EO; Ebenso, EE. Inhibitory action of methyl and phenyl thiosemicarbazone derivatives on the corrosion of mild steel in HCl. Mater. Chem. Phys 1995, 40, 87–93.

15. Babaqi, AS; El-Basiounyi, MS; Abdulla, RM. Kinetic study of corrosion and corrosion inhibition of aluminium in chloroacetic acids. Bull. Soc. Chim. Fr 1989, 3, 297–302.

16. Ateya, BG; El-Anadouli, BE; El-Nizamy, FMA. Corrosion Inhibition and adsorption behaviour of some thioamides on mild steel in sulfuric acid. Bull. Chem. Soc. Jp 1981, 54, 3157–3161.

17. Okafor, PC; Ebenso, EE; Ekpe, UJ. Inhibition of aluminium corrosion by some derivatives of thiosemicarbazone. Bull. Chem. Soc. Ethiopia 2004, 18, 181–192.

18. Quraishi, MA; Sardar, R; Khan, S. An investigation of the inhibitive capacity of synthesized thiosemicarbazides on the corrosion of carbon steel in acid solutions. Anti. Corros. Methods Mater 2008, 35, 60–65.

19. Ita, BI; Offiong, OE. Inhibition of steel corrosion in HCl by pyridoxal, 4-methylthiosemicarbazide, pyridoxal-(4-methyl thiosemicarbazone) and its Zn(II) complex. Mater. Chem. Phys 1997, 48, 164–169.

20. Ebenso, EE; Ekpe, UJ; Ita, BI; Offiong, OE; Ibok, UJ. Effect of molecular structure on the efficiency of amides and thiosemicarbazones used for corrosion inhibition of mild steel in HCl. Mater. Chem. Phys 1999, 60, 79–90.

21. Ita, BI; Offiong, OE. The study of inhibitory properties of benzoin, benzil, benzoin-(4-phenylthiosemicarbazone) and benzil-(4-phenylthiosemicarbazone) on the corrosion of mild steel in hydrochloric acid. Mater. Chem. Phys 2001, 70, 330–335.

22. Ekpe, UJ; Okafor, PC; Ebenso, EE; Offiong, OE; Ita, BI. Mutual effects of

TSC derivatives on the acidic corrosion of aluminium. Bull. Electrochem 2001, 17, 131–135.

23. Ebenso, EE; Okafor, PC; Offiong, OE; Ita, BI; Ibok, UJ; Ekpe, UJ. Comparative investigation into the kinetics of corrosion inhibition of aluminium alloy (AA 1060) in acidic medium. Bull. Electrochem 2001, 17, 459–464.

24. Kandemirli, F; Sagdinc, S. Theoretical study of corrosion inhibition of amides and thiosemicarbazones. Corros. Sci 2007,49, 2118–2130.

25. Eddy, NO; Odoemelam, SA; Odiongenyi, AO. Inhibitive, adsorption and synergistic studies on ethanol extract of Gnetum africana as green corrosion inhibitor for mild steel in H$_2$SO$_4$. Green Chem. Lett. Rev 2009, 2, 111–119.

26. Noor, EA. Potential of aqueous extract of Hibiscus sabdariffa leaves for inhibiting the corrosion of aluminium in alkaline solutions. J Appl Electrochem 2009, 39, 1465–1475.

27. Dehri, I; Ozcan, M. The effect of temperature on the corrosion of mild steel in acidic media in the presence of some sulphur–containing organic compounds. Mater. Chem. Phys 2008, 98, 316–323.

28. Eddy, NO; Odoemelam, SA; Odiongenyi, AO. Joint effect of halides and ethanol extract of Lasianthera Africana on the inhibition of the corrosion of mild steel in H$_2$SO$_4$. J Appl Electrochem 2009, 39, 849–857.

29. Quraishi, MA; Ahamad, I; Singh, AK; Shukla, SK; Lal, B; Singh, V. N-(Piperidinomethyl)-3-[(pyridylidene) amino] isatin: A new and effective acid corrosion inhibitor for mild steel. Mater. Chem. Phys 2008, 112, 1035–1039.

30. Ashassi-Sorkhabi, H; Shaabani, B; Seifzadeh, D. Effect of some pyrimidinic Schiff bases on the corrosion of mild steel in HCl solution. Electrochim. Acta 2005, 50, 3446–3452.

31. Tao, Z; Zhang, S; Li, W; Hou, B. Adsorption and Corrosion inhibition behaviour of mild steel by one derivative of benzoic–triazole in acidic solution. Ind. Eng. Chem. Res 2010, 49, 2593–2599.

32. Emregul, KC; Hayvali, M. Studies on the effect of a newly synthesized Schiff base compound from phenazone and vanillin on the corrosion of steel in 2 M HCl. Corros. Sci 2006, 48, 797–812.

33. Rodriguez-Valdez, LM; Villamisar, W; Casales, M; Gonzalez-Rodriguez, JG; Martinez-Villafane, A; Martinez, L; Glossman-Mitnik, D. Computational simulations of the molecular structure and corrosion properties of amidoethyl, aminoethyl and hydroxyethyl imidazolines.

Corros. Sci 2006, 48, 4053–4064.

34. Khaled, KF; Babić-Samardžija, K; Hackerman, N. Theoretical study of the structural effects of polymethylene amines on corrosion inhibition of iron in acid solutions. Electrochim. Acta 2005, 50, 2515–2520.

35. Bentiss, F; Lebrini, M; Lagren'ee, M; Traisnel, M; Elfarouk, A; Vezin, H. The influence of some new 2,5-disubstituted 1,3,4-thiadiazoles on the corrosion behaviour of mild steel in 1M HCl solution: AC impedance study and theoretical approach. Electrochim. Acta 2007, 52, 6865–6872.

36. Xia, S; Qiu, M; Yu, L; Liu, F; Zhao, H. Molecular dynamics and density functional theory study on relationship between structure of imidazoline derivatives and inhibition performance. Corros. Sci 2008, 50, 2021–2029.

37. Ebenso, EE; Arslan, T; Kandemirli, F; Caner, N; Love, I. Quantum chemical studies of some rhodanine azosulpha drugs as corrosion inhibitors for mild steel in acidic medium. Int. J. Quantum Chem 2010, 110, 1003–1018.

38. Arslan, T; Kandemirli, F; Ebenso, EE; Love, I; Alemu, H. Quantum chemical studies on the corrosion inhibition of some sulphonamides on mild steel in acidic medium. Corros. Sci 2009, 51, 35–47.

39. Hansch, C; Leo, A. Substituentz for Correlation Analysis in Chemistry and Biology; Wiley: New York, NY, USA, 1979.

40. Eddy, NO; Ebenso, EE. Quantum chemical studies on the inhibition potentials of some penicillin compounds for the corrosion of mild steel in 0.1 M HCl. J Mol Model 2010.

41. Gece, G. The use of quantum chemical methods in corrosion inhibitor studies. Corros. Sci 2008, 50, 2981–2992.

42. El Ashry, HE; El Nemr, A; Esawy, SA; Ragab, S. Corrosion inhibitors. Part II: Quantum chemical studies on the corrosion inhibitions of steel in acidic medium by some triazole, oxadiazole and thiadiazole derivatives. Electrochim. Acta 2006, 51, 3957–3968.

43. Eddy, NO; Ibok, UJ; Ebenso, EE; El Nemr, A; El Ashry, HE. Quantum chemical study of the inhibition of the corrosion of mild steel in H_2SO_4 by some antibiotics. J. Mol. Model 2009, 15, 1085–1092.

44. Karelson, M; Lobanov, VS. Quantum chemical descriptors in QSAR/QSPR studies. Chem. Rev 1996, 96, 1027–1043.

45. Cardoso, SP; Hollauer, E; Borges, LEP; Gomes, JA da CP. QSPR prediction analysis of corrosion inhibitors in hydrochloric acid on 22%-Cr stainless steel. J. Braz. Chem. Soc 2006, 17, 1241–1249.

46. Trainor, TP; Chaka, AM; Eng, PJ; Newville, M; Waychunas, GA;

Catalano, JG; Brown, GE, Jr. Structure and reactivity of hydrated heamatite (0001) surface. Surf. Sci 2004, 573, 204–224.

47. Vera, L; Guzman, M; Ortega-Luoni, YP. QSPR study of corrosion inhibitors; imidazolines. J. Chil. Chem. Soc 2006, 51, 1034–1039.

48. Khamis, E; El Ashry, ESH; Ibrahim, AK. Synergistic action of vinyl triphenylphosphonium bromide with various anions on corrosion of steel. Br. Corros. J 2000, 35, 150–154.

49. Eddy, NO; Ibok, UJ; Ebenso, EE. Adsorption, synergistic inhibitive effect and quantum chemical studies on ampicillin and halides for the corrosion of mild steel. J. Appl. Electrochem 2010, 40, 445–456.

50. Lukovits, I; Shaban, A; Kalman, E. Corrosion inhibitors: Quantitative structure activity relationships. Russ. J. Electrochem 2003, 39, 177–181.

51. Young, DC. Computational Chemistry, A Practical Guide for Applying Techniques to Real World Problems; Wiley Interscience: New York, NY, USA, 2001.

52. Fuentealba, P; Perez, P; Contreras, R. On the condensed Fukui functions. J. Chem. Phys 2000, 113, 2544–2551.

53. Stoyanov, SR; Gusarov, S; Kovalenko, A. Modelling of bitumen fragment adsorption on Cu^{2+} and Ag^+ exchanged zeolite nanoparticles. Mol. Simul 2008, 34, 943–951.

54. Stoyanov, SR; Gusarov, S; Kuznick, SM; Kovalenko, A. Theoretical modelling of zeolite nanoparticle surface acidity for heavy oil upgrading. J. Phys. Chem. C 2008, 112, 6794–6810.

55. Kittur, MIH; Majan Shetti, CS. Oleochemical: Part 1: Synthesis and biological evaluation of 1,4,6–oxadiazoles and 4H-1,2,4 –triazoles derived from long chain fatty acids. J. Oil Tech. Assoc 1984, 16, 49–54.

56. Ebenso, EE; Ibok, UJ; Ekpe, UJ; Umoren, SA; Jackson, E; Abiola, OK; Oforka, NC; Martinez, S. Corrosion inhibition studies of some plant extracts on aluminium in acidic medium. Trans. SAEST 2004, 39, 117–123.

57. Schmidt, MW; Baldridge, KK; Boatz, JA; Elbert, ST; Gordon, MS; Jensen, JH; Koseki, S; Matsunaga, N; Nguyen, KA; Su, SJ; Windus, TL; Dupus, M; Montgomery, JA. Games version 12. J Comp Chem 1993, 1347–1363.

58. Hyperchem Release 8.0.3 Hypercube Inc: Gainesville, FL, USA, 2002.

59. Steward, JP. Steward Computational Chemistry, 2009.

Chapter 14

CHEMICAL REACTIVITY AS DESCRIBED BY QUANTUM CHEMICAL METHODS

P. Geerlings and F. De Proft

Eenheid Algemene Chemie, Free University of Brussels (VUB), Pleinlaan 2,1050 Brussels,Belgium

ABSTRACT

Density Functional Theory is situated within the evolution of Quantum Chemistry as a facilitator of computations and a provider of new, chemical insights. The importance of the latter branch of DFT, conceptual DFT is highlighted following Parr's dictum "to calculate a molecule is not to understand it". An overview is given of the most important reactivity descriptors and the principles they are couched in. Examples are given on the evolution of the structure-property-wave function triangle which can be considered as the central paradigm of molecular quantum chemistry to (for many purposes) a structure-property-density triangle. Both kinetic as well as thermodynamic aspects can be included when further linking reactivity to the property vertex. In the field of organic chemistry, the ab initio calculation of functional group properties and their use in studies on acidity and basicity is discussed together with the use of DFT descriptors to study the kinetics of S_N2 reactions and the regioselectivity in Diels Alder reactions. Similarity in reactivity is illustrated via a study on peptide isosteres. In the field of inorganic chemistry non empirical studies of adsorption of small molecules in zeolite cages are discussed providing Henry constants and separation constants, the latter in remarkable good agreement with experiments. Possible refinements in a conceptual DFT context are presented. Finally an example from biochemistry is discussed: the influence of point mutations on the catalytic activity of subtilisin.

QUANTUM MECHANICS, QUANTUM CHEMISTRY, COMPUTATIONAL CHEMISTRY, DENSITY FUNCTIONAL THEORY: WHO IS WHO

From Quantum Mechanics to Quantum Chemistry and Computational Chemistry

The failure of classical physics (mechanics and electromagnetism) at the end of the 19th century led to the introduction of the Quantum Concept by Planck, Einstein, Bohr,... culminating in the birth of "modern" quantum mechanics around 1925 due to the work of Schrödinger, Heisenberg, Born, ... Schrödinger's equation occupied a central position in this new theory and, although later on complemented by its relativistic analogue by Dirac, stood the test of time and has been for now 75 years the central equation for the description both of the internal structure of atoms and molecules and their interactions. In his famous quote Dirac already in 1929 went so far to state [1] "The underlying physical laws necessary for the mathematical theory of a large part of physics and the whole of chemistry are thus completely known, and the difficulty is only that the exact application of these laws leads to equations too complicate to be soluble."

The step from Quantum Mechanics to Quantum chemistry can in principle be situated in the pioneering work by Heitler and London [2] on the hydrogen molecule in 1928 providing insight into, to quote Pauling, the Nature of the Chemical Bond [3]. However Quantum Chemistry is, at least in our opinion, more than the mere application of quantum mechanical principles to molecules and their interaction. In the years between 1930 and 1950 Pauling [3], Huckel [4], Coulson [5] indeed used quantum mechanical principles but combined them with their chemical intuition thereby gradually creating a new discipline, nowadays called Quantum Chemistry. The Valence Bond approach (after Heitler and London) was prominent in those days, to the detriment of Hund's and Mulliken's MO method [6]. A revolution was provoked by Roothaan's matrix formulation of the MO method in 1951 [7]. Its elegance together with the increasing computer power paved the way for the large scale introduction of the MO-LCAO method within the framework of the Hartree Fock Self Consistent Field (SCF) approach [8a] as excellently summarized in Pople's comprehensive treatise [8b].

The late seventies, eighties and nineties saw the development and/or the adaptation for systematic use of beyond SCF methods including (part of) electron correlation: Møller Plesset Perturbation Theory [9], the method of Configuration Interaction [10] and various types of Coupled Cluster Theory

[11]. The introduction of initially freely distributed, later on commercially available, computer programs which became more and more user friendly (cf Pople's GAUSSIAN series) [12] definitely promoted Quantum Chemistry, from a branch of Theoretical Chemistry almost exclusively reserved for "pure-sang" theoreticians and concentrating on diatomic and small polyatomic molecules, to a field also creating tools for non-specialists, in many other subfields of Chemistry (Inorganic, Organic, Biochemistry). The new subfield "Computational Chemistry" with P. Schleyer as a prominent figure [13] particularly stresses the "applied" aspects of Quantum Chemistry

The combination of conceptual and methodological improvements, and the growing performance of soft- and hardware led to an ever increasing accuracy in the treatment of problems of a given complexity. Orders of magnitude in the complexity.of problems that could be treated at a given level were gained. These evolutions are beautifully illustrated in Pople's two dimensional chart of Quantum Chemistry, given below in Figure 1 in a slightly adapted version [14].

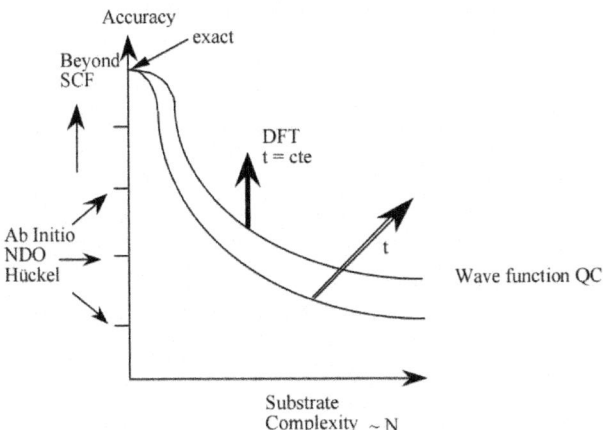

Figure 1: Accuracy versus complexity chart of Quantum Chemistry (After J.A.Pople).

A central paradigm in Quantum Chemistry is the Structure-Properties-Wave function triangle(Fig.2), Structure and Properties further determining reactivity. The third corner of the triangle is the ground state wave-function Ψ_0, or more generally also all excited state wavefunctions determining all properties of the system.

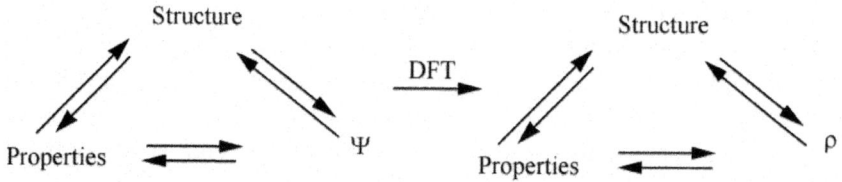

Figure 2: The central paradigm in Quantum Chemistry and its evolution upon the introduction of DFT.

Density Functional Theory Revolutionarized Quantum Chemistry from a Computational Point of View

A step of immense importance has been taken by Kohn and Hohenberg [15] in 1964. They proved that the information content in the electron density function $\rho(\underline{r})$, depending on only 3 variables, determines all ground state properties, thus replacing the crucial position of the complex wavefunction, Ψ, function of 4N variables (where N is the number of electrons).

Equation (1) clearly shows how much information in the wave function of a N-electron system is integrated out when passing to the electron density (x stands for a four vector containing three spatial coordinates r and one spin coordinates of an electron) [16]

$$\rho(\underline{r})= N\!\int \Psi^{*}\!\left(\underline{x},\ \underline{x}_2,\underline{x}_3\ ...,\ \underline{x}_N\right)\Psi\!\left(\underline{x},\ \underline{x}_2,\ \underline{x}_3,\ ...,\ \underline{x}_N\right)d\underline{s}\,d\underline{x}_2\,d\underline{x}_3\,...\,d\underline{x}_N$$

(1)

The problem of searching an optimal ρ instead of the much more complex optimal Ψ is most conveniently done within the Kohn-Sham formalism [17]) introducing orbitals φ_i, whose squares sum up to the electron density.

$$\rho = \sum_i |\varphi_i|^2$$

(2)

A variational procedure yields a pseudo-one electron equation, the analogue of the Hartree-Fock equations, which is written as

$$\left(-\frac{1}{2}\nabla_i^2 + v(\underline{r})+\int\frac{\rho(\underline{r})}{|\underline{r}-\underline{r}'|}d\underline{r}+v_{xc}(\underline{r})\right)\varphi_i = \varepsilon_i\varphi_i$$

(3)

Here, besides the electronic kinetic energy term $(-\frac{1}{2}\nabla_i^2)$, the nuclear attraction term $v(\ )r$ and the classical electronic repulsion term, the exchange correlation term $\upsilon_{xc}(\underline{r})$ appears whose form is actually unknown. One of the key features of present day DFT is the search for the best performing exchange correlation functionals [18]. Although this task is hampered by the lack of a unifying principle as present in wave-function theory (see e.g. Pople's Model Chemistry chart) [8b] impressive progress has been made in recent years among others via the so called hybrid functionals [19] which gained widespread use. Extensive testing of their capability in reproducing molecular properties has been performed [20, 21]. The whole of these efforts led to a methodology which affords the calculation of molecular ground state properties of high quality (in fact often way beyond SCF) at a much lower computational cost. Parr and Yang termed this branch of DFT "Computational DFT" [18]. A "témoignage par excellence" of the ever increasing importance of DFT is (the title of) Koch's book "A Chemist's Guide to DFT" [22] published in 2000 offering an overview of the performance of DFT for various properties to the practicing organic or inorganic chemist. As a result of this evolution the triangle is Fig. 2 can be adapted at one of its vertices, putting ρ (and its obtention via computational DFT) at equal footing with the wave-function Ψ.

DFT as a Provider of New Insights: Conceptual DFT

From Computational Chemistry to Chemical Insight

Both wave function Quantum Chemistry and Density Functional Theory, when being used to compute atomic and molecular properties, yield results which often and for most chemists are not always directly exploitable. The numbers they produce should in many cases be translated, or casted into a language or formalism pointing out their chemical relevance. As simply stated by Parr [23] "Accurate calculation is not synonymous with useful interpretation. To calculate a molecule is not to understand it". Quite often this translation involves terms going back to the early days of theoretical chemistry but still in use as a guideline for chemists in the interpretation of experimental data: hybridization, electronegativity, aromaticity, ...

A beautiful example in wave function quantum chemistry, dating from the sixties and seventies is the transformation of the Molecular Orbitals resulting from the Hartree-Fock equations, which are, usually delocalized over the entire molecule, to a set of localized Molecular Orbitals using a localization criterion [24]. The resulting MO picture is much closer to the Lewis picture of great use in organic chemistry (Figure 3), e.g. in the study of the electronic structure of bonds and its relation to spectroscopic properties. The relation between NMR coupling constants and the percentage of s character of the carbon atom hybrid involved in a CH bond [27], is a classical example, partly addressed in our own work [28]. Of importance in both branches (wave function and Density Functional Theory) is the visualization of the results (e.g. electron density or density difference plots (as in Figure 3), its extensive use also going hand in hand with hard- and software developments. Also in this area of obtaining a better chemical insight via Quantum Chemical calculations, a prominent role was played in recent years by DFT.

DFT as a source of Chemical Concepts

As already mentioned before, computational DFT is founded on a variational principle, more precisely for the energy functional

$$E = E[\rho] \qquad (4)$$

Looking for an optimal ρ, i.e. the one which minimizes E, is thereby subjected to the constraint that ρ should at all times integrate to N, the number of electrons.

$$\int \rho(\underline{r})d\underline{r} = N \qquad (5)$$

Within a variational calculation this constraint is introduced via the method of Lagrangian multipliers, yielding the variational condition

$$\delta[E - \mu\rho] = 0 \qquad (6)$$

where μ is the Lagrangian multiplier, a constant gaining its physical significance in the differential equation (the Euler equation) resulting from (6)

$$v(\underline{r}) + \frac{\delta F_{HK}}{\delta \rho} = \mu \qquad (7)$$

Here $v(\underline{r})$ is the external potential (i.e. due to the nuclei) and F_{HK} is the Hohenberg Kohn functional containing the electronic kinetic energy and the electron-electron interaction operators [29]. It has been Parr's impressive contribution to identify this abstract Lagrange multiplier as [30]

$$\mu = \left(\frac{\partial E}{\partial N}\right)_{V} = -\chi \qquad (8)$$

i.e. the derivative of the energy of the atom or molecule with respect to its number of electrons at constant external potential (i.e. identical nuclear charges and positions) (Figure 4). In this seminal paper, cited already more than 500 times, Parr thereby regained Iczkowski and Margrave's definition of electronegativity ($\chi = -\partial E / \partial N$) [31], Mulliken's 1934 definition can be considered as an approximation to it [32]. The Mulliken values, the arithmetic average of ionization energy (IE) and electron affinity (EA), were already shown before to correlate with the Pauling values [33] and received more and more importance in recent years on the basis of its simpler foundation. It can easily be seen that they correspond to the average slope of the E=E(N) curve at the N value considered.

$$\chi = \frac{1}{2}(IE + EA)$$

(9)

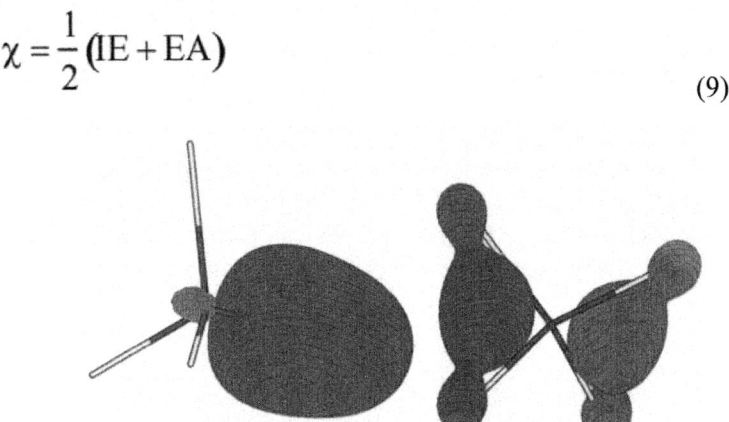

Figure 3: Delocalized Hartree Fock versus Localized Orbitals: one of the triply degenerate HOMO orbitals of methane versus a CH bond orbital. Electron density plot (Hartree Fock STO-3G/Boys localization procedure [25]) obtained with the software package [26]

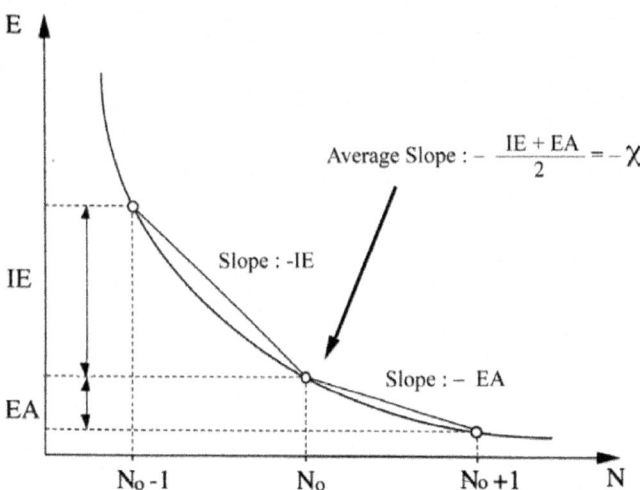

Figure 4: Atomic or molecular energy (E) versus number of electrons (N) at constant external potential: the modern definition of electronegativity.

In analogy with the thermodynamic potential

$$\mu_{\text{Therm}} = \left(\frac{\partial G}{\partial n} \right)_{P,T}$$

(10)

where G represents the Gibbs Free Energy function and n the number of moles, μ was termed electronic chemical potential which turns out to be the negative of the electronegativity.

Within a few years after Parr's contribution various other quantities representing the response of a system's energy to perturbation in its number of electrons and/or its external potential (cf the index v in 8), which both lie at the heart of chemistry, were published. They can nicely be ordered according to Nalewajski's charge sensitivity analysis [34] (Figure 5).

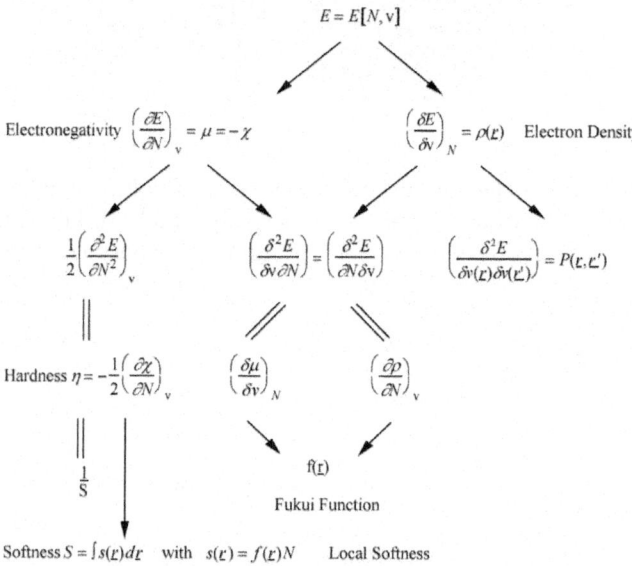

Figure 5: Nalewajski's Sensitivity Analysis: atomic and molecular properties as energy derivatives with respect to N and v.

Appearing in a natural way are -the chemical hardness η, an identification proposed by

Parr and Pearson [35] for the second derivative which respect to N, $\left(\frac{\partial^2 E}{\partial N^2} \right)_v$, and representing the resistance of a system to changes in its number of electrons.

The chemical softness S is naturally defined as the inverse of η

$$S = 1/2\eta$$

(11)

The analogue of equation (9) turned out to be

$$\eta = \frac{1}{2}(IE - EA)$$

$$(9')$$

- the Fukui function $f(\underline{r})$ [36], representing the change in electron density ρ at a given point r when the total number of electrons is changed, a generalization of Fukui's frontier orbital concept [37]

- a local version of S, $s(\underline{r})$, obtained by multiplying S and $f(\underline{r})$, the latter function distributing the local softness over various domains in space [38] (Other concepts introduced in this framework are reviewed in [39, 40])

In this way it is shown that DFT gave the possibility to sharply define concepts known for a long time in chemistry, but to which inadequate precision could be given to use them with confidence in quantitative studies.

The last 15 years showed growing importance of this branch of DFT, conceptual DFT, where these concepts were used as such or within the context of three important principles,

- Sanderson's electronegativity equalization principle [41, 42] stating that upon molecule formation, atoms (or more general arbitrary portions of space of the reactants) with initially different electronegativities $\chi_i^o (i = 1, ..., M)$ combine in such a way that their "atoms-in-molecule" electronegativities are equal. The corresponding value is termed the molecular electronegativity χ_M. Symbolically:

$$\chi_1^o, \chi_2^o, ..., \chi_M^o \longrightarrow \chi_1 = \chi_2 = ... = \chi_M$$

isolated atoms   molecule formation

Electron transfer thereby takes place from atoms with lower electronegativity to those with higher electronegativity, the latter reducing their χ value, the former increasing it (Fig.6)

- Pearson's Hard and Soft Acids and Bases Principle (HSAB) [43, 44] stating that Hard (Soft) Acids (electron pair acceptors), preferentially interact with hard (soft) Bases (electron pair donors).

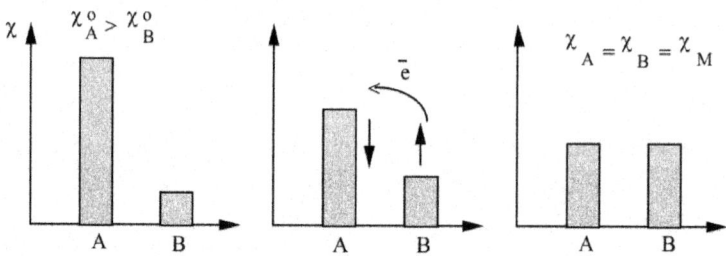

Figure 6: Sanderson's Electronegativity Equalization Principle.

Both principles were proven [30, 45] as was also the third one, the Maximum Hardness Principle, stating "that molecules try to arrange themselves to be as hard as possible" [44, 46]. In recent years our group was active in the development and/or use of DFT based concepts as such or within the context of the afore mentioned and other principles. Also performance testing was one of our objectives: setting standards for computational DFT in order that it can be used for a given type of problem with the same level of confidence as the combination of level and basis set in the case of Pople's model Chemistry for wave function theories. Studies were undertaken on IR frequencies and intensities, dipole and quadrupole moments, ionization energies and electron affinities and Molecular Electrostatic Potentials [20, 47,48,49].

THE STRUCTURE -PROPERTY (REACTIVITY)- ELECTRON DENSITY TRIANGLE: SOME EXAMPLES

Introduction

In this second part of the contribution examples are given on the role DFT studies, both conceptual and computational, can play in exploiting the structure, property -density triangle in Fig. 2 where we concentrate on properties directly related to reactivity (both seen from a thermodynamic and kinetic point of view). Examples will be taken essentially from our own work with reference to work of other groups if relevant to the discussion. As such this part is not aiming at completeness at all. The reader should consult other sources to have a complete overview of applications of conceptual DFT [39, 40a, 40b]. Illustrations will be taken from organic, inorganic and biochemistry.

Organic Chemistry

Group Properties and their use in Acidity and Basicity Studies

Functional groups are playing a fundamental role in rationalizing structure and reactivity, thus dictating transformations in synthetic chemistry [50], both in organic and inorganic chemistry.

An insight in the properties of these molecular building blocks is of utmost importance in the design of a rational chemistry. Whereas group electronegativity has already a longstanding history [33], the field of group softness and/or hardness is much less developed. Moreover a non-empirical uniform computational scheme obeying the working equations (9) and (9') was lacking in the period we started this work. We therefore presented a non-empirical computational scheme for group electronegativity, hardness and softness [51] for more than 30 functional groups

CH_3 ; CH_2CH_3; $CH=CH_2$; $C\equiv CH$; CHO ; $COCH_3$; $COOH$; $COCl$; $COOCH_3$; $CONH_2$; $C\equiv N$; NH_2 ; CH_2-NH_2 ; NO_2 ; OH ; CH_2OH ; OCH_3 ; F ; CH_2F ; CHF_2 ; CF_3 ; SiH_3 ; PH_2 ; SH ; CH_2SH ; SCH_3 ; Cl ; CH_2Cl ; $CHCl_2$; CCl_3.

Starting from the geometry the group usually adopts when being embedded in the molecule we calculated its η,S and χ values via (9) and (9'), considered as a radical, both at the Hartree Fock and CISD level using Pople's 6-31++G** basis [8b]. Figure 7 shows the correlation of the CISD calculated group electronegativities (showing a correlation coefficient r of 0.943 with the HF values for the same basis) with what is recently [52] considered as the most appropriate "experimental scale", the ^{13}C ^1JCC (ipso ortho) coupling constants in monosubstituted benzenes [53]. As can be seen the correlation fails for OCH_3 and SiH_3 for reasons that could not be detected. It is less convincing for groups containing triple bonds clearly to the higher demands for correctly describing electron correlation effects. Dropping these values a correlation coefficient r of 0.941 is obtained for the remaining groups. Typical trends to be observed are - the central atom effect

$$\chi_{CH_3} < \chi_{NH} < \chi_{OH} \quad cf. \ \chi_C < \chi_N < \chi_O$$

indicating that upon saturation of two different atoms with hydrogens the electronegativity of the resulting groups parallels that of the naked atoms.

- the second row effect indicating increasing electronegativity upon increasing s-character of the central atom of the group. For alkylgroups the intuitively expected decrease of χ upon increasing chain length or branching is found: Me: 5.12 ; Et: 4.42 ; n-Pr: 4.39 ; i-Pr: 3.86 (CISD values).

$$\chi_{CH_2CH_3} \; < \; \chi_{CH=CH_2} \; < \; \chi_{C+CH}$$

$$\chi_{CH_2NH_2} \; < \; \chi_{CH=NH_2} \; < \; \chi_{C+N}$$

Turning now to the hardness values, experimental scales are scarce, the best candidate being the corresponding radical hardness [54], although possible differences in geometry (e.g. for the CH_3 radical) indicate that this correlation should be looked upon with much reserve. Figure 8 shows a correlation, upon withdrawing the extremely hard CF_3 group as an outlier, of 0.926 for the 14 remaining cases for which experimental values are available. Typical trends which can be discussed are again (vide supra)

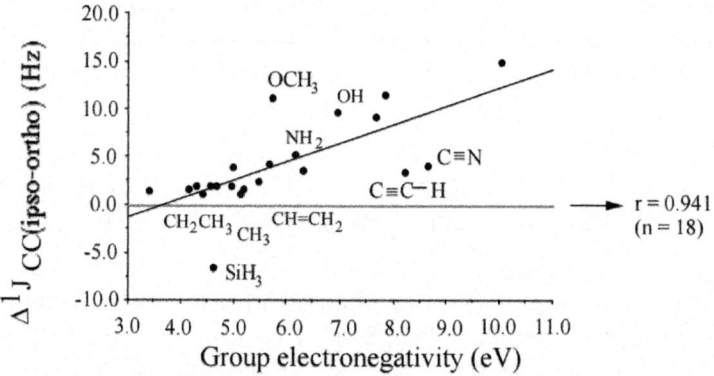

Figure 7: Calculated vs. Experimental Group Electronegativity values.

- the central atom effect

$$\chi_{CH_3} \; < \; \chi_{NH_2} \; > \; \eta_{OH} \; < \; \eta_F \quad \text{cf.} \quad \eta_C \; < \; \eta_N \; > \; \eta_O \; < \; \eta_F$$

- the second row effect (extremely important in forthcoming discussions but here already illustrating the opposite behaviour of χ and S)

$$\eta_{OH} \; > \; \eta_{SH}$$

$$\eta_{CCl_3} \; < \; \eta_{CH_3} \; < \; \eta_{CF_3}$$

$$\eta_{CH_3} \; > \; \eta_{SiH_3}$$

- the "volume" effect in alkyl groups, showing decreasing hardness (increasing softness) upon increasing chain length or branching. CISD η values illustrating this trend are

6.60 (H) ; 5.34 (Me) ; 4.96 (Et) ; 4.70 (n-Pr) ; 4.62 (i-Pr)

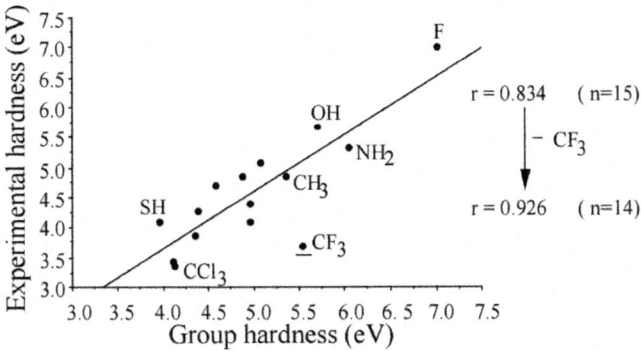

Figure 8: Calculated vs. Experimental Group Hardness values (see text).

As an example of the use of group properties we consider [55] the experimental acidity sequence of alkylalcohols showing an opposite behaviour in aqueous solution and gas phase [56] [57]. Whereas in aqueous solution the acidity decreases upon increasing carbon chain length and branching, the opposite behaviour is encountered in the gas phase. The former trend is usually traced back to the electron donating character of alkylgroups (the +I effect as well known in electrophilic aromatic substitutions on benzene [58]), whereas the latter tendency should imply an at first sight unexpected electron withdrawing character of alkyl groups.

We correlated the (gas phase) acidity quantified by the ΔG°_{acid} value with the calculated alkylgroup properties χ and η (all calculations were done at a uniform 6-31G* MP4 level). The sequences parallel those of the 6-31++G** CISD values in [51].

The values shown in Table 1 indicate the expected tendencies of increasing softness upon increasing chain length and branching accompanied by decreasing electronegativity, reflecting the traditional idea of an alkylgroup as electron donor (+I effect). It was tempting to have a more detailed look at the correlation between charge distribution, both in the acidic form ROH and the alkoxide anion RO-, and the acidity variation upon varying R. We therefore made use of χ and η values given in Table 1 with an electronegativity equalization scheme.

Working at functional group resolution and neglecting external perturbation higher order terms one obtains that upon embedding a functional group A with intrinsic χ_A^0 and η_A^0 into a molecule its electronegativity changes to χ_A with

$$\chi_A = \chi_A^0 - 2\eta_A^0 \Delta N_A$$

Table 1: 6-31G* MP4 alkylgroup properties: electronegativity χ (eV) and hardness η (eV)

	χ	η
Me	4.35	5.98
Et	3.76	5.45
n-Pr	3.68	5.17
i-Pr	3.44	5.04
n-Bu	3.57	5.05
t-Bu	3.30	4.19

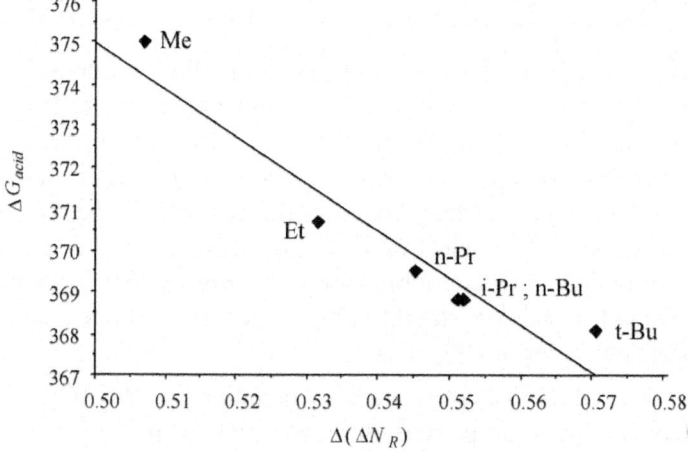

Figure 9: Experimental gas phase acidity of alkylalcohols (in kcal mol^{-1}) vs. $\Delta(\Delta N_R)$ (see text) (reprinted with permission by Pergamon/Elsevier-Reference [55]).

This equation is easily derived using a Taylor series expansion of the E = E[N,v] functional (cf. Fig.5) around a reference number of electrons and at constant external potential. Applying this relationship to both the acidic form of an alcohol ROH both for the R and OH groups and equalizing the electronegativities of both molecular building blocks, one gets for ΔN_{R_1} (the number of electrons transferred to the alkylgroup in the acidic form)

$$R\!:\!OH \qquad \Delta N_{R_1} = \frac{\chi_R^0 - \chi_{OH}^0}{2\left(\eta_R^0 + \eta_{OH}^0\right)} < 0 \tag{12}$$

In the case of the conjugate base RO$^-$ one obtains in a completely analogous way (taking into account again charge conservation, but now not at charge 0 but -1)

$$R{:}O^- \qquad \Delta N_{R_{II}} = \underbrace{\frac{\chi_R^0 - \chi_O^0}{2\left(\eta_R^0 + \eta_O^0\right)}}_{(a)} + \underbrace{\frac{\eta_O^0}{\eta_R^0 + \eta_O^0}}_{(b)} \qquad\qquad (13)$$

In (12) ΔN_{R_I} is negative as χ_{OH} is higher than χ_R^0 in the case of an alkylgroup. The intramolecular charge transfer is electronegativity dominated, as far as its direction is concerned, hardness playing a role in the amount of charge transferred. In the conjugate base form eqn.(13) indicates that a second, completely hardness-determined term, shows up which turns out to be larger in absolute value than the first term in all cases considered. This hardness dominated term consequently accounts for an electron transfer towards the alkyl group, the value being more important when the softness of the alkylgroup increases. Equating η^0 and χ^0 both for OH and O, the difference between $\Delta N_{R_{II}}$ and ΔN_{R_I}, $\Delta(\Delta N_R)$, can be approximately written as

$$\Delta(\Delta N_R)= \Delta N_{R_{II}} - \Delta N_{R_I} \approx \frac{\eta_O^0}{\eta_R^0 + \eta_O^0} \qquad\qquad (14)$$

indicating that the difference in electron transfer from R to the oxygen part of the alcohol upon deprotonation becomes more important when the alkylgroup is softer. Otherwise stated: $\Delta(\Delta N_R)$ is Rsoftness dominated. It may be therefore concluded that in order to correctly describe acid-base properties of molecules, the electronic properties of the charged form of the acid base equilibrium are of utmost importance. Figure 9 shows the excellent correlation between the experimental gas phase acidity and $\Delta(\Delta N_R)$ for the simplest alkylalcohols. In the present case alkylgroups act as electron acceptors; these electron withdrawing properties of alkylgroups were previously encountered in the literature in cases where alkylgroups are placed on negatively charged carbon atoms or reaction centers such as in α-alkylbenzyl carbanions [59].

We finally turned to the effect of solvent. We therefore recalculated group electronegativity and softness values, using a Self Consistent Reaction Field Model [60], introducing the solvent dielectric constant ε [61]. In Figure 10 we plot the $\Delta(\Delta N)$ quantity as a function of the Kirkwood function $\frac{\varepsilon-1}{2\varepsilon+1}$ related to the free energy of solvatation.

It is seen that an approximate linear relationship between $\Delta(\Delta N)$ and the Kirkwood function is found for the simplest alkylalcohols, with a crossing of the curves near the H_2O case. The gas phase acidity sequence is thereby inverted when passing to aqueous solution as found experimentally [62].

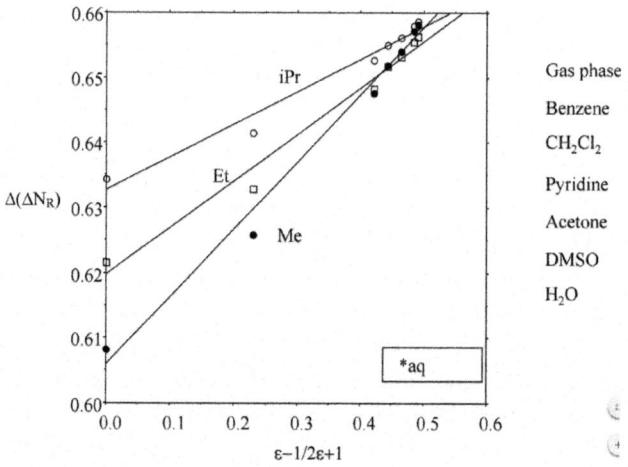

Figure 10: Plot of $\Delta(\Delta N_R)$ versus the Kirkwood function $\frac{\varepsilon-1}{2\varepsilon+1}$ (Reprinted with permission by the American Chemical Society - Reference [61]).

Kinetics of $S_N 2$ Reactions

The nucleophilic substitution reaction is one of the basic transformations in organic chemistry. It has been extensively studied both theoretically and experimentally, both in the gas phase and in solution [63].

In the case of gas phase reactions e.g. $X^- + CH_3Y \rightarrow Y^- + CH_3X$ a double well profile is present in the Reaction Profile (Fig. 11)

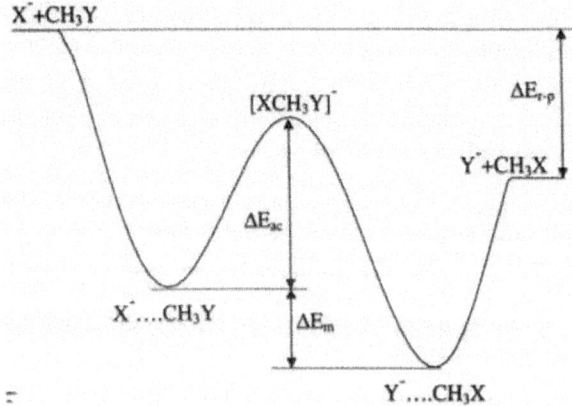

Figure 11: Reaction Profile for a $S_N 2$ reaction in the gas phase (Reprinted with permission by the American Chemical Society - Reference [64]).

We tried to interpret the reaction pathway via DFT based concepts (η, S ...) and principles (HSAB, MHP), using a variety of X, Y combinations [64]. 6-31+G* optimized structures were generated for reactants, products, transition states (TS) and ion-molecule complexes. A remarkable correlation is found between ΔE_m both with the calculated ΔE_{p-r} ($r^2 = 0.97$) and the experimental heat of reaction ($r^2 = 0.99$) indicating the relevance of ΔE_m. This quantity shows, e.g. in the case of the $X^- + CH_3F \rightarrow CH_3X + F^-$ reaction, a remarkable correlation with the difference in hardness between F and the group X (Figure 12). The hard F atom in CH3F can be expected also to harden its neighborhood, in casu the carbon atom thereby favoring an interaction with a harder X- according to the HSAB principle. One thereby regains Pearson's statement about S_N2 reactions that "when nucleophile and leaving group have similar hardnesses, reaction rates are relatively high". Gazquez [65] developed a formalism to relate the reaction energy and the activation energy to differences in hardness between reagents, products and TS. He showed that ΔE_{p-r} should be proportional to $\frac{1}{\Sigma S_r} - \frac{1}{\Sigma S_p}$, where ΣS_r and ΣS_p denote the softness sum for reactants and products, respectively

$$\Delta E_{p-r} \sim \frac{1}{\Sigma S_r} - \frac{1}{\Sigma S_p} \tag{15}$$

It is clearly seen that an exothermic reaction yields

$$\Sigma S_r > \Sigma S_p \tag{16}$$

i.e. products which are harder than reagents, thus recovering the MHP. It should however be remarked that the stringent conditions to be satisfied in the proof of the MHP (constancy of both the external and chemical potential) put severe restrictions on situations in which the principle is applied. For a recent detailed discussion we refer to Chandra and Uchimaru [67]. When applying (15) to the ion-reagent and product complexes one expects a relationship

$$\Delta E_m \approx \frac{1}{S_{prod \atop complex}} - \frac{1}{S_{reag \atop complex}} \sim \frac{1}{\alpha_{prod \atop complex}} - \frac{1}{\alpha_{reag \atop complex}} \approx \Delta E_{r,\alpha} \tag{17}$$

where the proportionality between softness and polarizability (α) [66] has been exploited. In Fig. 11 we observe that $\Delta E_{r,\alpha}$ is always negative (evolution from a complex with lower hardness to one with higher hardness) and that ΔE_m correlates with $\Delta E_{r,\alpha}$ (separate correlations for hard and soft R groups have been drawn). Turning now to kinetic aspects the central barrier can be regarded as providing the activation energy ΔE_{ac}. Adopting a similar ansatz as before in eq. (17) we looked for two correlations between ΔE_{ac} and the difference $\frac{1}{\alpha} - \frac{1}{\alpha_{TS}}$

. Again two correlations can be drawn, one for soft groups, one for hard groups as seen in Fig. 14.

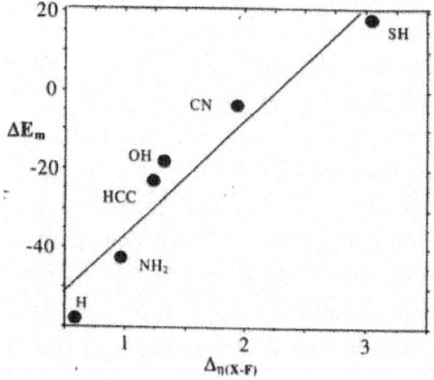

Figure 12: Calculated reaction energy ΔE_m (in kcal mol^{-1}) as a function of the group hardness difference between the X group and fluorine, $\Delta\eta_{X\text{-}F}$ (Reprinted by permission by the American Chemical Society - Reference [64]).

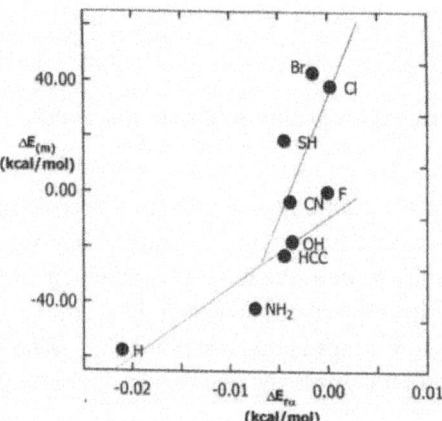

Figure 13: Calculated reaction energy ΔE_m (in kcal mol^{-1}) as a function of the reaction energy $\Delta E_{r,\alpha}$ (in a.u.) obtained via the Gazquez approach. Separate linear correlations for hard (X = H, NH$_2$, OH, F) and soft (X = HCC, CN, SH, Cl) nucleophiles are shown (Reprinted by permission by the American Chemical Society - Reference [64]).

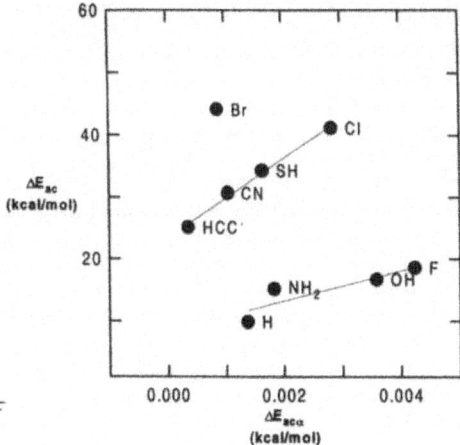

Figure 14: Calculated values of the central barrier energies ΔE_{ac} (kcal mol^{-1}) as a function of the central barrier $\Delta E_{ac'\alpha}$ value (in a.u.) obtained via the Gazquez approach. Separate linear correlations for hard and soft nucleophiles are shown as in Figure 12. (Reprinted by permission by the American Chemical Society - Reference [64]).

These recent results show that the use of DFT based concepts such as softness might shed further light on the reactivity within the context of both the HSAB and MHP principles. In the next paragraph the HSAB principle is further exploited in the study of regioselectivity in organic reactions.

Regioselectivity in Organic Reactions: the HSAB Principle

As stated above the HSAB principle has been given theoretical support by Parr and Pearson, later on by Gázquez. Based on the global properties of the reacting systems (acid (A) and base (B)) with global softness values S_A and S_B it was shown, based among others on the idea of equalization of electronic chemical potentials (cf. §1.3.2), that the largest interaction energy for constant SA occurs when S_B equals S_A thereby regaining the HSAB principle [68] . If the interaction sites in A and B are specified (say k in A and l in B) it was shown by Gazquez and the present author [68] that the demand

$$S_A = S_B \qquad (18a)$$

is converted into an analogous equation at local level , i.e.

$$s_{A,k} = s_{B,l} \qquad (18b)$$

where $s_{A,k}$ and $s_{B,l}$ are the local softness values at sites k and l.

Obviously this expression is ready for use in the study of regioselectivity problems. As an example (for further applications see [69]) we briefly discuss the regioselectivity of Diels Alder reactions using the softness matching (18b) approach in at local level [68b].

The predominance of orthoregioisomers in the cycloaddition of 1-substituted dienes 1 and asymmetrical dienophiles 2 cannot be explained by electronic effects, as replacement of an electron donating substituent by an electron-attracting one does not alter the regioselectivity as discussed by Anh and coworkers [70].

In the double local-local approach of the HSAB principle we considered the local softness resemblance of the termini combinations 1 and 1', 4 and 2', as compared to the 1-2', 4-1' ones. The latter combination yields the meta product, the former the paracycloadduct, considered (cf sterical hindrance) as the contrathermodynamic reaction product.

In the cases R=-Me, -OMe, -COOH, -CN, -NH$_3$, -NMe$_2$, -NEt$_2$, and -OEt and R' = -COOH, - COOMe, -CN, -CHO, -NO$_2$ and -COMe (ensuring in most cases a Normal Electronic Demand reaction type, the dienophile being the electrophilic partner and the diene the nucleophilic one), we calculated local softness values s_1^-, s_1^+, s_2^+, and s_4^- from 3-21G optimized structures at the same level* .

In order to look for a simultaneous fulfillment of the local HSAB principle at both termini the following local softness similarly indicators were evaluated.

$$S_{ortho} = \left(s_1^- - s_{1'}^+\right)^2 + \left(s_4^- - s_{2'}^+\right)^2$$

$$S_{meta} = \left(s_1^- - s_{2'}^+\right)^2 + \left(s_4^- - s_{1'}^+\right)^2 \tag{19}$$

Upon analysis of the Frontier Molecular Orbital-energies, the diene could always be identified as the electron donating system and the dienophile acting as electron acceptor.

It is seen that in the 8×6=48 cases studied, corresponding to all R and R' combinations, S_{ortho} is always smaller than S_{meta} except when a CN substituent is present either in the diene or the dienophile. However the DFT related reactivity parameter for the CN group was shown to be highly sensitive to correlation effects. Moreover the idea when writing down equations (19) is based on the hypothesis of a reaction with both couples of termini reacting at the same rate (synchronicity). Concertedness however is not a synonym to synchronicity prompting us to look for the smallest of the four quadratic forms in (19) which in almost all cases turns out to be the $(s_4^- - s_2^{'+})^2$ term.

This result points into the direction of the C_4-C_2' bond forming faster than the C_1-C_1' bond, in obvious agreement with the demand of equal softness of interacting termini as there are most remote from R and R'. The asynchronicity in the mechanism suggested on this basis is confirmed by Houk's transition state calculations [71] where in all cases with R and R'≠H asymmetric transition states were found with the C_4-C_2' distance being invariably shorter than C_1-C_1'. This HSAB study performed at the locallocal level provides an answer to Anh's long pending hypothesis [70] stating that "it is likely that the first bond would link the softest centers". Important work broadening the scope of this type of approach has been delivered in recent years among others by Nguyen and coworkers, partly in collaboration with the authors, including free radical additions to olefins [72], hydration of cumulenes [73], 2+1 cycloadditions between isocyanide and heteronuclear dipolarophiles [74], cycloaddition reactions between a 1,3-dipole and a dipolarophile [75] and excited state [2+2] photocycloaddition reactions of carbonyls [76]. On the other hand Ponti [77] presented a generalization of the ansatz via eqn (19).

Similarity in Reactivity

Similarity in a fundamental concept in chemistry and pharmacology. In the design of drugs for example one supposes that molecules with similar structures will also exhibit similar biological or physiological activities [78]. The rapid development of computational techniques in recent years has enabled a systematic investigation of the similarity concept suited for quantitative studies of molecular activity [79]. Several similarity indices have been proposed among which the Carbo Quantum Molecular Similarity Index ZAB has played a prominent role [80]. In simplest form it can be written as

$$Z_{AB}^\rho = \frac{\int \rho_A(\underline{r})\rho_B(\underline{r})d\underline{r}}{\left[\int \rho_A^2(\underline{r})d\underline{r} \cdot \int \rho_B^2(\underline{r})d\underline{r}\right]^{1/2}}$$

(20)

or introducing the shape factor $\sigma(\underline{r})$ [81]

$$\rho(\underline{r}) = \sigma(\underline{r})N$$

$$Z^\rho_{AB} = \frac{\int \sigma_A(\underline{r})\sigma_B(\underline{r})d\underline{r}}{\left[\int \sigma^2_A(\underline{r})d\underline{r} \cdot \int \sigma^2_B(\underline{r})d\underline{r}\right]^{1/2}}$$

(21)

indicating that the similarity index only depends on the shape of the density distribution and not on its extent. To obtain a more reactivity related similarity index we proposed to replace the electron density in eq. (21) by the local softness s( )r (cf. § 1.3.2.) [82] yielding the following expression

$$Z^s_{AB} = \frac{\int f_A(\underline{r})f_B(\underline{r})d\underline{r}}{\left[\int f^2_A(\underline{r})d\underline{r} \int f^2_B(\underline{r})d\underline{r}\right]^{1/2}}$$

(22)

where the local softness has been filtered out.

The index (22) has been tested via similarity analysis of peptide isosteres first using the Carbo and Hodgkin Richards indices [82a]. In the field of peptidomimetics analogues of the peptide bond [83] are looked for, showing however resistance to nucleophilic attack. Functional groups proposed in the vast literature in this domain are CH=CH, CF-CH, CH_2-CH_2, CH_2-S, $COCH_2$, CH_2-NH etc. We tested the similarity of a model system CH_3-CO-NH-CH_3 with its analogues on the basis of the similarity indices involving ρ (20) and s (22). DFT has been used here also as a computational tool using the B3PW91 exchange correlation potential [19a] [84] combined with a 6-311G* basis set [8b]. Table 2 shows that the sequences for Zρ and Zs are not identical. The highest density similarity is observed for transbutene, which is not unexpected in view of the trans structure of the peptide bond and the resulting matching of the terminal chains. However, turning to the softness similarity, the Z fluorinated alkene structure shows higher resemblance with the amide bond due to the similarity in polarity with the carboxyl group. The potential use of C=C-F as a peptidomimetric is in accordance with the results of [85].

Notice that the softness similarity has been calculated on the basis of an average of $s^+(\underline{r})$ and $s^-(\underline{r})$. The high similarity index thus represents a similarity in an average interaction of the model system, including terminating groups, with the surroundings, whereas (not shown) the similarity for a nucleophilic attack is very low (0.016 for the Z^{s+}_{AB} index) [82]. Summarizing, a combined search for optimal Z^ρ_{AB} (identical shape) - if necessary with a different type of index, such as Hodgkin's one [82][86] including extent - and optimal Z^s_{AB} s might be a valuable approach not only in this case but in rational drug design as a whole. In a more recent study [82b] one of the remaining problems in this approach, the orientational and translational dependence of Z, was avoided by introducing the concept of an autocorrelation function of the property considered.

Table 2: Similarity indices Z_{AB}^{ρ} and Z_{AB}^{s} for peptide isosteres of CH_3-CO-NH-CH_3 (B3PW91/ 6- 311G* calculation). The centers of the central bond were brought to coincidence

- CO-NH -	Z_{AB}^{ρ}	Z_{AB}^{s}
- CH=CH - (trans)	0.582	0.576
- CH=CH - (cis)	0.506	0.501
- CF=CH - (Z)	0.418	0.635
- CH2-CH2 -	0.466	0.495
- CH2-S -	0.034	0.337
- CO-CH2 -	0.531	0.108
- CH2-NH -	0.412	0.559

Inorganic Chemistry: Zeolites

Zeolites are alumino-silicates consisting of SiO_4 and AlO_4 tetrahedra linked to each other by their corner oxygens. The alumino silicate structure is negatively charged due to the isomorphic subsitution of silicon by aluminum [87]. This negative charge is balanced by exchangeable cations. When protons are introduced as counterions, the zeolite becomes a Brφ nsted acid, the protons being positioned on an oxygen atom connecting an aluminum and a silicon atom. These bridging hydroxyls [88] are at the origin of the acid catalysis application of zeolites [89]. Another aspect of the structure of zeolites is the occurrence of large vacant interconnected spaces forming long, wide channels of varying size depending on the type considered allowing the crystal to act as a molecular sieve [90]. For many years [91] the concepts of electronegativity, softness, hardness .... were exploited by us in the study of the acidity of bridging hydroxyl in zeolites, of utmost importance in their catalytic properties (for a review see [92]).

Due to space limitations and because some fundamental aspects of this kind of study are already present in § 2.2.1. on the acidity of alcohols, we concentrate in this paragraph on a recently aborded topic especially highlighting the possibilities of Computational Chemistry, switching at the end however again to Conceptual DFT: adsorption in zeolites.

The problem we recently addressed [93, 94, 95] was the selectivity of adsorption of gases [96] for which up to now no parameter free ab initio quantum chemical studies were performed yet [90]. Below we summarize the results for the interaction of small molecules (such as N_2, O_2, CO), with a NaY faujasite type zeolite, more specifically with the α cage.

At sufficiently low pressure adsorption is governed by Henry's law [96]

$$q = Kp \tag{23}$$

where q is the amount of substance per unit volume in the adsorbed phase and p the pressure. Henry's constant K can be written as [97]

$$K = \frac{BI}{aRT} \tag{24}$$

where B is the number of cavities in which adsorption can take place per unit mass of zeolite. The quantity "a" equals 1 for a monoatomic gas, 4π for a linear molecule and $8\pi^2$ for a non linear molecule. The configuration integral I is defined as

$$I = \int e^{-E(\underline{r},\underline{\Phi})/RT} \, d\underline{r}d\underline{\Phi} \tag{25}$$

where E represents the interaction energy of the adsorbing molecule with the zeolite cage at position r and with orientation Φ. The integration is performed over the volume V of the supercage.

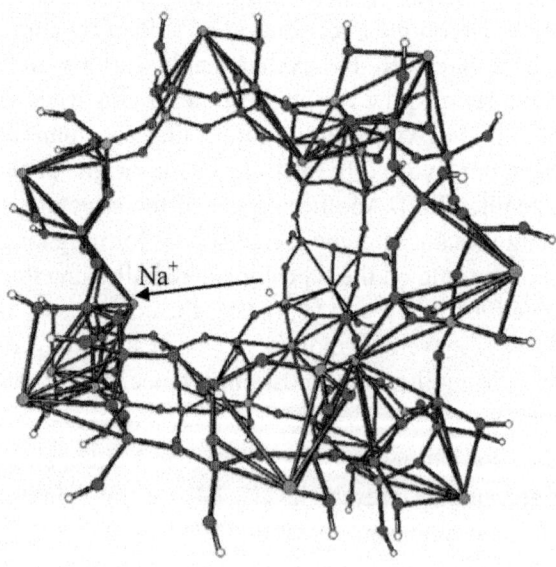

Figure 15: Cluster used to model the large cage of faujasite Y (Reprinted with permission by J. Wiley - Reference [94]).

The cluster representing the supercage and its nearest environment was chosen to be "as large as possible" (within computational limits) and with an Si/Al ratio of 3. The 232 atoms cluster, involving OH groups as terminators is shown in Figure 15, where also four cationic adsorption sites of type II, the ones "active" in the conditions that we consider are present, one of them being indicated.

An embedded cluster method was then adopted, considering the molecule in the field generated by the cage atom charges, the latter being obtained in a single run at Hartree Fock STO-3G or 3-21G level. The points at which the interaction energy is evaluated are selected by constructing a cubic grid with a grid distance of 0.50 Å in each direction and considering the center of each cube generated in this way. The expression (25) is thereby approximated as

$$I \approx \frac{4\pi}{3} \Sigma_i \left(e^{-E_{x,i}/RT} + e^{-E_{y,i}/RT} + e^{-E_{z,i}/RT} \right) \Delta V_i$$

$$(26)$$

where the three terms account for three mutually perpendicular orientations of the molecule, each one given an equal weight to simplify the orientational integration $d\Phi$. ΔV_i equals 0.125 Å3 .

A supplementary condition was introduced imposing a minimal distance between an atom of the adsorbing molecule and a cage atom (cf. the treatment of the repulsive term in the potential, for details see [94]). At these small distances E becomes positive and its contribution to I, via the exponential function, very small. This procedure leads to the consideration of not less than ± 3000 cage-points necessitating an in-depth search for an optimal quality/ cost ratio in the procedure for the interaction energy evaluation. The optimal procedure, from a quality/cost ratio point of view, turned out the following one [94,95,98]. The starting point was the energy expansion of a molecule in a non-uniform electric field [99].

$$E(F_\alpha, F_{\alpha\beta}, F_{\alpha\beta\gamma}, \ldots) = E - \mu_\alpha F_\alpha - \frac{1}{3}\Theta_{\alpha\beta}F_{\alpha\beta} - \frac{1}{15}\Omega_{\alpha\beta\gamma}F_{\alpha\beta\gamma}$$

$$- \frac{1}{105}\Phi_{\alpha\beta\gamma\delta}F_{\alpha\beta\gamma\delta} - \frac{1}{2}\alpha_\alpha F_\alpha F_\beta - \frac{1}{6}\beta_{\alpha\beta\gamma}F_\alpha F_\beta F_\gamma$$

$$(27)$$

where F_α, $F_{\alpha\beta}$, ...represent the field, field gradient, ... components whereas μ_α, $\Theta_{\alpha\beta}$, $\Omega_{\alpha\beta\gamma}$,... stand for the dipole, quadrupole, octadecapole, ... components and α_α, $\beta_{\alpha\beta}$, ... are the dipole polarizability and the first hyperpolarizability. In the case of a neutral molecule of the $C_{\infty,v}$ type (CO) the expression simplifies to (the molecular axis is taken to be the z-axis)

$$E(q, R, \theta) = E^o + \sum_i \mu_z q_i R_i^{-2} \cos\theta_i + \sum_i \Theta_{zz} q_i R_i^{-3} (3\cos^2\theta_i - 1)/2$$

$$+ \sum_i \Omega_{zzz} q_i R_i^{-4} (5\cos^3\theta_i - 3\cos\theta_i)/2$$

$$+ \sum_i \Phi_{zzzz} q_i R_i^{-5} (35\cos^4\theta_i - 30\cos^2\theta_i + 3)/8$$

$$- \sum_i E_i^2 (\alpha_{zz}\cos^2\theta_i + \alpha_{xx}\sin^2\theta_i)/2 \tag{28}$$

E_i represents the electric field at the origin of the molecule due to the surrounding point charges. In $D_{\infty,h}$ cases (N_2, O_2, CO_2, ...) terms in μ and Ω vanish.

The quality of this approach in dependent on the number of terms retained in the expressions (27) and (28) and the quality of the multi-pole moments, polarizabilities ... These were calculated with DFT methodology (B3LYP functional [19]) combined with Dunning's extremely large basis sets (the augmented -correlation consistent and polarized - valence - quadruple or quintuple basis sets (AVQZ or AV5Z) [100][101]).

This methodology was proved by us to yield multipole moments which are in excellent agreement with experiment [98] [102].

In Table 3 the main results are summarized for N_2, O_2, CO, CO_2, C_2H_2. Besides Henry constants, also separation constants α (ratio of two Henry constants) and the isosteric heats of adsorption are tabulated. The latter values were obtained via the Van't Hoff equation

$$\frac{\partial \ln K}{\partial (1/T)} = -\frac{\Delta H^o}{R} \tag{29}$$

for which K was calculated in the temperature interval between 260 and 340K with an increment of 10K and using a linear regression.

Table 3: Henry constants, separation constants and isosteric heats of adsorption on NaY: comparison between theory and experiment [94][98]

	N_2	O_2	CO	CO_2	C_2H_2
K	3.320	1.830	9.951	63.33	196.8
K_{exp}	31.4	15.4	85		
ΔH^o	-12.9	-7.9	-18	-27	-29
ΔH_{exp}	-14	-9.4	-20	-37	

	N_2/O_2	CO/N_2	CO_2/CO
α	1.81	2.99	6.36
α_{exp}	2.04	2.70	

As a whole these results yield K values which are systematically one order of magnitude too small but showing correct sequences. The order of magnitude should be considered within the correct context: a uniform underestimation of the interaction energy of only 1.4 kcal mol-1 already yields an order of magnitude underestimation of K. This extremely high sensitivity is of course due to the sensitivity of the potential in the repulsion part, plugged into an exponential in the configuration integral. The ratio K_{exp}/K is almost constant with values of 9.45, 8.41, 8.54 respectively. As a consequence the separation constants are in excellent agreement with experiment. The isosteric heats of adsorption also show very good agreement in absolute value and reproduce the experimental sequence. In our opinion these investigations pave the way for future studies involving more complex molecules where the lines drawn here, together with increasing computer hard and software and DFT based methodology, may finally yield to calculations which will be of great use in the design of zeolites for well defined purposes (e.g. gas separation).

An alternative for the interaction energy evaluation including electrostatic and polarization effects was proposed in a conceptual DFT context. Using a DFT perturbational approach we obtained the following expression for the interaction energy

$$\Delta E = \sum_i q_i V(\underline{R}_i) + S \sum_i \sum_j q_i q_j \left(\int \frac{f(\underline{r})}{|\underline{r} - \underline{R}_i|} d\underline{r} \int \frac{f(\underline{r}')}{|\underline{r}' - \underline{R}_j|} d\underline{r}' \right) - \int \frac{f(\underline{r})}{|\underline{r} - \underline{R}_i||\underline{r} - \underline{R}_j|} d\underline{r} \tag{30}$$

The first term corresponds to the electrostatic interaction $V(\underline{R}_i)$ as given by the molecular electrostatic potential at position $\underline{R}_i$ where charge q_i is located, multiplied by this charge. The second term is the polarization term [103] in which both the total softness, S, and the fukui function, $f(\underline{r})$, appear. Further work to evaluate in an efficient way the integral in the second term is in progress, the basic requirements for the evaluation of the interaction energy being the molecular electrostatic potential and fukui function $(V(\underline{r})$ and $f(\underline{r}))$ and the total softness nowadays ready available at a high precision level.

Conceptual DFT can thus be exploited in the study of adsorption behaviour of zeolites, as also especially witnessed in the work by Chatterjee and coworkers using the condensed Fukui function and local softness to estimate and rationalize the interaction energy of several small molecules with a zeolitic framework [104-106]. An attempt was made to explain selective permeation of these molecules [105] and the choice of the best template for a particular zeolite synthesis by estimating the reactivity of the templating molecule [106].

Biochemistry: Influence of Point Mutations on the Catalytic Activity of Subtilisin

As an example of our work in biochemistry we summarize very briefly recent studies in which the embedded cluster approach presented in § 2.3 was used to study the active sites in enzymes, more precisely the catalytic triad in subtilisin [107] and Ribonuclease T$_1$ [108].

The methodology followed in these studies shows similarities with part of the zeolite adsorption studies (high level quantum chemical calculation on that part of the system at which the active site is situated and a point charge environment for the residues farther away). The results of the subtilisin study are given as an example.

Subtilisin is a bacterial enzyme belonging to the class of the serine proteases characterized by a catalytical apparatus consisting of three amino acid residues, serine, histidine and aspartate: the catalytic triad (Asp32 - His64 - Ser221 for subtilisin) (Fig. 16). In the rate determining step of the hydrolysis reaction (see Fig. 17) the hydroxyl proton of Ser221 is transferred to the N$_{\varepsilon2}$ atom of His64. Simultaneously a nucleophilic attack by the hydroxyl oxygen of Ser221 at the carbon atom of the scissile peptide bond occurs. The role of the aspartate residue is to enhance the nucleophilicity of Ser221 due to the electric field of the charged aspartate side chain and to provide electrostatic stabilization of the tetrahedral intermediate.

The role of the catalytic triad amino acids was studied by Carter and Wells: both single and double alanine substitutions of these residues led to a lowering of the catalytic rate constant k$_{cat}$ [109]. Russell and Fersht studied the effect on k$_{cat}$ for mutations occurring outside the catalytic triad [110].

These effects were studied by placing the catalytic triad into an environment of ChelpG point charges representing all atoms of the amino acids within a 15 Å sphere around His64, and obtained from ab initio calculations on isolated amino acids; the structural data on the wild type enzyme were taken from X-ray diffraction studies [111], mutations were carried out in computero. The nucleophilicity of the Ser221 oxygen was investigated using local softness and (models for) the local hardness: the charge on the oxygen atom and the MEP.

Local softness turned out to be not successful when correlating so − with kcat values in line with Fersht's statement that the nucleophilic attack of the serine on the substrate can be considered as an attack on a hard nucleophilic center [112]. We now further concentrate on local hardness descriptors. When comparing the wild type enzyme and the His64 Ala mutant with, respectively, the Asp32 Ala and Asp32 Ala: His64 Ala mutants, Table 4 shows a less negative charge and, much more pronounced, a less negative MEP value, indicating that the interaction of the catalytic serine with an electrophile is less advantageous in the aspartate mutants. These results are in agreement with experiments pointing out an enhancement of the serine nucleophlicity by the aspartate residue [113].

Table 4: Serine oxygen charge (q_0) (au) and MEP minimum around this oxygen atom $(V(\underline{R})_{min})$ (in kcalmol^{-1}) calculated for the wild-type and mutant enzymes (3-21G Hartree Fock with ChelpG charges)

	q_O	$V(\underline{R})_{Min}$
Wild Type	-0.7719	-121.72
Asp32 Ala	-0.7440	-75.64
His64 Ala	-0.7665	-113.13
Asp32 Ala : His64 Ala	-0.7487	-67.92

A more quantitative picture is obtained when studying the atomic charges and MEP vs. the experimental k_{cat} values [109] for the enzymes when the mutations were performed in the environment of the catalytic triad (Table 5). A correlation coefficient of 0.927 was obtained in the case of the charges (Figure 18).

Table 5: Experimental k_{cat} values (s-1) and calculated charges q_0 (au) and MEP minimum $(V(\underline{R})_{min})$ (kcal mol^{-1}) for the serine oxygen for wild-type, aspartate 99 and glutamate 156 mutant enzymes

	k_{cat}	q_O	$V(\underline{R})_{min}$
1. Wild Type	57	-0.7719	-121.72
2. Asp99 Ser	45	-0.7610	-121.33
3. Asp99 Lys	30	-0.7581	-121.30
4. Glu156 Ser	55	-0.7700	-122.96
5. Glu156 Lys	79	-0.7748	-124.57

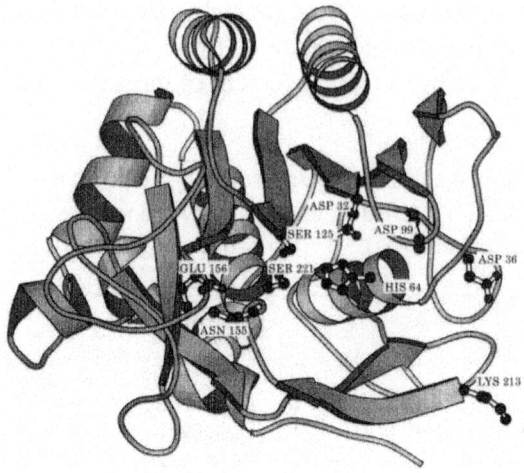

Figure 16: Schematic drawing of subtilisin. Side chains of the residues of importance in the discussion are shown explicitly (Reprinted with permission by Academic Press, Reference [107]).

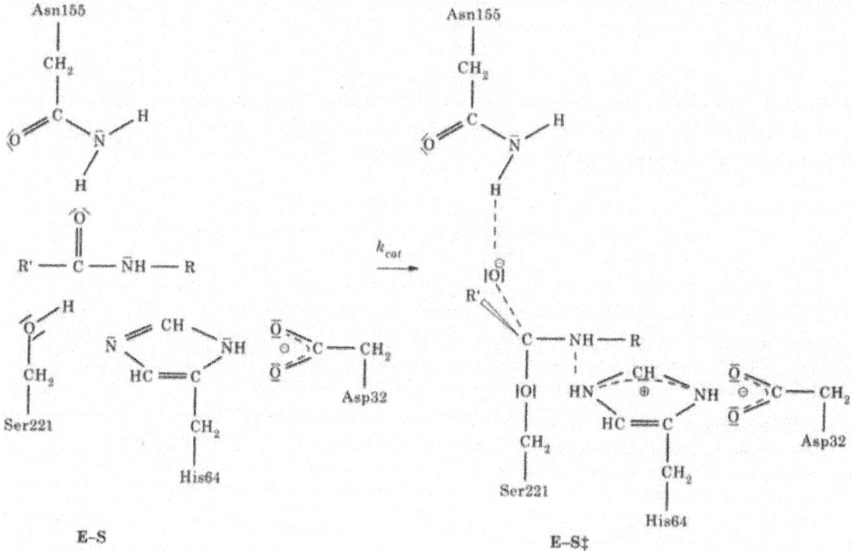

Figure 17: Schematic representation of the enzymatic reaction of serine proteases; E-S is the enzyme substrate complex, E-S$^{\neq}$ the tetrahedral reaction intermediate (Reprinted with permission by Academic Press - Reference [107]).

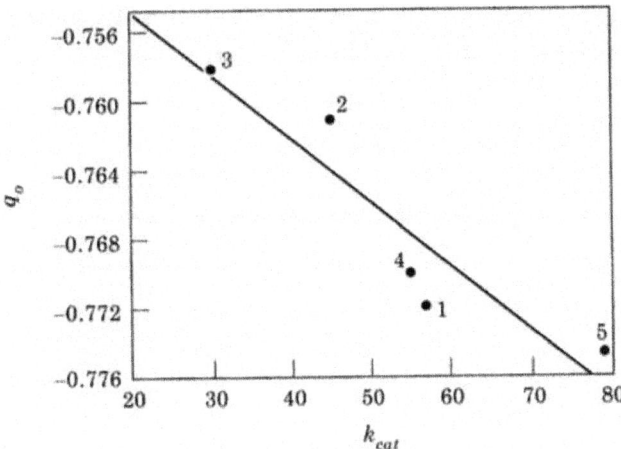

Figure 18: Atomic charge on the serine oxygen atom q_0 vs. experimental k_{cat} (see Table 5).

CONCLUSIONS

In this contribution it is seen that Quantum Mechanics gave birth to new branches in chemistry, Quantum Chemistry and Computational Chemistry, which nowadays provide the theoretical/conceptual and computational framework for studying "real" chemical problems, i.e. systems whose size promotes them to valuable models for the actual reaction partners, their environment, and their interactions. Within this context Density Functional is a source of increasing importance both for concepts and computational techniques.

Application to reactivity problems (considered within a broad context including kinetic as well as thermodynamic aspects) in organic, inorganic and biochemistry shows that present day quantum chemistry becomes a priceless tool to compute and understand phenomena in various traditional subfields of chemistry. Especially the DFT ansatz provides an ideal situation to bridge the gap between chemistry and physics in the years to come [114].

ACKNOWLEDGEMENTS

Besides his coauthor PG wants to thank his past and present Ph.D. students and postdoctoral associates W. Langenaeker, A. Baeten, K. Choho, B. Safi, G. Boon, S. Damoun, F. Tielens, whose work was summarized in this contribution. Special thanks to Professor F. Mendez (Mexico) for the collaboration on organic reactivity during his stay in Brussels, to Professors W. Mortier (ExxonMobil-KU Leuven), R. Schoonheydt (KU Leuven) and G. Baron (VUB) for the longstanding collaboration on zeolites and Professors L. Wyns and J. Steyaert (VUB) for backing the quantum biochemical research.

REFERENCES

1. Dirac, P.A.M. Proc. Roy.Soc.(London), 1929,123, 714.

2. Heitler, W.; London, F. Z. Phys. 1927,44, 455.

3. Pauling,L. The Nature of the Chemical Bond, Third Edition, Cornell University Press, Ithaca, 1960.

4. For a classic and authoritative account of Hückel's approach and its applications see A. Streitwieser, A. Molecular Orbital Theory for Organic Chemistry; John Wiley, 1961.

5. Mc Weeny, R. Coulson's Valence; Third Edition, Oxford, 1979.

6. (a) Hund, F. Z. Phys., 1928, 51, 759.

7. (b) Mulliken, R. S. Phys Rev., 1928, 32, 186.

8. Roothaan, C. C. J. Rev. Mod. Phys., 1951, 23, 69.

9. (a) For an overview see Hartree, D.R. The Calculation of Atomic Structures, John Wiley, New York, 1957.

10. (b) Hehre, W. J.; Radom, L.; Schleyer ,P. v. R.; Pople, J. A. Ab Initio Molecular Orbital Theory, Wiley, New York, 1986.

11. MØller, C.; Plesset, M.S. Phys. Rev. 1934, 46, 618.

12. Shavitt, I. The Method of Configuration Interaction in Modern Theoretical Chemistry, Vol.3, Methods of Electronic Structure Theory, H.F.Schaefer III Editor, Plenum Press, New York, 1977, p.189.

13. Bartlett, R.J. J. Phys. Chem., 1989, 93, 1697.

14. Pople, J.A., et al., GAUSSIAN 98, 1998, and previous releases (GAUSSIAN 94, GAUSSIAN 92, ..., GAUSSIAN 70), Gaussian Inc., Pittsburgh A.

15. The Encyclopedia of Computational Chemistry, Schleyer P. v. Rague; Allinger, N. L.; Clark,T.; Gasteiger, J.; Kollman, P. A.; Schaefer, H. F.; Schreiner, P. R. (Eds.), John Wiley & Sons Ltd., Chichester 1998.

16. Pople, J.A. J. Chem. Phys., 1966, 43, S229.

17. Hohenberg,P.; Kohn, W. Phys. Rev. B ,1964, 136, 864.

18. Mc Weeny,R.; Sutcliffe, B.T. Methods of Molecular Quantum Mechanics, Academic Press, London, 1969, Chapter 4.

19. Kohn W.; Sham, L. Phys. Rev. A, 1965, 140, 1133.

20. Parr,R.G.; Yang,W. Ann. Rev. Phys. Chem., 1995, 46, 701.

21. (a)Becke, A.D. J. Chem. Phys, 1993, 98, 5648.

22. (b) Lee, C.; Yang, W.; Parr, R.G. Phys. Rev. B, 1998, 37, 785.

23. Geerlings, P.; De Proft,F.; Langenaeker,W. Adv. Quantum Chem., 1999, 33, 303.

24. Kohn, W.; Becke, A.; Parr, R.G. J. Phys. Chem., 1996, 100, 12974.

25. Koch,W.; Holthausen, M.C. "A Chemistry's Guide to Density Functional Theory", Wiley-VCH, Weinheim, 2000.

26. Parr,R.G. in "Density Functional Methods in Physics", Dreizler R.M.; da Providencia,J., Editors, Plenum, 1985, p. 141.

27. England, W.; Salmon,L.S.; Rüdenberg,K. Fortschr. Chem. Forsch., 1971, 23, 3 and references therein

28. Foster, M.; Boys, S.F. Rev. Mod. Phys. 1960, 32, 300.

29. Schaftenaar, G.; Noordik, J.H. "Molden: a pre- and post-processing program for molecular and electronic structures", J. Comp.-Aided Mol. Design, 2000, 14, 123.

30. Müller, N.; Pritchard, D.E. J. Chem. Phys., 1959, 31, 768.

31. Figeys, H. P.; Geerlings, P.; Raeymaekers, P.; Van Lommen, G.; Defay, N.; Tetrahedron , 1975, 31, 1731.

32. Parr, R.G.; Yang,W. Density Functional Theory of Atoms and Molecules, Oxford, 1989, Chapter

33. Parr, R.G.; Donelly, R.A.; Levy, M.; Palke,W.E. J. Chem. Phys.,1978, 68, 3801.

34. Iczkowski, R.P.; Margrave, J.L. J. Am. Chem. Soc. 1961, 83, 3547.

35. Mulliken, R.S. J. Chem. Phys., 1934, 2, 782.

36. For a series of papers covering various aspects of electronegativity see: Structure and Bonding, Vol.66, Sen, K.D.; JØrgensen, C.K., Editors, Springer Verlag, Berlin, 1987.

37. Nalewajski, R.F.; Parr, R.G. J. Chem. Phys. 1982, 77, 399.

38. Parr, R.G.; Pearson, R.G. J. Am. Chem. Soc., 1983, 105, 7512.

39. Parr, R.G.; Yang,W. J. Am. Chem. Soc., 1984, 106, 4049.

40. Fukui, K.; Yonezawa, T.; Shinghu, H. J. Chem. Phys. 1952, 20, 722.

41. Yang, W.; Parr, R.G. Proc. Natl. Acad. Sci.. 1985, 82, 6723.

42. Chermette, H. J. Comp. Chem., 1999, 20 ,129.

43. a) De Proft, F.; Geerlings, P. Chem. Rev., 2001, 101, 1451.

44. b) Geerlings, P.; De Proft, F.; Langenaeker, W. Chem. Rev. , submitted.

45. Sanderson, R.T. Science, 1955, 121, 207.

46. Sanderson, R.T. Polar Covalence, Academic Press, New York, 1983.

47. Pearson, R.G. J. Am. Chem. Soc., 1963, 85, 3533.

48. Pearson, R.G. Chemical Hardness, J. Wiley, New York, 1997.

49. Chattaraj, P.K.; Lee, P.K.; Parr, R.G. J. Am. Chem. Soc., 1991, 113, 1855.

50. Parr, R.G.; Chattaraj, P.K. J. Am. Chem. Soc., 1991, 113, 1854.

51. De Proft, F.; Martin, J.M.L.; Geerlings, P. Chem. Phys. Lett., 1996, 250, 393; 1996, 256, 400.

52. De Proft, F.; Geerlings, P. J. Chem. Phys., 1997, 106, 3270.

53. De Oliveira, G.; Martin,J.M.L.; De Proft, F.; Geerlings, P. Phys. Rev., 1999, A60, 1034.

54. See for example the impressive series The Chemistry of Functional Groups, Patai, S. Editor, Interscience Publishers, London.

55. De Proft, F.; Langenaeker, W.; Geerlings, P. J. Phys. Chem., 1993, 97, 1826.

56. Datta, D.; Nabakishwar, S.S. J. Phys. Chem., 1990, 94, 2184.

57. Wray, V.; Ernst, L.; Luna,T.; Jakobsen, H.-J. J. Magn. Res., 1980,40, 55.

58. Pearson, R.G. J. Am. Chem. Soc.,1998, 110, 7684.

59. De Proft, F.; Langenaeker,W.; Geerlings, P. Tetrahedron , 1995, 55, 4021.

60. Lias, S.G.; Bartmess, J.-E.; Liebman, J.F.; Holmes, J.L.; Levin, R.D.; Mallard, W.G. J. Phys. Chem. Ref. Data, 1988, 17 (Suppl 1).

61. Bartmess, J.E.; Scott, J.A.; Mc Iver Jr., R.T. J. Am. Chem. Soc., 1979, 101, 6056.

62. Taylor, R. Electrophilic Aromatic Substitution, J. Wiley, New York, 1990.

63. Vanermen, G.; Toppet, S.; Van Beylen, M.; Geerlings,P. J. Chem. Soc., Perkin Transactions 2, 1986, 699.

64. Wiberg, K.; Keith, T.A.; Frisch, M.J.; Muacko, M.; J. Phys. Chem., 1995, 99, 9072.

65. Safi, B.; Choho, K.; De Proft, F.; Geerlings P. J. Phys. Chem., 1998, A102, 5253.

66. Isaacs, N. Physical Organic Chemistry, Longman Scientific and Technical, Singapore, 1995.

67. Shaik, S.S.; Schlegel, H.B.; Wolfe, S. Theoretical Aspect of Physical Organic Chemistry, John Wiley, 1992.

68. Safi, B.; Choho, K.; Geerlings, P. J. Phys. Chem, , 2001, A105, 591.

69. Gazquez, J.L. J. Phys. Chem., 1997, A101, 8967.

70. Politzer, P. J. Chem. Phys., 1987, 86, 1072.

71. Chandra, A. K. ; Uchimaru, T. J. Phys. Chem. A, 2001, 105, 3578.

72. (a) Gazquez, J.L.; Mendez, F. J. Phys. Chem., 1994, 98, 459.

73. (b) Damoun, S.; Van de Woude, G.; Mendez, F.; Geerlings, P. J. Phys. Chem.,1997, A101, 886.

74. Geerlings, P.; De Proft, F. Int. J. Quant. Chem., 2000, 80, 227.

75. Eisenstein, O.; Lefour, J.M.; Anh, N.T.; Hudson, R.F. Tetrahedron, 1977, 33, 523.

76. K.N. Houk, K.N.; Li,Y.; Evanseck., J.T. Angew. Chem. Int. Ed. Engl., 1992, 31, 682.

77. Chandra, A. K. ; Nguyen, M. T. J. Chem. Soc. Perkin Trans. 2, 1997, 1415.

78. (a) Raspoet, G. ; Nguyen, M. T. ; McGarraghy, M. ; Hegarty, A. F. J. Org. Chem., 1998, 63, 6867.

79. (b) Nguyen, M. T. ; Raspoet, G. Can. J. Chem. 1999, 77, 817.

80. (c) Nguyen, M. T. ; Raspoet, G. ; Vanquickenborne, L. G. J. Phys. Org. Chem., 2000, 13, 46.

81. (a) Chandra, A. K. ; Geerlings, P. ; Nguyen, M. T. J. Org. Chem., 1997, 62, 6417 .

82. (b) Nguyen, L. T. ; Le, T. N. ; De Proft, F. ; Chandra, A. K. ; Langenaeker, W. ; Nguyen, M. T. ; Geerlings, P. J. Am. Chem. Soc., 1999, 121, 5992.

83. (a) Chandra, A. K. ; Nguyen, M. T. J. Comput. Chem., 1998, 19, 195.

84. (b) Nguyen, L. T. ; De Proft, F. ; Chandra, A. K. ; Uchimaru, T. ; Nguyen, M. T. ; Geerlings, P. J. Org. Chem. 2001, 66, 6096.

85. (c) Le, T. N. ; Nguyen, L. T. ; Chandra, A. K. ; De Proft, F. ; Geerlings, P. ; Nguyen, M. T. J. Chem. Soc. Perkin Trans. 2 1999, 1249.

86. (d) Chandra, A. K. ; Uchimaru, T. ; Nguyen, M. T. J. Chem. Soc. Perkin Trans. 2 1999, 2117.

87. (e) Chandra, A. K. ; Nguyen, M. T. J. Phys. Chem. A 1998, 102, 6181.

88. Sengupta, D. ; Chandra, A. K. ; Nguyen, M. T. J. Org. Chem. 1997, 62, 6404.

89. Ponti, A. J. Phys. Chem. A 2000, 104, 8843.

90. Rouvray, D.H. Top. Curr. Chem., 1995, 173, 2 .

91. Mishra, P.C.; Kumar, A. Theor. Comput. Chem., 1996, 3, 257.

92. Carbo, M.; Arnau, M.; Leyda, L. Int. J. Quant. Chem., 1980, 17, 1185.

93. Parr, R.G.; Bartolotti, L.J. J. Phys. Chem., 1983, 87, 2810.

94. a) Boon, G.; De Proft, F.; Langenaeker,W.; Geerlings, P. Chem. Phys.

Lett., 1996, 295, 122.

95. b) Boon, G.; Langenaeker, W.; De Proft, F.; De Winter, H.; Tollenaere, J.P.; Geerlings, P. J. Phys. Chem.A, 2001, 105, 8805.

96. Spatola, A.F. in Chemistry and Biochemistry of Amino Acids, Peptides and Proteins, vol. 7, Weinstein, B. Editor, Marcel Dekker, New York, 1983, p. 267.

97. Perdew,J.P.; Wang, Y.; Phys. Rev. B, 1992, 45, 13244.

98. (a) Allmendiger, T.; Felder, E.; Hungerbühler, A. Tetr. Lett. , 1990, 7301.

99. (b) Bartlett, P.A.; Otake, A. J. Org. Chem., 1995, 60, 3107.

100. Hodgkin, E.E.; Richards, W. G. Int. J. Q. Chem., Quantum Biology Symp., 1987, 14, 1051.

101. Breck, D.W. Zeolite Molecular Sieves: Structure, Chemistry and Use, John Wiley, Canada, 1974.

102. Uytterhoeven, J.B.; Christner, L.G.; Hall, W.K. J. Phys. Chem., 1965, 69, 2117.

103. Van Bekkum, H.; Flanigen, E.M.; Jansen, J.C. Introduction to Zeolite Science and Practice, Elsevier, Amsterdam, 1991.

104. Sherman, J.D.; Proc. Natl. Acad. Sci. USA, 1999, 96, 3471.

105. Van Genechten, K.; Mortier, W.J.; Geerlings, P. J. Chem. Phys. , 1987, 86, 5063.

106. Langenaeker, W.; De Proft,F.; Geerlings, P. Recent Developments in Physical Chemistry, Vol. 2, Transworld Research Network, Triviandum, India, 1998,1219.

107. Peirs, J.C.; De Proft, F.; Baron, G.; Van Alsenoy, C.; Geerlings, P. Chem. Comm., 1997, 531.

108. Tielens, F.; Langenaeker, W.; Ocakoglu, A. R.; Geerlings, P. J. Comp. Chem., 2000, 21, 909.

109. Tielens, F. ; Geerlings, P. J. Mol. Catal., 2001, A166, 175.

110. Ruthven, D. M. Principles of Adsorption and Adsorption Processes, John Wiley, Canada, 1984.

111. Kiselev, A. V. Pure Appl. Chem., 1980, 52, 2161.

112. (a) Tielens, F.; Geerlings, P. Int. J. Quant. Chem. 2001, 84, 58.

113. (b) Tielens, F. ; Geerlings, P. Chem. Phys. Lett. 2002, 354, 474.

114. Buckingham, A.D. in Intermolecular Interactions, Pullman, B. Editor, Wiley, 1988.

115. Dunning Jr., T.II. J. Chem. Phys., 1989, 90, 1007.

116. Kendall, R.A.; Dunning Jr., T.H.; Harrison, R.J. J. Chem. Phys., 1992, 96, 6796.

117. De Proft, F.; Tielens, F.; Geerlings, P. J. Mol. Struct. (Theochem), 2000, 506, 1.

118. Langenaeker, W.; De Proft, F.; Tielens, F.; Geerlings, P. Chem. Phys. Lett., 1998, 288, 628.

119. Chatterjee, A. ; Iwasaki, T. ; Ebina, T. J. Phys. Chem. A 1999, 103, 2489.

120. Chatterjee, A. ; Iwasaki, T. J. Phys. Chem. A 1999, 103, 9857.

121. Chatterjee, A. ; Iwasaki, T. J. Phys. Chem. A 2001, 105, 6187.

122. Baeten, A.; Maes, D. ; Geerlings, P. J. Theoret. Biol., 1998, 195, 27.

123. Mignon, P.; Loverix, S.; Van Houtven, S.; Steyaert, J.; Geerlings, P. in preparation.

124. Carter, P.; Wells, J.A. Nature, 1988, 332, 565.

125. Russell, A.J.; Fersht, A.R J. Mol. Biol., 1987, 193, 803.

126. Bott, R.; Vetsch, M.; Kossiakoff, A.; Graycar, T.; Katz, B.; Power, S. J. Biol. Chem., 1988, 263, 7895 .

127. Fersht, A. Enzyme Structure and Mechanism, New York, W.H. Freeman and Co., 1985.

128. Wells, J.A.; Estell, D.A. TIBS, 1988, 13, 291 (1988).

129. Geerlings, P.; De Proft, F.; Langenaeker, W. Editors, Density Functional Theory: a Bridge between Chemistry and Physics, Proceedings of a Two Day Symposium at the VUB, May 14-15, 1998, VUB-Press, Brussels, 1999.

CITATION

CHAPTER 1

Khusenov, M. , Dushanov, E. and Kholmurodov, K. (2014) Molecular Dynamics Simulations of the DNA-CNT Interaction Process: Hybrid Quantum Chemistry Potential and Classical Trajectory Approach. Journal of Modern Physics, 5, 137-144. doi: 10.4236/jmp.2014.54023.

CHAPTER 2

R. Morales and C. Hernández, "Cyanopolyynes as Organic Molecular Wires in the Interstellar Medium,"International Journal of Astronomy and Astrophysics, Vol. 2 No. 4, 2012, pp. 230-235. doi:10.4236/ijaa.2012.24030.

CHAPTER 3

Masoud Saravi and Seyedeh-Razieh Mirrajei, Numerical Solution of Linear Ordinary Differential Equations in Quantum Chemistry by Spectral Method, ISBN 978-953-51-0372-1.

CHAPTER 4

Nelson Henrique Morgon, Composite Method Employing Pseudopotential at CCSD(T) Level, ISBN 978-953-51-0372-1.

CHAPTER 5

Aline Thaís Bruni and Vitor Barbanti Pereira Leite (2012). Quantum Chemistry and Chemometrics Applied to Conformational Analysis, Quantum Chemistry - Molecules for Innovations, Dr. Tomofumi Tada (Ed.), ISBN: 978-953-51-0372-1, InTech, DOI: 10.5772/34994.

CHAPTER 6

Irena Kratochvilova, Charge Carrier Mobility in Phthalocyanines: Experiment and Quantum Chemical Calculations, ISBN 978-953-51-0372-1.

CHAPTER 7

Ramakrishnan, R. et al. Quantum chemistry structures and properties of 134 kilo molecules. Sci. Data 1:140022 doi: 10.1038/sdata.2014.22 (2014).

CHAPTER 8

Babbush, R., Love, P.J. & Aspuru-Guzik, A. Adiabatic Quantum Simulation of Quantum Chemistry. Sci. Rep. 4, 6603; DOI:10.1038/srep06603 (2014).

CHAPTER 9

Yung, M.-H. et al. From transistor to trapped-ion computers for quantum chemistry. Sci. Rep. 4, 3589; DOI:10.1038/srep03589 (2014).

CHAPTER 10

Wymore T, Brooks CL (2012) From Molecular Phylogenetics to Quantum Chemistry: Discovering Enzyme Design Principles through Computation. Computational and Structural Biotechnology Journal. 2 (3): e201209018. doi: http://dx.doi.org/10.5936/csbj.201209018

CHAPTER 11

Yulia Monakhova and Bernd Schneider. The Intramolecular Diels-Alder Reaction of Diarylheptanoids — Quantum Chemical Calculation of Structural Features Favoring the Formation of Phenylphenalenones. DOI: 10.3390/molecules19045231

CHAPTER 12

Chang-Yu Hsieh and Raymond Kapral. Correlation Functions in Open Quantum-Classical Systems. doi:10.3390/e16010200.

CHAPTER 13

Eno E. Ebenso, David A. Isabirye and Nnabuk O. Eddy. Adsorption and quantum chemical studies on the inhibition potentials of some thiosemicarbazides for the corrosion of mild steel in acidic medium. doi:10.3390/ijms11062473

CHAPTER 14

P. Geerlings and F. De Proft. Chemical reactivity as described by quantum chemical methods. doi:10.3390/i3040276

INDEX